Fluid Preservation

Fluid Preservation

A Comprehensive Reference

John E. Simmons

ROWMAN & LITTLEFIELD
Lanham • Boulder • New York • Toronto • Plymouth, UK

Published by Rowman & Littlefield
4501 Forbes Boulevard, Suite 200, Lanham, Maryland 20706
www.rowman.com

10 Thornbury Road, Plymouth PL6 7PP, United Kingdom

British Library Cataloguing in Publication Information Available

Library of Congress Cataloging-in-Publication Data
Simmons, John E.
Fluid preservation : a comprehensive reference / John E. Simmons.
pages cm
Includes bibliographical references and index.
ISBN 978-1-4422-2965-5 (cloth : alk. paper) — ISBN 978-1-4422-2966-2 (electronic)
1. Liquids—Storage. 2. Liquids—Storage—History. 3. Fluids—Storage. 4. Fluids—Storage—History. 5. Chemicals—Storage. 6. Canning and preserving. I. Title.
QC147.S56 2014
570.75'2—dc23

2013051343

™ The paper used in this publication meets the minimum requirements of American National Standard for Information Sciences—Permanence of Paper for Printed Library Materials, ANSI/NISO Z39.48-1992.

Printed in the United States of America

Contents

List of Figures

List of Tables

Preface

Fluid preservation refers to the use of chemicals in liquid form in which organic materials are submerged and intended to be kept in good condition indefinitely. Most fluid-preserved specimens are animals or plants in alcohol-, aldehyde-, or glycol-based preservatives, but fluid preservation also includes acids, oils, and other chemicals and objects (e.g., some fossils and minerals are stored in oils, some anatomical preparations are stored in acid solutions). Although ethyl alcohol has been in use as a preservative for more than 350 years, and the practice of preservation of organic materials by other means is much older, our knowledge of the chemistry of fluid preservation is limited, and many of the chemicals, techniques, and practices currently in use do not prolong the useful life of specimens as they should.

The way most specimens are collected, prepared, and preserved has long been communicated mostly by oral tradition, despite the rigor applied to most scientific techniques and procedures. While oral tradition includes some good techniques, it also promotes many damaging or inadequate preservation practices that are based on incorrect information or faulty assumptions. Most people who work with collections assume that the chemicals, techniques, and collection care practices (collectively called preservation technology) in the oral tradition were developed by someone wiser than themselves, or at least that preservation technology is based on experimentation or years of trial-and-error efforts—in fact this is rarely the case. The basis of the vast majority of fluid preservation technology is anecdotal at best, and very few techniques or chemicals have been evaluated over the long term. Most fluid preservation techniques were discovered to work accidentally, and evaluated only over a short time frame, if at all. The situation is made more complicated by the lack of good preservation histories for most preserved specimens—many assumptions are made about how specimens were euthanized, fixed, and preserved in the past, but in most cases, these are only guesses. It is rare to find useful details of field preservation techniques recorded in collectors' notes or museum catalogs. Just because a particular technique was known to have been used by a particular individual once does not mean it was used consistently.

We need to subject the practices and materials used in fluid preservation and in managing fluid-preserved collections to careful questioning and analysis, particularly accelerated aging tests that will give us a better idea of whether or not what we are doing really does prolong the useful life of our specimens. A number of the most basic questions are difficult to answer based on existing data, such as what strength of alcohol provides the best compromise between protection and dehydration of the specimens, how long should specimens remain in a fixative, is formaldehyde fixation necessary for long-term preservation, does glycerin added to preservatives really protect specimens from dehydration and keep them flexible, or are there fixatives as good as formaldehyde and preservatives as good as alcohol that are safer to use?

ABOUT THIS REFERENCE

Fluid Preservation: A Comprehensive Reference is aimed at anyone interested in fluid preservation, including people who work in natural history museums, medical museums, anatomical museums, art museums, and nature centers, as well as researchers, students, teachers, and historians of science. My purpose in writing this book is to summarize information from the literature and oral tradition on fluid preservation to help people take better care of their collections, to make specimens more useful, to provide a basis for future preservation research, to encourage the development of better preservation technologies, and to provide a context for diagnosing and solving preservation problems. The book begins with a history of fluid preservation, reviews the chemistry of fluid fixation and preservation, summarizes best practices, and considers fluid preservation in historic and cultural contexts.

A few words are necessary about what is not included in this book. I have not included techniques that make short-term use of fixatives or preservatives for specimens that are intended to be stored dry (e.g., for arthropods prior to pinning, as described by Gibb and Oseto [2006]), unless the processes or chemicals involved shed some light on other fluid preservation practices. The preparation of anatomical and histological specimens is addressed for preparations intended to be kept long term, but not techniques intended for immediate examination or diagnosis. Embalming techniques and fluids are not included except when embalming overlaps with true fluid preservation because embalmed bodies are rarely kept submerged in fluid preservatives. The collecting and euthanizing of live specimens are not discussed except for those instances in which they affect specimen preservation (e.g., for many invertebrates, relaxation and euthanizing techniques are integral to the preservation process); collecting and euthanizing techniques have been reviewed elsewhere in the literature (e.g., Simmons 2002, and others).

The book is divided into three parts. Part 1 contains six chapters, beginning with the history of fluid preservation, the surprising variety of fluid preservatives that have been used over the last 350 years, and a summary of early instructions for preserving specimens. Chapter 2 covers fixation, formaldehyde chemistry, and other fixatives. The topics covered in chapter 3 include preservation, preserving DNA, alcohol and other preservatives, botanical, anatomical, and histological preservation, and the search for alternative preservatives. Chapter 4 summarizes the physical and chemical effects that fluid fixatives and preservatives have on specimens. In chapter 5, best practice guidelines are provided for managing fluid preserved collections, identifying fluids, and evaluating fluid quality. The history of specimen containers and why they fail is addressed, as well as storage environments, working with old containers and old specimens, and health and safety issues. Chapter 5 concludes with a short summary of best practices. Chapter 6 considers fluid preservation in popular culture, art, music, literature, and film.

Part 2 contains more than one thousand references from diverse sources related to fluid preservation, divided into two bibliographies—references cited in the text and tables, and references I reviewed but did not cite.

Part 3 is composed of twenty-three reference tables. Table 1 is an alphabetical list of published fluid preservation techniques, without comments as to their efficacy for long-term preservation. Table 2 provides a chronology of the introduction of techniques and chemicals, and table 8 provides a list of chemicals known to have been used in fluid preservation and when they were first introduced. Subsequent tables provide information on proprietary chemicals, narcotizing agents, anatomical preparations, the effects of preservatives on specimens, containers and labels, rehydration, and best practices.

HOW TO USE THIS REFERENCE

I started compiling a bibliography on fluid preservation in 1987 and quickly discovered that there is an astounding amount we don't know about the subject, and at the same time, a fascinating literature devoted to it. This combination of frustration and discovery led to the writing of this book because of the need for a single, comprehensive review of what is known (and not known) about fluid preservation. This book is not intended to be a how-to guide for specific preservation techniques, but rather is an overview of procedures and chemicals that we can use to care for fluid preserved specimens and improve preservation technology. Anyone interested in the step-by-step details of specific techniques should refer to the original sources, which are provided in the bibliography. The reader is cautioned to keep in mind that many techniques and chemicals were in use before they were mentioned in the literature, and that the use of some techniques and chemicals was never recorded at all, which prevents the historic chronology from being as comprehensive as we might wish it to be.

I have been fascinated by collecting and collections for as long as I can remember. By the time I was in the eighth grade, my collecting had reached the point that I had my own natural history museum in a corner of the basement; by age twenty, I was living in a bamboo hut in the Amazon forest, collecting and preserving amphibians and reptiles for a major university natural history museum. Other than a year spent as a zoo keeper taking care of live animals, I have worked with fluid-preserved collections throughout my career. Along the way, the more I have learned about fluid preservation, the more fascinating the subject has become.

One cold, snowy morning in February 2010, Julianne Snider and I met Erik Åhlander at the Naturhistoriska Riksmuseet (Swedish Museum of Natural History) in Stockholm. Erik escorted us through a maze of hallways and down to the depths of collection storage and at last, into a large vault behind a massive steel door. There, on the shelves, we saw jar after jar of fabulous fluid-preserved specimens, the collection known as the *Museum Adolphi Friderici*, once part of the private cabinet of curiosities of King Adolf Fredrik of Sweden (1710–1771). It isn't just the age of these specimens that makes them special—many of them were illustrated in the great *Thesaurus* of Albertus Seba (published in four volumes between 1734 and 1765), and were described in a catalog of the king's collection written by Carl Linnaeus (who studied them between 1751 and 1754), making them type specimens in his new system of classification of the biodiversity of the planet. We were able to examine firsthand the hatchling crocodiles, Surinam toads, and a particularly significant armadillo that had been suspended in time for three hundred years, floating in their jars of alcohol (see Simmons and Snider 2012). I hope that readers of this book can share this sense of awe and wonder at the connection with nature and history that we have through fluid-preserved specimens.

Please join me each year on the fourth of June, and raise a glass of your favorite spirits to honor the spirit of William Croone who, in 1662, introduced the marvel of fluid preservation to the world.

Acknowledgments

Over the many years that I have worked on this project, I have received the help of numerous friends and colleagues from around the world. I do not have the space to acknowledge everyone by name, but I would like to mention those individuals who have contributed critical or extremely obscure citations, cultural references, and invaluable advice, including Andrew Bentley, Brent Brock, Bob Child, Anna Dhody, Dante Fenolio, Linda S. Ford, Catharine A. Hawks, Marinus Hoogmoed, Susan Jewett, Liesl Jonker, Todd Kelly, Brad Kemp, Edward Kowalski, James B. Ladonski, William Lamar, Tor Limbo, Sandra C. Lindstrom, Sofia Galarza Liu, William López-Rosas, Elizabeth Merritt, Peter Mudde, Yaneth Muñoz-Saba, Giovanni Onore, Lars Österdahl, Timothy Paine, Jane Pickering, Ann Pinzl, Jeff Plant, Cindy Ramotnik, Roberto Rodriguez, Stephen P. Rogers, Steffan Roth, Sally Y. Shelton, Ron Skylstad, Julianne Snider, Sylvia Suarez, and Arnold Suzumoto.

I thank Sandra Jaime Silva for inviting me to supervise her outstanding redesign and renovation of the Colección de Anatomía at Universidad El Bosque in Bogotá, Colombia, and for sharing what she learned during the project as she applied many of the principles and practices recommended in this volume.

I am extremely grateful to several people for many in-depth discussions on fluid preservation, including Andrew Bentley, Julian Carter, Oliver Crimmen, Catharine A. Hawks (who taught me how to ask the fundamental questions, and why they should be asked), Simon Moore, Robert W. Waller (who taught me to how to think about fluid preservation from a chemical perspective), and Andries van Dam. I thank Tracy Hicks for introducing me to fluid preservation from a completely different perspective—as art—and inviting me to collaborate with him on several fascinating projects involving creative uses of fluid-preserved objects.

I thank Chris Collins for inviting me to participate in the Cloth Makers Foundation Expert Workshop on Benchmark Standards for the Preservation of Wet Collections at the Natural History Museum in London, and my fellow panelists for their lively discussions and questions—Andrew Bentley, Julian Carter, Chris Collins, Oliver Crimmen, Simon Moore, Birger Neuhaus, Dirk Neumann, and Andries van Dam.

Joachim Händel generously sent me a copy of his comprehensive book, *Makroskopische Präparations-technik: Wirbellose. Leitfaden für das Sammeln, Präparieren und Konservieren* (Piechocki and Händel 2007). Ann Pinzl kindly shared her extensive unpublished research on botanical fluid preservation. Dirk Neumann was particularly helpful in providing information, comments, and references, and in reviewing earlier drafts of portions of the text.

Kristian Murphy Gregersen deserves a very special acknowledgment and much gratitude for sharing his extensive unpublished research and in-depth knowledge on the history of formaldehyde.

Simon J. Moore has been extraordinarily generous in sharing not only his vast knowledge of fluid preservation, but for allowing me to participate in his most excellent and highly recommended fluid preservation course at the Horniman Museum in London, in which I had the opportunity to put into practice many techniques and procedures I had only read about.

I thank my editor at Rowman and Littlefield Publishers, Charles Harmon, for taking a chance on this book and for his much appreciated diligence throughout the writing and production process. I cannot imagine getting this project to completion without his help.

Julianne Snider helped select and prepare images for this book and made numerous suggestions to improve the content and formatting. More significantly, this project would never have been completed without her advice, encouragement, and unending support.

A project of this complexity (involving the review of more than one thousand references) could not have been accomplished without access to excellent library resources and assistance from dedicated professional librarians. The reference librarian who first taught me how to find information in the scientific literature was Ruth Fauhl, a beloved librarian at the University of Kansas when I was an undergraduate. While doing the early research for this project, I was fortunate to have the friendly and able assistance of two other knowledgeable and extremely helpful librarians at the University of Kansas, Chanette Kirby and Bayless Harsh. I started this project well before the days of electronic access to published materials, much less the easy access provided now through the Internet (even today, however, much of the literature is still not available electronically). We must not lose sight of the fact that the electronic resources we now access so easily are available only because professional librarians organize vast amounts of information and make it accessible—and keep the system functioning and expanding. Therefore, this book is dedicated to the fine professional librarians at the University of Kansas and in libraries everywhere, who are an underappreciated but absolutely irreplaceable resource, fundamental to successful scholarship.

I

FLUID-PRESERVATION TECHNIQUES AND COLLECTIONS

1

The History of Fluid Preservation

> Who does not know the vast importance which our illustrious Cuvier attributed in the progress of the natural system, to him who first conceived the idea of preserving objects in alcohol?
>
> (Gannal and Harlan 1840)

FLUID PRESERVATION IN THE ANCIENT WORLD

The idea of preserving organic material by submerging it in a fluid is very old. Vinegar (acetic acid) has long been used by humans to preserve food as it is sufficiently acidic to afford protection against bacteria, and many cultures figured out how to store food long-term in sealed containers of vinegar, honey, oil, or brine (Shephard 2000). A concentration of about 15 percent acetic acid will stop bacterial growth, but most vinegar contains only 4–5 percent acetic acid, so salt and spices containing volatile oils were usually added to the vinegar to strengthen its preservative properties (Anderson 1925). Herodotus (ca. 484–425 BCE) reported that the Egyptians consumed "wine . . . made of barley" and ate fish that had been preserved by being "pickled in brine" (Grene 1987, 163–164). Greek and Roman writings mention the preservation of fish products using vinegar (Anderson 1925).

Before the Egyptians had perfected mummification, the Babylonians figured out one of the secrets of successful long-term preservation—keep oxygen away from the tissues, which they accomplished by submerging bodies in honey (Grene 1987; Ransome 1937). According to tradition, submersion in honey is how the body of Alexander the Great (356–323 BCE) was preserved for the two-year journey back to Egypt after his death in Babylon in 323 BCE (Ransome 1937). The bodies of the Spartan king, Agesipolis (400–360 BCE), and the philosopher Democritus

(460–370 BCE) were also preserved in honey (Ransome 1937). As part of the mummification process, the Egyptians sometimes preserved the important organs of the body in oil in canopic jars (El Mahdy 1989). The important organs did not include the brain, which they thought (as did Aristotle) was only good for producing mucous (French 1994).

HISTORY OF ETHYL ALCOHOL

Ethyl alcohol is one of the oldest known organic chemicals, long accessible to humans from natural fermentation processes (Monick 1968), and later from controlled fermentation processes used to produce alcoholic beverages. Archaeological evidence indicates that the production of beer (fermented grain) and mead (fermented honey) preceded the production of wine (fermented grapes) (Amerine 1964). The production of beer and wine is depicted on Sumerian pottery from 4200 BCE. There is chemical evidence for beer production from the Godin Tepe site in the Zagros Mountains of Iran (Michel et al. 1992) in the form of carbonized barley and calcium oxalate deposits (a by-product of brewing barley beer) in the grooves of beer jars. However, the oldest evidence for the preparation of a fermented alcohol beverage dates to at least nine thousand years ago in China (McGovern 2009).

The problem with fermentation is that the process stops when alcohol concentration reaches about 12 percent, because the alcohol that is produced kills the yeast (Nitz 1989). To produce alcohol stronger than 12 percent, the brew has to be rectified, or purified. Rectification is the process of purifying alcohol by repeatedly, or fractionally, distilling it to remove water. Alchemy, the source of modern science, was concerned in part with finding a way to transform base metals into precious metals. In their search for the most pure form of each substance, alchemists developed various processes of purification and distillation. Tradition attributes the invention of distillation to an alchemist best known as Miriam, the sister of Moses (she was later afflicted with leprosy for speaking out against Moses's marriage to an Ethiopian), although the precise time and place of the origin of distillation is unknown (Forbes 1948). There is archaeological evidence from around 3500 to 3000 BCE of the use of extraction pots in the Mesopotamian region (Brock 1993; Levey 1960). Extraction pots were made with a double-rim trough that was percolated with holes. When the pot was heated, the steam condensed in the lid and dripped back into the pot, producing a progressively more rectified distillate. The pots were used by herbalists and perfume makers, and later adopted by alchemists, but how strong an ethanol solution might have been made in an extraction pot is not known. The Chinese distilled a beverage made from rice beer as early as 800 BC, and arrack has long been distilled in the East Indies from sugarcane, coconut flower sap, fruit, or rice (Shipman and Thomas 1991). Raymond Lully, or Ramón Lull (1235?–1315), a late thirteenth-century Spanish philosopher, wrote that distilled spirits were "An element newly revealed to man but hid from antiquity because the human race was then too young to need this beverage destined to revive the energies of modern decrepitude" (Lichine 1989, 188).

By about 50 CE, Pliny (23–79) was aware of the presence of a combustible substance in wine, which he described as "wine which can be ignited" (Monick 1968, 7), or in another translation, "There is now no wine known that ranks higher than the Falernian; it is the only one, too, among all the wines that takes fire on the application of flame" (Bostock and Riley 1855, 240). To catch fire means that the wine was at least 50 percent ethyl alcohol, as that is the minimum strength at which alcohol will burn. The wine may have been similar to brandy, which was first distilled from wine "in crook-necked pot stills" (Lichine 1989, 188) that were designed for continual reheating (and thus continual rectification) of the contents. The idea behind making brandy was to reduce the bulk of wines for shipping; once the wine arrived at its destination, the consumer could add water to reconstitute it—but the consumers preferred the stronger brandy (Lichine 1989). The word brandy comes from the Dutch word *brandewijn*, which means wine that burns (brandy is typically 50 percent or more ethyl alcohol).

The word distillation comes from *distillare*, which means "to drip off" or "to drop off" (Forbes 1948). The technique is based on the difference in the boiling points of alcohol (78.5°C, or 173.3°F) and water (100°C, or 212°F). When heated to between 78.5°C and 100°C, an alcoholic liquid produces a vapor that can be condensed. The condensate will have a higher alcohol concentration than the original liquid (Shipman and Thomas 1991). Distillation has a high heat requirement, so to be effective the process requires a lot of fuel (Shipman and Thomas 1991), and the earliest stills left so much water in the solution that the alcohol would not burn (hence it was less than 50 percent ethyl alcohol). Success came from two innovations. The Arabian alchemists made a stronger wine by adding salts or tartar (potassium carbonate) to absorb some of the water, based on the philosophical principle that a substance of one nature can be used to purify a substance of another nature (Leicester 1956; Partington 1989). The earliest known account of how to produce strong alcohol says to mix a pure and very strong wine with three parts salt, and then heat it (Crombie 1952; Leicester 1956). In the first century CE, Alexandrian chemists developed a better still head that more efficiently collected and cooled the distillate so it could be redistilled (Brock 1993). More sophisticated apparati followed that allowed for continuous distillations (Brock 1993). For example, better cooling systems were developed in Europe in the twelfth and thirteenth centuries (Forbes 1948). Taddeo Alderotti (1223–1303) developed a still head with a three-foot-long water-cooled tube that produced 90 percent pure alcohol (Crombie 1952; Holmyard 1957; Leicester 1956).

Hieronymus Brunschwygk (1450–ca. 1512) published the first edition of his *Little Book of Distillation* (*Liber de arte distillandi simplicia et composita*) in 1500, describing steam distillation, the use of a water bath to heat the substance, and an air-cooled conical alembic to condense it. The various editions of Brunschwygk's book stimulated much interest in these techniques and the development of many different sorts of distilling apparati, some of which are still in use today (Leicester 1956;

figure 1.1), such as the pot still and the continuous still. A pot still is an enclosed vessel, heated from underneath or by steam coils inside, with a cylindrical bulb at the top that connects to a tube that collects and cools the vapor that is produced (Shipman and Thomas 1991). A continuous still is a tall cylinder containing a series of perforated plates that collect the water-rich vapor while allowing the alcohol-rich vapor to pass through. In effect, the perforated plates function like a series of small pot stills (Shipman and Thomas 1991).

By 1606, the word rectified was being used in English in reference to its chemical meaning, "Purified or refined by renewed distillation; redistilled." The actual quotation in which rectified first appeared in this sense in English is, "This humour . . . doth resemble the rectified animal aquavita" (Oxford English Dictionary).

Alcohol and water together do not form a true solution, but rather a binary azeotrope in which each substance keeps its own properties (e.g., the vapor pressure of alcohol and water are different, which explains why alcohol evaporates from a mixture of alcohol and water faster than the water evaporates). Distillation can increase the concentration of ethyl alcohol only to the binary azeotrope of 95.6 percent. For ethanol of greater purity, it is necessary to use chemical dehydration, which is why absolute alcohol (100 percent ethyl alcohol) is much more expensive than 95.6 percent alcohol, and why the latter has become the standard solution for museum use.

Alcohol strength is often given as a percent proof, which was first used in English in 1705, with one hundred proof defined as a concentration of eleven parts of alcohol and ten parts of water, which would permit the ignition of gunpowder (Nitz 1989). Proof is now defined as a mixture of alcohol and water with a specific gravity of 0.91984, containing 0.495 by weight or 0.5727 by volume absolute alcohol

Figure 1.1. A pot still, alembic (still head), and cooling coils (in barrel on right).

(Oxford English Dictionary), or twice the volume percent of ethanol in a sample measured at 15.56°C (Waller and Strang 1996).

ORIGIN OF THE NAME ALCOHOL

One of the curious footnotes in the history of alcohol is its many names, most of which come from alchemy. The alchemist and monk Jean de Roquetaillade (also known as Johannes de Rupescissa or John of Rupescissa, ca. 1310–1366) thought that alcohol was the supreme remedy against corruption, and named it as the fifth element, *quintesseare* (Leicester 1956).

The first distillation of alcohol produces *aqua ardens* or water that burns (Brock 1993), about 60 percent ethyl alcohol (Crombie 1952). The second distillation makes *aqua vitae*, or water of life, which is about 96 percent ethyl alcohol (Crombie 1952). The word whiskey is derived from a Gaelic word (*uisgebeatha*, or whiskybae) that means water of life (Lichine 1989). Its first recorded use in English was in a book published in 1715, which noted that "Whiskie shall put our brains in rage" (Oxford English Dictionary). In the sixteenth century, the Swiss alchemist Paracelsus (Philippus Aureolus Theophrastus Bombastus von Hohenheim, 1493–1541) derived the name of alcohol from *al kohl*, an Arabic word referring to a fine black powder used for eye shadow (Crosland 2004; Leicester 1956). Kohl is a paste made of black antimony sulfide—also known as stibnite, Sb_2S_3—in fat (Crosland 2004). In alchemy, each substance was reduced to its most pure form (Monick 1968), so a finely ground powder was considered to be an essence, or spirit, of an element, hence *al kohl* became alcohol in reference to the essence, or spirit of wine. The first recorded uses of the word (then spelled alcohole) in English (the first in 1615, and next in 1626) were by Roger Bacon (1214–1294) in reference to the fine powder (Oxford English Dictionary).

GLASS CONTAINERS

Glass blowing originated in Syria about 1000 BCE (Derry and Williams 1960). By the seventeenth century, glass was made all over Europe, but it was very expensive. In 1615, the British navy needed more ships, so the crown issued a proclamation that gave the navy priority use of timber for shipbuilding, forbidding its use in glass furnaces (Derry and Williams 1960). Glass furnaces had to switch to coal, which burned much hotter than wood. In 1675, George Ravenscroft (1632–1683), who had learned glassmaking in Italy, developed a mixture of potash, lead oxide, and calcinated flints that could be fired in these hotter ovens to produce an inexpensive, clear flint glass with fewer bubbles (Derry and Williams 1960). This was a huge advance for fluid preservation—affordable, clear glass, and you could look through the container to see what was inside.

THE DISCOVERY OF PRESERVATION OF SPECIMENS IN ETHYL ALCOHOL

At least one author has attributed the discovery of ethanol preservation to Roger Bacon (1214–1294), but without providing a citation for this attribution (Penrith 1971). Some authors cite the diary of Elias Ashmole (1617–1692) (e.g., Down 1989; Reid 1994) as evidence that Ashmole had in his possession a fetus preserved in alcohol as early as 1660. What the diary actually says (the entry is for October 12, 1660) is that:

> This morning I showed the King the young children which Dr. Warner had preserved. The one was a male infant about 4 months, who was cut out of a woman's belly in Covent Garden (she was dying of a consumption) and had been (now four years past) luted up in a glass, and preserved by a liquor of his preparation from putrefaccon, the flesh not so much as rumpled, but plump as it was when taken out of ye wombe. The other was 2 girls joined together by the breast and belly (which monster was borne about the king's coming in), they were dryed and preserved with spices. (Gunther 1927, 67)

This "liquor" could mean any number of fluids—including vinegar, turpentine, or brine—that were in use for preservation at the time. The word *liquor* can refer to a liquid of any kind, alcoholic or nonalcoholic (e.g., liquor was used in 1604 to refer to liquid mercury, according to the Oxford English Dictionary). As late as 1890 a taxidermy text referred to temporary fluid preservation of skins by immersing them "in a barrel of brine, or what is technically called 'liquor'—in fact, parts of alum and salt dissolved in water, in the proportion of 6 lbs. of alum and 3 lbs. of salt, sea salt if possible" (Ward 1890, 27). The word *luted* in Ashmole's description refers to the sealant on the container, which was probably wax, tar, or a similarly viscous substance. Furthermore, had the fetus been preserved in alcohol for four years, it would probably have been dehydrated, not "plump."

Several authors have made or repeated the claim that Robert Boyle (1627–1691) developed the process of preservation in alcohol, but this is not supported by evidence in the literature. Boyle was certainly interested in finding ways to preserve animal tissues, particularly human body parts, an interest that began while he was at Oxford between 1655 and 1668 (Kaplan 1993), but prior to 1662, Boyle refers to the use of brine solutions as preservatives (Reid 1994). In 1664, in the second part of *Some Considerations Touching the Usefulness of Experimental Natural Philosophy*, Boyle wrote:

> [I]t cannot but be a great help to the Student of Anatomy to be able to preserve the parts of human Bodies, and those of other Animals, especially such Monsters as are of a very singular or instructive Fabrick, so long that he may have recourse to them at pleasure, and contemplate each of them so often and so considerately, till he have taken sufficient notice of the shape, situation, connection, etc. of the Vessel, Bone, or other part, and firmly impressed an idea of it upon his memory. (Boyle 1664, 21)

Boyle goes on to discuss various ways of preserving animal tissue by dehydration. Kaplan (1993, 51) states that "Boyle was able to develop a preservative based on turpentine that was useful in preserving small specimens," however, this was not a fluid preservative, but rather turpentine condensed by evaporation into a gum that was then used to coat specimens. Boyle also reported on preparations in oil and in spirits of wine that he observed but did not claim to have prepared himself:

> But I must not insist on theses Fancies, but rather adde, That I have known an *Embrio*, wherein the parts have been very perfectly delineated and distinguishable, preserved unputrified for several Yeares; and I think it still continues so, by being seasonably an artificially embalmed with Oyl (if I much misremember not) of Spikes. And I have elsewhere seen a large *Embrio*, which after having been preserved many Years, by means of another Liquor (whose Composition I doe as yet but guesse at) did, when I saw it, appear with such an admirable Entirenesse, Plumpnesse, and Freshnesse, as if it were but newly dead: And that which concurs to make me hope that some nobler way may yet be found out, for the preservation of dead Bodies, is that I am not convinc'd that nothing can powerfully resist Putrefaction in such Bodies, but things that are either Saline and Corrosive, or else Hot, nor that the Embalming Substances cannot be effectually applied, without ripping open the Body to be preserv'd by them. . . . And I shall anon tell You, that I have seen a Liquor, which without being at all either acid or caustic, is in some bodies far more effectual against Putrefaction, than any of the corrosive Spirits of Nitre, Vitriol, Salt, etc. and than any of the other Saline Liquors that are yet in use." (Boyle 1664, 24–25)

Oyl of spikes refers to an essential oil made with lavender flowers. Many experiments in preservation were being carried out at this time—for example, Bolnest (1672) referred to various nonalcoholic preservatives such as salt and other chemicals in water as spirit or well rectified spirit, and a physician named Francis Hume was experimenting with the preservation of fish and beef in limewater in the mid-1700s—the fish remained in good condition for seven weeks, the beef did not (Clephane 1753).

Some authors have claimed that Boyle's description of preserving avian embryos (Boyle 1666) is the first account of fluid preservation, but there are older published accounts of preservation in alcohol. Yet another author remarked, without providing a citation for the evidence, that:

> It will thus be realised what an incalculable advance was made in museum technique by Robert Boyle's experiment, carried out on William Croune's suggestion, of preserving parts of, or even whole, animals in spirit. Boyle's specimens of "a linnet and a little snake, preserved already four months, entrails and all, without any change in colour, in some spirit of wine," placed in the repository of the Royal Society at Gresham College in 1663, were still in existence a hundred and fifty years later. (Dobson 1970, xii)

The role of "William Croune" will be discussed next. Dobson was quoting the *Proceedings of the Royal Society* for January 20, 1664. The full quotation is as follows:

> [Mr. Boyle] presented the society with a linnet and a little snake, preserved already four months, entrails and all, without any change of the colour, in some spirit of wine made

> after a peculiar manner. Both the glasses, wherein they were kept, were ordered to be sealed up with wax, and put into the repository.
>
> Dr. CLARKE mentioned, that there was an intention of preserving the whole body of a man with spirit of wine in a vessel well glased; suggesting the usefulness of this method of preserving, that by this means there might be had in readiness for occasions an eye, hand, muscles, larynx, &c.
>
> Dr. MERRET affirmed, that he had preserved the very entrails of a cat for twelve years past, though they were somewhat shrunk.
>
> Mr. BOYLE observed, that he had a liquor compounded of spirit of wine and a little oil of turpentine, whereby not only the bodies of animals or the parts thereof might be preserved, but also it might be known what it is that comes away; which being such, that it will not mingle with the oil, falls to the bottom, and is found there.
>
> He offered to preserve a hand and a larynx. (Birch 1968, 374)

In fact, the linnet and the snake referred to were preserved in 1664 (Cole 1944), as recorded in *Musaeum Regalis Societatis, or a Catalogue and Description of the Natural and Artificial Rarities Belonging to the Royal Society and Preserved at Gresham Colledge*, published by Nehemiah Grew (1641–1712) in 1681. In this detailed catalog, Grew (1681, 48) describes "A snake preserved in Spirit of Wine"; "A young LINET which being first embowel'd, hath been preserved found and entire, in rectified Spirit of Wine, for the space of 17 years. Given by the Honourable Mr. *Boyl.* Who, so far as I know, was the first that made trial of preserving Animals this way. An Experiment of much use. As for the preserving of all sorts of Worms, Caterpillars, and other soft-Insects in their natural bulk and shape, which otherwise shrink up, so as nothing can be observed of their parts after they are dead. So also to keep the Guts, or other soft parts of Animals, fit for often repeated Inspections. And had the Kings or Physitians of *Egypt* thought on't, in my Opinion, it had been a much better way of making an everlasting Mummy" (Grew 1681, 58); and "A young CHICKEN emboweled and put into rectified Oil of Turpentine, at the same time, with the *Linet*, and preserved found; Only there is a little sediment at the bottom of the Glass" (Grew 1681, 58). As the specimens had been in the collection for 17 years, this would mean they were preserved in 1664 or shortly before.

The first recorded mention of the use of alcohol for fluid preservation was on June 4, 1662. A physician, William Croone (1633–1684) (his name was also spelled Croune), appeared before the Royal Society of London and "produced two embryos of puppy-dogs, which he had kept eight days, and were put in spirit in a glass-vial sealed hermetically" (Birch 1968; Cole 1944). Croone was born in London and received his M.D. degree in 1662; he was mentioned several times in the diary of Samuel Pepys (Payne, Wilson, and Hartley 1960). Croone believed that muscular motions were produced by "vital spirits" (an idea from Decartes) that flowed along the nerves and mixed with the spirit of the blood (Debus 1968). After Croone's presentation of the preserved embryos to the society, the *Transactions* report that "Mr. Boyle promised to make the like experiment in rabbets." The next mention of the technique was when another member brought in an "abortive human foetus, kept in spirit well rectified." In 1664, Boyle brought to the Royal Society the specimens of a linnet and snake which he had preserved in spirit of wine four months previously (Cole 1944), and in 1665, the society records noted that Boyle showed "the head of a *colt*" which he "Caused . . . to be put into a Vessel, and covered with *Spirit of Wine*, thereby chiefly intending, to give good example, together with a proof, that by the help of the said Spirit . . . the parts of *Animals*, and even *Monsters*, may in *Summer* it self be preserved long enough, to afford *Anatomists* the opportunity of examining them" (Boyle 1665).

Although Boyle took an interest in the use of alcohol as a preservative, his speculations on the topic in *Some Considerations Touching the Usefulness of Experimental Natural Philosophy* were clearly based on the work of others:

> Nor were it amisse that diligent Tryal were made what use might be made of Spirit of Wine, for the preservation of a humane Body. For this Liquor being very limpid, and not greasie, leaves a clear prospect of the Bodies immers'd in it, and though it doe not fret them, as Brine, and other sharp things, commonly imploy'd to preserve flesh, are wont to do, yet it hath a notable Balsamic faculty, and powerfully resists Putrefaction, not onely in living Bodies (in which, though but outwardly applied, it hath been found of late one of the potentest Remedies against Gangrenes) but also in dead ones. And I remember that I have sometimes preserv'd in it some very soft parts of a Body for many Moneths) and perhaps I might had done it for divers Years, had I had opportunity) without finding that the consistence or shape was lost, much lesse, that they were either putrified or dried up. (Boyle 1664, 25–26)

Nevertheless, Boyle later experimented with fluid preservation, and based on his experience, recommended changing the alcohol solution after the initial preservation of the specimen, redistilling the preservative when it became diluted or discolored, and testing the strength of ethyl alcohol by soaking a strip of paper or cloth in the solution and then trying to light it (Boas 1958).

The idea of fluid preservation caught on quickly, but the use of the technique was limited by the expense of the alcohol and the glass containers. As late as 1813, one anatomical preparation manual lamented that, "Preparations of almost every part are occasionally kept in spirits, unless their size renders it impracticable, more especially diseased parts; as by this mode they undergo less change of appearance than by any other method of preservation, and consequently give the best idea of the natural or diseased appearance; but the expensiveness of the glass and spirits is a great inducement to the making of so many dry preparations" (Pole 1813, 163). The preservation methods recommended were not always very refined. For example, in 1748, René-Antoine Ferchault de Réaumur (1683–1757) stated that, "There is no great Skill required for putting one or several [specimens] into a Vessel full of Spirit of Wine, or very strong Brandy. It has been usual for a long time to make use of those Liquors with Success for preserving the Flesh of dead Animals" (Réaumur 1748, 307). There are records of the preservation of a human fetus in 1665 by T. Coxe, of insects in alcohol by 1670 by Jan Swammerdam (1637–1680), and the body of a child being kept for several years in alcohol, reported in 1678 (Cole 1944). The most distinctive and best preserved specimens from this time were those prepared in Holland by Frederik Ruysch (1638–1731), a surgeon and anatomist in Amsterdam. Ruysch distilled his own alcohol from barley, producing about 67 percent ethanol from this process (Singer 1921). Although most of Ruysch's techniques were closely guarded secrets, it is thought that some of his preservative solutions included black pepper, cardamom, and cloves mixed with the alcohol (Cole 1944). Ruysch perfected the methods of vascular injection that revolutionized the quality of fluid preserved specimens by providing a better means of distributing preservative in the tissues (Singer 1921). Ruysch was able to preserve soft brain tissue by a combination of injection of the veins and arteries with wax and vermilion and the use of his own alcohol recipe, and made a few brief mentions of his techniques in the catalog of his collection, his *Thesaurus Anatomicus* (published between 1727 and 1744) and in letters to colleagues (Cole 1921; Gere 2003). Ruysch sold a collection of anatomical preparations to Peter the Great (1682–1725) for his Kunstkamera in 1717 (Peter the Great had first seen the Ruysch collection in 1697 when he visited Amsterdam). Many of the specimens survived war and other afflictions and can still be viewed at the Anthropological and Ethnographic Museum in St. Petersburg, Russia (Humphries 2003).

One very famous collector of the era was the Englishman James Petiver (ca. 1663–1718), who solicited specimens from many travelers around the world. He published a sheet of instructions called "Brief Directions for the Easie Making and Preserving Collections of all Natural Curiosities for James Petiver Fellow of the Royal Society of London" that was widely distributed to travelers and sailors (Stearns 1953, 363). Petiver advised that "All small Animals, as Beasts, Birds, Fishes, Serpents, Lizards, and other Fleshy Bodies capable of corruption, are certainly preserved in Rack, Rum, Brandy, or any other Spirits; but when these are not easily to be had, a strong Pickle, or Brine of Seawater may serve" (Stearns 1953, 363). *Rack* is a shortened form of *arrack*, an alcoholic beverage distilled from either the juice of coconut palm or a mash of rice and molasses. Even with these recommendations, some of Petiver's collectors improvised fluid preservatives of their own. John Lawson, collecting in North Carolina in 1710, wrote to Petiver to explain that his "bottles are too small & too narrow mouths for some large snakes, etc. I can no ways preserve snakes, Lizards & small birds but in Spirits. I had a curious liquor of Mr. Fettiplace Bellers that preserved bird-skins very well it is all gone I have made a liquor of my own which does it reasonably well wch. is aloes, myrrh, allom & tobacco steept in rum. I take his to be something of that nature but it gives them a shining varnish wch. mine does not pray get his recipe & speak to him to send a quantity" (Lefler 1967, 270). When the artist and naturalist Maria Sybilla Merian (1647–1717) returned to Europe from a two-year stay in Surinam, she wrote to a physician to explain that:

> When I was in that country I painted and described the larvae and caterpillars as well as the kind of food and habits; but everything I did not need to paint [there] I brought with me, such as the butterflies and beetles and everything which I could steep in brandy and everything which I could press I am now painting the same way as I did when I was in Germany, but everything on vellum in large format with the plants and creatures life size. (Owens 2007, 148)

A good number of the published recipes for fluid preservatives from the earliest days onward call for additives to the alcohol, particularly alum and mercuric chloride, but including a wide variety of acids and salts, and other substances (tables 1 and 2). The additives were assumed to make the alcohol a better preservative. At the time, it was difficult to assess the strength of alcohol (other than seeing if it would burn, which indicated a strength of about 50 percent or greater), so many of the additives were probably necessary to compensate for low concentrations of ethanol that were available. In addition, many of the additives had been used as preservatives before the discovery of the preservative properties of ethanol, so it was natural to combine them in an effort to produce a better preserving fluid.

In 1753, Carl Linnaeus (1707–1778) published *Instructo musei rerum naturalium*, which included a number of preservation techniques (such as attaching dried fish to paper sheets in a manner similar to how plants are mounted in herbaria), including the recommendation to add sugar to alcohol for preserving lower vertebrates (Holm 1957). Linnaeus also provided a detailed account of how to seal glass cylinders of specimens by using cork stoppers sealed with tragacanth (a gum extracted from the legume *Astragalus gummifer*), a sealant so successful that several of the jars sealed by Linnaeus are still extant (Holm 1957).

In a description of his travels in Guiana, published in 1769, Edward Bancroft (1744–1821) described some techniques then used for fluid preservation, referring to a technique "since published in one of the monthly Magazines; and the author, if I mistake not, thought it an important discovery, which, it seems, he obtained in *Paris*, tho' not without great difficulty" (Bancroft 1769, 184). Bancroft is most probably referring either to a publication by Réaumur (1748) or a review of it (Bancroft refers to Réaumur by name a few pages later). It is interesting to note that Bancroft recommended the use of alcohol for both specimens that would later be dried (birds) and those that will remain in fluid (snakes):

> The method of doing this in *Guiana*, is to put the Bird, which is to be preserved, in a vessel, and cover him with High Wines, or the first Running of the Distillation of Rum. In this Spirit he is suffered to remain for twenty four or forty-eight hours, or longer, according to his size, til it has penetrated thro' every part of the body. When this is done, the Bird is taken out, and his feathers, which are no ways changed by this immersion, are placed smooth and regular. He is then put into a machine, made for the purpose, among a number of others, and its head, feet, wings, tail &c. are placed exactly agreeable to life. In this position they are all placed in an oven, very moderately heated, where they are slowly dried, and will ever after retain their natural position, without danger of putrefaction. This method might perhaps in *England* be deemed expensive, as the great duty

> on Spirits has raised their price to an enormous height; but in a country where Rum is sold for ten pence sterling per gallon, the case is far different. (Bancroft 1769, 184–185)
>
> The method which I use in preserving these Animals, may perhaps not be unworthy of a communication. When the Snake is killed, it must first be washed clean, and freed from all filth and nastiness; then it is to be put into a glass of a proper size, the tail first, and afterwards the rest of the body, winding it in spiral ascending circles, and disposing the back, which is always the most beautiful, outwardly. A thread, connected to a small glass bead, is, by the help of a needle, to be passed thro' the upper jaw from within outwardly, and then thro' the cork of the bottle where it must be fastened: by this means the head will be drawn into a natural posture, and the mouth kept open by the bead, whereby the teeth, &c. will be discovered: the glass is then to be filled with rum, and the cork sealed down, to prevent its exhalation. A label, containing the name and properties of the Snake, is then to be affixed to the wax over the cork; and in this manner the Snake will make a beautiful appearance, and may be thus preserved a great number of years; nor will the spirits impair or change the lustre of its colours." (Bancroft 1769, 218–220)

Not everyone was convinced that fluid preservation was an acceptable means of preserving specimens. Tessor Samuel Kuckahn (misspelled as T. S. Kuckhan in the *Philosophical Transactions*), in describing various means of preserving birds, stated that:

> A second method of preserving birds is, by immerging them in spirits, and if the barely keeping the carcase of birds from putrefaction is all that is required, I must own this method is an effectual one, and congratulate the naturalists on the facility with which they may now procure foreign birds. We have nothing to do but send by sailors bound to different parts of the world, a few kegs of spirits, and we shall be sure of birds enough preserved in this manner; but then what becomes of their proportions, attitudes, graces, and in short, of every thing that gives them life and motion? (Kuckhan 1771, 304–305)

Two such birds collected by the naturalists on one of Captain Cook's voyages and given to the Hunterian collection in 1792 are among the oldest extant fluid-preserved bird specimens. They were examined and dissected when rediscovered, and were described as being in "remarkably good condition" after 190 years of preservation (Burton 1969, 389).

By the early 1800s, the practice of preservation in spirit of wine was well enough known for many people to have an opinion about it, including those who had never done it, which led to the criticism directed at the surgeon who preserved the body of Admiral Nelson. After the Battle of Trafalgar in October 1805, the body of Admiral Lord Nelson (1758–1805) was preserved in fluid for the long trip back to England. The ship's surgeon, William Beatty (1773–1842), later published an account of the event, explaining the procedure—Nelson's body, clad only in a shirt, was put in a large cask that was then filled with brandy and lashed to the deck. The cask had two apertures, one at the bottom for drawing off brandy and another at the top for adding fresh brandy as needed. As it happened,

> On the 24th there was a disengagement of air from the Body to such a degree, that the sentinel became alarmed on seeing the head of the cask raised: he therefore applied to the Officers, who were under the necessity of having the cask spiled to give the air a discharge. After this, no considerable collection of air took place. The spirit was drawn off once, and the cask filled again, before the arrival of the Victory at Gibraltar (on the 28th of October): where spirit of wine was procured; and the cask, shewing a deficit produced by the Body's absorbing a considerable quantity of the brandy, was then filled up with it. . . . When the Victory had proceeded some weeks on her voyage, adverse winds and tempestuous weather having prolonged the passage much beyond the period that is generally expected, it was thought proper to draw off the spirit from the cask containing Lord NELSON'S Body, and renew it; and this was done twice. On these occasions brandy was used in the proportion of two-thirds to one of spirit of wine. (Beatty 1807, 63–64)

Once the body arrived in England, Beatty was criticized in the British press for not preserving the body in rum, which was commonly believed to be a better fluid preservative than brandy. In his defense, Beatty argued that:

> Brandy was recommended by the Surgeon in preference to rum, of which spirit also there was plenty on board. The circumstance is here noticed, because a very general but erroneous opinion was found to prevail on the Victory's arrival in England, that rum preserves the dead body from decay much longer and more perfectly than any other spirit, and ought therefore to have been used: but the fact is quite the reverse, for there are several kinds of spirit much better for that purpose than rum: and as their appropriateness in this respect arises from their degree of strength, on which alone their antiseptic quality depends, brandy is superior. Spirit of wine, however, is certainly by far the best, when it can be procured. (Beatty 1807, 62)

EARLY INSTRUCTIONS FOR PRESERVING SPECIMENS IN FLUIDS

Initially, fluid preservation was a laboratory procedure, but quickly became a means for preservation of specimens in the field as well. By the early 1800s the published recommendations for alcohol-based preserving fluids usually included additives (table 2). Most of the people who were preserving specimens in the field were not naturalists but sailors or other travelers who were acquiring specimens to sell to collectors in Europe—for example, Albertus Seba (1665–1736) is not known to have done any fieldwork himself, but purchased his specimens from sailors at the docks or from commercial dealers, then rebottled his specimens for his museum. Throughout the 1800s the published instructions for fluid preservation tended to either specify just beverage alcohol (easy to obtain and use) or called for spirits of wine with various chemical additives.

The years following Croone's 1662 demonstration were a time of great experimentation with fluid preservation, with numerous individuals making recommendations for their own particular concoctions. In 1688, Blanckaert (in *Tractatus de Balsamatoione*) recommended three different liquors—salt water with alum, oil of turpentine, and rectified spirit of wine with a little sal ammoniac (ammonium chloride) (Edwards and Edwards 1959). As confidence in the preservative powers of alcohol-based liquids grew, people were willing to invest more in the system. In addition to reporting the linnet that had been preserved for seventeen years (discussed earlier), Grew also described a human fetus: "The skin hath been kept white and smooth for so long a time scil: above 15 years, by being included with rectified Spirit of Wine in a Cylindrical Glass to the middle of which the Foetus is poised, by means of a bubble of an inch diameter, the Neck whereof is fastened to the Anus of the Foetus by a wire" (quoted in Edwards and Edwards 1959, 12).

Specific instructions for field preservation of specimens were not long in coming. In 1691, John Woodward published *Brief Instructions for Making Observations* which recommended drying specimens "unless you rather think fit to put some of the more rare, curious, and tender, into small Jars, filled with Rum, Brandy, or Spirit of Wine, which will keep them extremely well" (Edwards and Edwards 1959, 12). While writing a catalog of the museum in the Jardin des Plantes, Daubenton mentioned that although spirit of wine was used because it was plentiful, all fermented liquors seemed to be equally effective, certainly better than salt or alum, and he reported success with rum and observed that specimens sent from the Americas in undistilled rum were often yellowed. Daubenton claimed to know Ruysch's secret formula (mix 1 oz. 6 dr black pepper, 1/2 oz. small cardamomium seeds, and 1/2 oz. of cloves in 12 lbs. of spirit of wine; add 2 oz. of camphor, then distill the fluid down to 11 lbs. 3 oz. and dilute before use) (Edwards and Edwards 1959).

Most specimens were preserved in fluid in the field simply by dropping them—alive or dead—into a wooden cask or keg (or later into metal containers) of some form of alcohol. Glass vessels were rarely taken into the field for several reasons—glass was expensive and fragile; it was difficult to get an effective seal on glass containers (see the following discussion); and most collecting was done by individuals for whom collecting was not a major occupation, thus they did not carry much preserving equipment with them. Although there are some accounts of some glass vials and cylinders being used for field preparation, until the mid-1800s jars had to be sealed with bladders and disks made of wood, cork, or tin. After John Mason invented the screw-on zinc jar lid for threaded jars in 1858, glass containers became much more commonly used in the field.

In reference to containers for specimens, in 1748, Réaumur wrote that:

> This vessel may be a jar of glass, if it is only intended for receiving small birds; one may contain a great number of them, which you may put in at different times, accordingly as you get them, till it is quite full. Wooden barrels however are preferable to jars, as they are not liable to break in long journeys; there are to be had very small ones for smaller birds, and some large enough for others of the tallest size. The barrel is to have a hole large enough for passing the birds through: this hole can be no other than the bung widened, it will even be better placed in one of the heads. (Réaumur 1748, 309).

Réaumur went on to say that once bird specimens had been soaking in alcohol for eight days to six weeks (depending on size), the specimens could be removed from the fluid and packed in a box in chaff or straw for shipment. Réaumur had previously (1746) described two methods for sealing jars of specimens. If the jar had a stopper, mercury or thickened nut oil could be put into the alcohol and the container inverted to allow the additive to settle around the stopper. Réaumur also mentioned receiving a jar with a lid that formed a gutter around the neck for the application of mercury or oil. If the container did not have a stopper, Réaumur recommended a layer of thickened nut oil spread under the bladder (Edwards and Edwards 1959). Dumberton suggested sealing stoppered jars with tallow or a mastic (Edwards and Edwards 1959). Nicola (1771) recommended tying on the bladder, covering it with a layer of putty, then turning the container upside-down and setting it into a cup of melted tallow, or coating the neck of the container with oil and using an oil-soaked cork along with putty and a bladder (Edwards and Edwards 1959). In 1607, Sir Hugh Plat devised a system for the British navy for sealing jars of preserved food with a layer of olive oil (Shephard 2000). A description of how to preserve specimens written in 1773 recommended putting specimens in a cask of rum (or any other spirits) for preservation and shipment (Walker 1811).

A well-known guide for field collection and preservation first published in 1817 (*The Naturalist's Pocket Book*) recommended that small specimens of quadrupeds could be preserved in a mixture of alcohol and alum:

> The solution of alum should be made by pouring one quart of boiling water on eight ounces of alum and when cool, the water should be poured off, as some water will not hold that quantity in solution; and if a larger quantity be dissolved at any one time than is required, the water may be evaporated either over a fire, or by placing the solution within the influence of the sun, and the alum will be deposited in crystals, which only require being burnt over a common fire to be fit for using again. (Graves 1817, 54)

Graves recommended preparing burnt alum "by burning common alum over a fire till it looses its transparency" (1817, 55–56). His specific instructions for preserving large turtle and tortoise specimens included opening the skin at the throat, removing as many internal organs and as much muscle mass as possible, then submerging the specimen "in a strong solution of burnt Allum for a few hours, after which, it should be removed into a jar, and completely covered with the same kind of preparation as directed for preserving Quadrupeds in spirits, the feet and tail may also be preserved in the same way" (Graves 1817, 152 153). Smaller specimens "may be preserved entire, for these it will be proper to mix one half of pure spirit, with an equal quantity of Allum; the latter to be made by pouring three pints of boiling water over one pound of burnt Allum, which shou'd be suffered to cool previous to mixing it with the spirit. Frogs, the smaller kinds of lizards and Serpents, are best preserved in these last-mentioned preparations, as likewise the eggs of such as are not covered with a hard shell" (Graves 1817, 153). Fishes may be preserved "as for lizards" (Graves

1817, 184) and "vermes" and shell-less invertebrates in a mixture of one-third spirits and two-thirds burnt alum. Concerning quadrupeds in general, Graves wrote that:

> As it may be more convenient, at the time of procuring many of the smaller species of Quadrupeds, to preserve them entire, till a more convenient opportunity offers for stuffing them, they can be safely put into glass or earthen jars, or small casks filled one third spirit of wine, arrack, rum, or other spirit, and two thirds of a strong solution of burnt alum, care should be taken not to use, if avoidable, coloured spirit of any kind, as it frequently happens that when coloured, it will leave a stain on the lighter parts of the skin or fur, that cannot be removed. (Graves 1817, 54)

Instructions for field preservation extracted from a French publication and reprinted in Edinburgh in 1828 (Bory de Saint-Vincent 1828) recommend a variety of fluids, including brine, vinegar, oil, and acid solutions, with a comment that while alcohol is the most efficient fluid preservative, it causes alterations of colors and hardens tissues. The author notes that "It will be kept in mind, that alcohol becomes milky when diluted with common instead of distilled water" and recommends diluting alcohol to 16°–22° Baumé for preservation, and adding camphor (the Baumé scale is an expression of specific gravity such that 16°–22° Baumé = 0.959 to 0.921 specific gravity, or about 59–47 percent ethyl alcohol in water). Bory de Saint-Vincent recommended cleaning specimens before preservation, and cautioned that square glass bottles are a better shape for traveling than round bottles as they are easier to pack. Fishes should be packed in small casks with iron hoops, with an opening in one end that is six inches in diameter; specimens should be wrapped in cloth and have a wooden tag attached before they are inserted into the cask. When the cask is full, cotton or flax should be stuffed in to prevent the specimens from moving, then the cask should be sealed. In warm climates, Bory de Saint-Vincent recommended suspending the specimens in the preservative by cork floats to allow the preservative to penetrate from all sides. Incisions should be made in the abdomens of large animals. Vessels containing specimens should be sealed by soaking a cork in a mixture of three parts wax and one part tallow, then covered with a mastic of four parts pitch, one part sulfur, and one-half part tallow, melted together. Alternatively, a round glass plate may be placed over the container, then covered with a piece of oiled parchment, then lead, then parchment soaked in oil colored with lampblack and tied on with a cord (a cork stopper may be substituted for the glass plate). An alternative sealant can be made by melting common resin and yellow wax together, then adding red ochre a little at a time before boiling for seven or eight minutes, then adding turpentine; this solution is to be brushed over the cork and covered with a piece of linen.

A manual on specimen preservation written in 1831 stated that reptile and fish specimens should be "transported in spirit; we know of no better method than that proposed by M. Dufresne:—Wrap the fish in a cloth, and sew him up; fill a cask two-thirds with any sort of spirit, choosing that which is not too highly coloured,—its strength should be equal to 14 or 15 degrees of Baumé's aerometer: at the bottom of the cask put the largest fish, and decrease in size till you have filled it; the cask should be headed and hooped tight, and done over with pitch" (Anon. 1831, 79–80).

In 1852 Spencer Fullerton Baird (1823–1887) instructed those interested in preparing specimens for the U.S. National Museum that "The collector should have a small keg, jar, tin box, or other suitable vessel, partially filled with liquor, into which specimens may be thrown as collected. They should be alive, or as near it as possible when this is done, as besides the speedy and little painful death, the animal will be more apt to keep sound" (Baird 1852, 13). A few years later Templeton Hoffman Bean (1846–1916) noted that "Oak kegs, holding about 10 gallons each and provided with iron hoops, are capital containers for large fishes and they will stand the wear and tear of railway travel better than most other receptacles" (Bean 1881, 237).

In 1831, Leonhard Hess Stejneger (1851–1943) published directions for collecting and preserving specimens that recommended a one-gallon copper collecting can and a chest of copper tanks to fill with alcohol to preserve specimens in the field and then ship them home. In a review of pamphlets issued to collectors by the American Museum of Natural History, Myers (2000) reported that the earliest version he found (from 1919) recommended shipping specimens in metal cans (five-gallon square kerosene cans); later versions still recommended cans in addition to jars. For example, "For shipping or transportation, after the specimens have been in preservative for two weeks, the packages should be removed from the liquid, packed snugly in a can . . . the cans should then be sealed with solder" (Myers 2000, figure 43). The 1921 version of the *Handbook of Instructions for Collectors* issued in several editions by the British Museum (Natural History) in London stated that:

> The collector should provide himself with the best and strongest alcohol procurable. This can afterwards be diluted with water or with weaker spirit, until it is reduced to the degree of strength required in each case. As difficulty is often experienced in obtaining spirit while traveling, the collector is advised to take a sufficient supply with him. If an excise permit can be obtained, "industrial methylated spirit" may be recommended on account of its greater cheapness; but specimens do not keep so well in this liquid as in pure spirit, and it should not be used for the preservation of particularly valuable specimens, or of such as are likely to form the subject of anatomical investigation, in cases where pure spirit can be employed. In default of spirit the collector may use arrack, brandy, rum, or any other spirituous liquid which he can procure, provided it possesses the requisite strength. Any spirit which takes fire immediately on the application of a light, without having been previously warmed, is strong enough to be used for the preservation of animals. Spirit in which specimens are finally packed for transmission should contain 74–80 per cent. by volume of alcohol, or about 30–40° over Proof.
>
> It is sometimes found advisable to mix an emetic, crushed colocynths or other disagreeable ingredient with the spirit, in order to deter pilferers from drinking it. (Anon. 1921, 46–47)

The same publication later states that formaldehyde is not recommended as a preservative for amphibians, reptiles, or fish with soft and delicate skin (Anon. 1921).

LATER INSTRUCTIONS FOR PRESERVING SPECIMENS IN FLUIDS

Although quite a number of publications exist that describe various ways to preserve specimens in fluid (table 1), there are few detailed accounts of how these practices were actually carried out in the field by collectors. Joseph Richard Slevin (1881–1957) of the California Academy of Sciences wrote that "The specimens, having been thoroughly hardened and preserved, must be packed in jars, cans, or tanks, for storage or shipment" (Slevin 1927, 242). Similarly, a British Museum (Natural History) pamphlet of instructions for collectors recommended glass jars as well as "metal tanks, fitted with screw covers and encased in boxes" as being "more economical of space than boxes of jars and are stronger." The pamphlet further advises that "Well-galvinised milk cans of 3 to 10 gallons capacity are perhaps the most convenient of all field-containers" (Anon. 1953, 6). Improvements in the construction of metal containers is reflected in a later statement in the same publication that "Wooden casks are unsuitable, especially in hot countries, for the transport of specimens in spirit; they should not be employed except in cases of necessity or for packing large dry or salted specimens" (Anon. 1953, 10).

Clifford Hillhouse Pope (1899–1974) collected in China between 1921 and 1926 for the American Museum of Natural History Central Asiatic Expeditions (Adler 1989). In 1940, Pope published a popular account of his fieldwork, including a description of his working habits, writing that "My own ambition was to bring back nothing but flawless specimens" (Pope 1940, 184), while lamenting that "The preparation of specimens in the field required training and experience, although it was not highly skilled work" (Pope 1940, 188). To accomplish his goal, Pope hired one Chinese assistant to skin mammals and another to serve as his "injector" of fluid-preserved specimens. Pope provided the following detailed description of how his specimens were prepared:

> Only the largest reptiles were skinned: monitor lizards, large pythons, and other snakes more than five feet long. A snake was excoriated by making a short incision along the throat, severing the neck inside, and peeling the skin back to the tail, which, like the head, was left whole and attached to the skin. The body was discarded. Snakes not prepared in this way were preserved whole by injections of formaldehyde into the body cavity and tail and under the skin along each side. Very small individuals did not require the subcutaneous injections. Lizards in general were treated like snakes of medium or small size. All elongated reptiles were fitted into their containers immediately after preservation and before formaldehyde had made them hard and stiff.
>
> Turtles were easily preserved by inserting the injection needle into the body through one of the exposed fleshy parts, but of all reptiles no kind is really so hard to pack. Place turtles as you will, much space remains vacant. Relatively little is known about these creatures because most collectors simplify transportation by saving only their dried shells. Snakes, on the other hand, can be packed like so much rope, the small specimens filling the space inside the big ones.

> Frogs and salamanders were easily preserved because they have moist skin in life which, after death, is readily penetrated by formaldehyde. Only the large frogs had to be injected although I found it wise to see that a little preservative entered the body cavities of those medium in size. Fish more than a few inches long made better specimens when injected as thoroughly as amphibians, which they resemble in having permeable skin.
>
> The chief implement used in reptile and amphibian preparation was a large ear syringe with needles of all sizes. Rubber gloves are indispensible for many workers because formaldehyde hardens and cracks the skin, producing in some an almost incurable irritation of the skin. The worst pain that I suffered was not by some dread Chinese malady, but by lack of caution in the handling of formaldehyde. No bad effects may be felt from a single day's work with this preservative; after the third or fourth successive one, however, the skin cracks appear into which the liquid flows, causing sharp pain. The scalpels and scissors used on the mammals also served in the preparation of the cold-blooded animals. . . .
>
> The "wet" reptiles, amphibians, and fishes were first carefully packed in five-gallon Standard Oil tins. These were nailed in wooden boxes by twos after the tops had been soldered in place to keep the surplus formaldehyde from leaking out. (Pope 1940, 190–192)

FLUID-PRESERVED COLLECTIONS

Preserving the specimens in fluid was one thing, maintaining them was something else. Thomas Pole (1753–1829) noted in 1790 that "It is found to be attended with no small difficulty, to enclose wet preparations in glasses, so as to prevent effectually the evaporation of the spirits, which occasions very considerable trouble, and no small expense to keep a large anatomical collection in good order" (Pole 1790, 259). As noted earlier, there were many variations in sealing glass cylinders, but the basic technique was to cover the mouth of the jar with one or two pig or sheep bladders, which were then varnished and heavily coated with wax, lead, glaziers putty, or other substances, including bitumen and various jar cements (figure 1.2). Sometimes the bladders were put over a cork, glass plate, waxed paper, or piece of lead or tin foil. If done right, it was a fairly effective seal, except for the fact that it had to be destroyed to retrieve the specimen. After a discussion of commonly used techniques to bladder glass cylinders, Parsons's (1831) recommendation was:

> to cover the edge or rim of the glass, with fine soft Glazier's putty; then cover the mouth completely with a piece of flat common window glass, cut to the exact circumference of the rim of the vessel it is designed to cover; the putty should be laid on with great smoothness, so as to guard against any air-holes; the surfaces of the glass to come in contact with the putty, should be previously rubbed with a little boiled linseed oil, the glass cover should then be carefully applied, over this may be stretched a bladder or two, and bound as before described, covering the bulge of the vessel: when perfectly dry, the edge of the bladder round the bulge, should be cut even with a knife, and the bladder

> covered with a black varnish, to make it more secure, defend it from wet, and give it a neater appearance; or the glass vessels may be made with covers, fitted on with putty. (Parsons 1831, 122)

Parsons reported that he made his varnish by mixing lamp black with copel (oil varnish). However, acid-based preservatives will decompose the putty and the oil of turpentine will soften the putty, so instead "A lute may in such cases be made of finely powdered, and dry brick dust, four parts; common rosin, three parts; yellow wax with sufficient spirits of turpentine to mix it, one part" (Parsons 1831, 123). Other suggested sealants were made of gum mastic; chalk and egg white; or a mixture of common resin, red ochre, yellow wax, and oil of turpentine.

The traditional jar sealing method in the Netherlands was to use a cork stopper or "chiselled plate of schist," covered by a pig or sheep bladder, varnished, and then coated with a wax, such as beeswax, carnauba wax, or a mixture of colophonium and chalkpowder (van Dam 1997).

If done right, a bladder closure could be very effective—some of the fluid-preserved specimens featured in Seba's *Thesaurus* (1734) are still extant, having been carefully preserved in their original jars (sealed with distinctive red wax) in the collection of Lidth de Jeude and acquired in 1867 by the British Museum (Thomas 1892). On the other hand, many specimens were lost due to failures

Figure 1.2. Glass jar sealed with a bladder (note rupture in bladder).

of the containers or closures, particularly during the long and difficult journey from the field to the museum. For example, a student of Linnaeus, Peter Forsskal (1732–1763), collected fish in alcohol as naturalist on a Danish expedition to the Red Sea from 1761 to 1767. Forsskal and three other expedition members died on the trip. Unfortunately, there was a delay of two years before Forsskal's fishes could be shipped back to Denmark, and by then, all the specimens had been destroyed because the alcohol had evaporated (Davis 1993). The explorer and naturalist Alexander von Humboldt (1769–1859), recounting his 1799–1804 trip to the New World, remarked that "Sad experience taught us but too late, that from the sultry humidity of the climate, and the frequent falls of the beasts of burden, we could preserve neither the skins of animals hastily prepared, nor the fishes and reptiles placed in phials filled with alcohol" (von Humboldt 1889). Similarly, Manuel Almargo y Vega (1834–1895), a member of the 1864–1865 Comisión Científica del Pacífico, wrote in his diary that "Bien es verdad que la adquisición de los objetos de historia natural era sumamente difícil a causa de los escabroso del suelo y de la abundancia de las lluvias; pero si la adquisición era difícil, la conservación lo era más aún, pues la grande humedad que había nos obligada a inventar medios artificiales para secar nuestras plantas y aves disecadas" (It is true that the acquisition of objects of natural history was extremely difficult because of the rocky soil and the abundance of rainfall; but if the acquisition was difficult, the conservation was even more, as high humidity forced us to invent artificial means to dry our plants and stuffed birds) (Cabodevilla 1998, 103).

In his 1849 text on embalming, Jean Nicolas Gannal (1791–1852) summed up the available fluid preservatives by writing that:

> Alcoholic liquors are most generally used for the preservation of animal substances, if they are more costly, they are liable to fewer objections. Brandy, rum, tafea, are colored by a resinous substance which clouds their transparency, and which is liable to be deposited. The alcohol of cherries, of grain, of cider, or of wine, is preferred at present, which can be procured well rectified with distilled water, so as to obtain alcohol very limpid, marking from 22° to 30° of Baumè's areometre. (Gannal and Harlan 1840, 167). [22° to 30° Baumè is equivalent to 92 percent to 88 percent ethyl alcohol in water]

OTHER FLUID PRESERVATIVES

As mentioned previously, ethyl alcohol was not the only fluid preservative in use—brine and vinegar were quite common. In 1663, one year after Croone showed the first fluid-preserved dog embryos before the Royal Society, Robert Hooke (1635–1703) demonstrated "leeches in vinegar" to the same group (Huxley 2007, 101). Around 1753, in his *Instructio musei rerum naturalium*, Linnaeus recommended adding some sugar to the alcohol for the preservation of lower vertebrates (Holm 1957). In 1749, the Paris museum was still preserving specimens in rum (Cole 1944). In his 1840 book, Gannal suggested as alternatives to alcohol (which he considered to be very expensive) a solution of water (2 lbs.), alcohol (1 lb.), and aluminum sulfate (6 oz.); a solution proposed by the English naturalist George Graves consisting of alum (8 oz.), common water (1 lb.) and alcohol (1/2 lb.); or a solution promoted by Abbe Manesse of alum (1 lb.), nitre (1 lb.), sea salt (1 lb.), common water (4 lb.), and alcohol (1 lb.), with the caution that "all these liquors are inferior to spirits of wine, inasmuch as they are liable to freeze" (Gannal 1840, 180). Gannal reported that Chaussier recommended the use of deuto-chloride of mercury (mercuric bichloride) in distilled water (Gannal and Harlan 1840, 166).

As late as 1884 it was reported that Albert Günther (1830–1914) at the British Museum was using 56 percent ethanol to preserve fishes. Because of the cost of ethanol, many other preservatives were also in use, including Goadby's solution (composed of salt, alum or arsenic, mercuric chloride, and water) (Browne 1884). A chapter on "wet preparations" in a book on anatomical preparation techniques published in the early nineteenth century included as options alcohol; alcoholic solutions of a metallic, alkaline, or earthy salt; aqueous salt solutions; oils (particularly oil of turpentine); acids; acids and alcohol; alcohol and oils; and ammonia (Parsons 1831). Small amounts of elemental mercury were found in two alcoholic specimens of lizards collected and preserved prior to 1901 (Simmons et al. 2007). The mercury, presumably added to the preservative as mercuric chloride, had caused the specimens to noticeably darken. Mercuric chloride may be reduced to elemental mercury by the action of resistant heterotrophic bacteria, interactions with proteins, by reduction and oxidation reactions, or by interaction with an organic solvent such as ethanol.

As noted previously, vinegar has been used as a preservative, particularly for food, since at least the time of the Romans (Shephard 2000), and was a commonly used means of preservation in sixteenth-century England; the word pickle means to preserve in vinegar (Shephard 2000). Industrial methylated spirits (IMS), which is currently composed of ethyl alcohol denatured with methyl alcohol, went into duty-free production in Great Britain in 1855 (Derry and Williams 1960). According to Borodin (1930), during the mid-twentieth century there were two forms of denatured alcohol available under British regulations. Mineralized methylated spirits (100 gallons was composed of 90 percent ethyl alcohol, 9.5 parts wood-naphtha, 0.5 parts crude pyridine, 3/8 gallons mineral naphtha, and 1/4 ounce methyl violet dye) was sold for fuel and cleansing purposes. Industrial methylated spirits (95 percent ethanol and 5 percent wood tar) was sold for arts and industry purposes.

SUMMARY

Preservation of organic materials in fluids has a long and often obscure history. Table 1 provides a compilation of published instructions for preserving organic materials in fluids; table 2 lists some of the most significant advances in the preservation of natural history specimens in fluid, beginning with Croone's appearance before the Royal Society with his two preserved dog embryos in 1662. Two things that should be kept in mind while reviewing this history are the preponderance of chemicals used for preservation and the lack of rigorous testing of any of them. Because so many different chemical substances have been used outright or added to alcohol over the centuries, it is wise to use caution when working with older fluid-preserved specimens. Many of the chemicals reported in the literature are dangerous or downright toxic. The preservation of invertebrate specimens and anatomical preparations seem to have inspired a particularly diverse array of mixtures, brews, and solutions, as have (to a lesser extent) botanists trying to maintain the color of plants, but the bottom line is that any fluid-preserved specimen of unknown preservation history should be treated as potentially hazardous.

What is lacking in the literature (and in the practice of fluid preservation) are well-designed controlled studies that demonstrate which fluid preservatives really do work best for which organisms. Even such common practices as adding a little glycerin to alcohol preservatives (based on the assumption that should the preservative evaporate, the glycerin will form a coating on the specimens that will protect them longer from dehydration) have never been tested—we have, instead, proceeded for centuries to rely on oral tradition and anecdotal evidence to prepare the specimens that are then used in rigorous scientific work, with little attention to how preservation may have altered the specimens. Although the majority of the mixtures and additives that have been proposed since 1662 probably do not really improve the preservative qualities of ethyl alcohol, a few do merit further research, particularly the use of light oils (especially the proposal detailed in Chamberlin [1925]).

2

Fixation

Broadly speaking, fixation and preservation are processes used to prevent postmortem changes in tissues; beyond that, fixation is difficult to define. As one author stated, "the major objective of fixation has been to maintain excellent morphological features" (Eltoum et al. 2001a, 173). A standard histological definition of tissue fixation is "to differentiate the solid phase of the protoplasm from the aqueous phase, to change cell parts into materials that will remain insoluble during subsequent treatment . . . to protect cells from distortion and shrinkage . . . to improve the staining potential of tissue parts and to alter their refractive indices for better visibility" (Presnell and Schreibman 1997, 17). Another definition is that fixation is a process that preserves tissues "against microbial activity, osmotic damage and autolysis and should ensure, even after removal of the fixative, that the structure of the tissue remains an accurate representation of the structure of the living tissue" (Jones 1976, 155). From a microtechnique point of view, Galigher and Kozloff (1971) describe fixation as a process to prevent postmortem changes, produce optical differences (by raising the relative refractive indices of various structures), and prepare tissue for staining. Fixation is sometimes used as a synonym for hardening, which means that alcohol and other chemicals that do not form cross-links are considered fixatives.

In the post-formaldehyde histological literature, the consensus seems to be that true fixation means that covalent bonds, called cross-links, are formed to link the molecules composing the tissue (Fox and Benton 1987; Pearse 1980; Stoddard 1989). Stoddard (1989) differentiated between true fixatives (chemicals that form cross-links, such as aldehydes) and pseudo-fixatives (chemicals that harden tissue but do not form cross-links, such as alcohols). Aldehydes, and particularly formaldehyde, are used as fixatives because they form strong bonds (Stoddard 1989), but these bonds do great damage to DNA—by definition, all fixatives damage DNA. Stoddard's pseudo-fixative agents include methanol, ethanol, acetone, chloroform, and acetic acid. Pseudo-fixatives and denaturing agents do not form cross-links but unwind and disorder proteins, and alter the patterns of hydrogen bonding in tissues by removing water (Stoddard 1989). Many previously popular pseudo-fixatives were rapidly abandoned once formaldehyde became available. For example, "Alone, alcohol has only limited application as a fixing fluid" although methyl alcohol in concentrations of 80–100 percent is a good fixative for blood and tissue films and smears. Absolute or 100 percent alcohol penetrates tissues rather slowly (Lillie 1954, 35–36).

Preservation is not the same thing as fixation—fixation was developed as a histological technique. One of the earliest methods recorded for hardening tissues for histological study was used by Marcelo Malpighi (1628–1694) who, in 1666, boiled kidneys and brains, sliced them in thin sections, and coated the cut surface with ink to study the cortical blood vessels (Bracegirdle 1978; Jones 2001). About this same time, Robert Hooke (1635–1703) mounted specimens using olive oil as a hardener (Jones 2001). Antony van Leeuwenhoek (1632–1723) used spirits of wine (alcohol) to harden some of his specimens (Dobell 1932). One of the problems in treating tissues for histological study is that "Almost any method of fixation induces shrinkage, swelling, and hardening of tissues" (Eltoum et al. 2001a, 173), so histologists were always searching for better fixatives. In fact, the history of histology is filled with accounts of discoveries of new ways to stabilize, harden, fix, and stain tissues for study, and it is this quest that led to the discovery of the fixative powers of formaldehyde in 1893. Formaldehyde has been described as "the ideal fixative. However, in this context as in many others, the search for ideality can be equated in terms of reality with the search for the Philosophers Stone" (Jones 1973, 2). Eltoum et al. (2001b) cautioned that there is no single best fixative as different chemicals (as well as under- and over-fixation) affect immunochemistry, effectiveness of stains, and other histological processes. Galigher and Kozloff (1971, 64) include in their list of fixatives "alcohol; acetic acid; formalin; mercuric chloride; potassium dichromate (potassium bichromate); picric acid; nitric acid, and osmium tetroxide ('osmic acid')" and add that "a number of other chemicals are used, though much less frequently."

In any case, a major change in fluid preservation came about with the introduction of formaldehyde fixation.

FORMALDEHYDE

Aleksandr Butlerov (1828–1886) discovered formaldehyde (HCHO) in 1858, detecting it by its odor as he attempted the synthesis of methylene glycol (Seymour and Kauffman 1992). It wasn't until 1868 that August Wilhelm von Hofmann (1818–1892; his name is sometimes incorrectly spelled Hoffmann) figured out how to prepare an aqueous solution of formaldehyde by passing a mixture of methanol vapors and air over a heated platinum spiral (Hofmann 1867; Walker 1964).

The commercial manufacture of formaldehyde began in Germany shortly after the first process to mass-produce it was patented in 1889 (Fox et al. 1985; Seymour and

Kauffman 1992), first at the Mercklin and Lösekann chemical company in Seelze, near Hanover (Walker 1964). The chemical company Meister, Lucius and Brünig (in Höchst am Main) had begun producing formaldehyde for commercial use by 1891. Eager to find new uses for the chemical, company representatives asked a young physician named Ferdinand Blum (1865–1959) to test formaldehyde's antiseptic properties (Blum 1893a; Fox et al. 1985). Blum diluted the aqueous formaldehyde 1:9 with water and tested it on various bacterial species. In 1893, Blum noted that he had accidentally hardened the epidermis of his own fingers, thus discovering that formaldehyde was a good fixative (Blum 1893b). Ferdinand Blum and his father, Isaak Blum (1833–1903; publishing under the name J. Blum), published several key papers (F. Blum 1894a, 1894b, 1896, 1897; J. Blum 1894) describing the action of formaldehyde as an antiseptic and a fixative, using a 4 percent solution (the standard 1:9 laboratory dilution). As stated by Fox and Benton (1987, 199), "Laboratories have been using a 4 percent solution ever since, not because of any scientific evidence, but because that was the dilution of the commercial product used by Dr. Blum when he fixed his fingers." Blum's account of the discovery of the preservative properties of formaldehyde has been translated as follows:

> This slow and certain disinfectant appears to rest on a peculiar transformation of the organic material, by which the tissues . . . are changed from their . . . soft state into a harder and more resistant modification. I made this observation first on my own fingers when the epidermis became completely hardened during my work with formaldehyde. Afterwards I observed that a mouse, infected with anthrax, which had been opened and placed overnight in a formaldehyde solution, felt like a spirit preparation after a short time." (Edwards and Edwards 1959, 65–66)

However, in an unpublished manuscript, Kristian M. Gregersen (2008) has pointed out that Ferdinand Blum was not the first to propose that formaldehyde might be useful as a preservative. Gregersen's research revealed that in 1889, a French chemist named Jean-Auguste Trillat (1861–1944) applied for a patent in France and Germany on a method to produce formaldehyde, and on October 9, 1891, applied for a French patent on the use of formaldehyde as an antiseptic, a polymerizing agent, and speculated that formaldehyde could be used to preserve meat, fish, cadavers, and plants. In 1892, Trillat published a paper on formaldehyde as an antiseptic, further speculating that it could be used to preserve food (animal and plant tissues) (Trillat 1892). Gregersen's detailed research into the history of formaldehyde identified several other overlooked papers regarding the antiseptic actions of formaldehyde that included comments about it hardening skin (e.g., Stahl 1893), prior to Ferdinand Blum's 1893 publication. Nevertheless, most of the literature on the history of formaldehyde gives Ferdinand Blum full credit for the discovery without reference to prior publications by other workers.

In October 1893, Isaak Blum experimented with 10 percent and 20 percent formaldehyde as a preservative for a fish, a lizard, a snake, a frog, a white mouse, snails, fruits, flowers, and a banana (J. Blum 1894). Not long afterward, formaldehyde was tested and advocated as a botanical fixative (Penzig 1894). The elder Blum later preserved human embryos and a near-term human fetus (J. Blum 1894; Kenyon 1895).

ORIGIN OF THE NAMES FORMALDEHYDE, FORMOL, AND FORMALIN

When formaldehyde first went into commercial production, various companies called it by different names. English chemical companies originally called the new substance formic aldehyde. One German company invented the name formol; another company coined formalin, which was the name adopted by the American manufacturers (Fox et al. 1985). However, there was some dispute about the spelling as the word formol came into common use. For example, Baker (1944) refers to a "formal-calcium" solution used in his work on the Golgi element, noting that "The word is often spelled formol, but the standard abbreviation for aldehydes is -al (e.g., chloral, furfural). The ending in -ol is best reserved for alcohols and phenols. There is no more reason for writing formol than there would be for writing phenal" (Baker 1944, 7). Ferdinand Blum also weighed in on the use of the common name (Blum 1896, 1897), as did Kenyon (1897). Stroud (1897) opined that "the term *Formaldehyde* is a cumbersome one" and that "*Formalin* is a term which has no meaning" (Stroud 1897a, 92), preferring the use of "*Formal*." Nevertheless, formol prevailed in common usage. Formaldehyde is also known as methanol formic aldehyde (Browning 1965).

The concentration of commercial formaldehyde is often misreported or misunderstood; for example, one publication referred to it as "a solution of 33 percent to 50 percent gaseous formaldehyde in water with a small amount of methanol as a stabilizer" (Burroughs et al. 2006, 49). In common use, formol and formalin may refer to either a saturated solution of formaldehyde gas in water, or a dilution of the saturated solution—some authors restrict the use of the term formaldehyde to the gas, and use formol or formalin for a solution of the gas in water. A saturated solution of formaldehyde gas in water may be listed as either a 40 percent aqueous solution or a 37 percent aqueous solution—both are, in fact, the same thing. Producers used to give the concentration in grams per 100 cc of aqueous solution (weight/volume), although most producers now use grams per 100 grams of solution (weight/weight). Both the 40 percent and 37 percent aqueous solution are 100 ml of formaldehyde in 900 ml of water on a volume basis, which is the concentration that results from bubbling formaldehyde gas through water until no more will go into solution. When one part formaldehyde is diluted with nine parts water, the result is a solution commonly called "10 percent formalin," which is really 3.7 or 4.0 percent formaldehyde gas in water.

COMMERCIAL FORMALDEHYDE

The most commonly used method of producing formaldehyde commercially is by oxidizing methanol, but sometimes methods based on the oxidation of hydrocarbon gases are used (Steedman 1976b). The oxidation of methanol produces formaldehyde containing traces of methanol and formic acid; hydrocarbon oxidation produces formaldehyde containing traces of other aldehydes, alcohols, ketones, fatty acids, and other chemicals that are removed through other chemical processes.

Formaldehyde undergoes virtually spontaneous condensation reactions with itself when stored in a concentrated form. To prevent polymerization and the formation of formic acid, 8 percent methanol is generally added to commercial formaldehyde in summer, 13 percent in winter (Fox and Benton 1987; Seymour and Kauffman 1992). When a commercial formaldehyde solution is diluted with water, the protective effect of the methanol is lost, and the pH of formaldehyde solutions will rapidly drop, which is why it is important to buffer formaldehyde solutions (Fox and Benton 1987).

The recommended storage temperature for formaldehyde is in the range of 10–25°C, with the caution that paraformaldehyde haze and deposits will form in the –5 to +5°C range (Steedman 1976b):

> A lowering of temperature will reduce the amount of formaldehyde and increase the proportion of paraformaldehyde. Storage in cold conditions, or in refrigerators, may therefore lead to deposits of this polymer. The first indication of its presence is a slight turbidity of the solution, which is then followed by a white deposit which sinks to the bottom of the container. Warming the solution, or adding small amounts of NaOH will assist the depolymerization of the paraformaldehyde and after this treatment a clear solution of formaldehyde will be produced. (Steedman 1976b, 104)

FORMALDEHYDE AS A FIXATIVE AND PRESERVATIVE

The use of formaldehyde quickly became popular with many workers. In 1896, Blum reported that more than fifty papers had already been published on formaldehyde as a fixative (Blum 1896). Nevertheless, the use of formaldehyde began generating controversy almost from the first. In a report of a discussion at a medical society meeting in 1897, a Dr. Mandelbaum claimed he could maintain good colors in anatomical specimens by exposing fresh tissues to formalin vapor for twenty-four hours in a large museum jar, followed by submergence in 95 percent alcohol for twenty-four to forty-eight hours and storage in a strong solution of potassium acetate in one part glycerin and one part water. Mandelbaum was quoted as stating that "The formalin at first bleaches the specimen slightly, but the alcohol restores the color and brings out the blood perfectly. After this the specimen does not undergo any further change" (Anon. 1897, 248). Dr. J. S. Ely then claimed that he had used this technique for intestinal tissue but that the tissues began to deteriorate in the glycerin, which also rendered the specimens transparent, and that although the specimens looked good at first, within a year or two they would all fade, and Dr. Larkin claimed that the potassium acetate made the specimens brittle.

Formaldehyde went into limited production in the United States in 1901, but it wasn't until 1904 that regular commercial production got underway at the Heyden Chemical Company (Garfield, New Jersey) and Perth Amboy Chemical Works (Philadelphia, Pennsylvania) (Walker 1964). However, we know that formaldehyde (imported from Germany) was available in the United States by 1894, because Alleger (1894) and Fish (1895) reported on their fixation experiments with it. Fish

(1895) recorded that formaldehyde could be purchased for US$2.00/gallon, twice as high as the price in Germany, but much less than the cost of alcohol.

One of the first reports on the use of formaldehyde as a botanical fixative was presented at a meeting of the Brooklyn Botanical Club in 1894 (Newcombe 1894). Other early reports include a short article in *Natural Science* in 1895 (Anon. 1896) and notes by Hornell (1895), Kellicott (1896), Parker and Floyd (1896), and Redenbaugh (1895). Lucas (1896) reported on a presentation by Smithsonian biologist Leonhard Stejneger at the annual meeting of the Biological Society of Washington on March 21, 1896, entitled "Formalin in the Field," during which Stejneger showed formaldehyde-preserved specimens of plants, insects, fishes, and reptiles. In his talk, Stejneger mentioned that formaldehyde was inexpensive and compact to carry in the field, and had "the property of preserving specimens which could not be kept in alcohol, or could not be kept in such good condition" (Lucas 1896, 604). Gregersen's (2008) research provides a detailed list and analysis of reports on early formaldehyde use around the world.

Formaldehyde was quickly adopted by the American funeral industry, and between 1914 and 1939, the production of formaldehyde in the United States increased by sixteen-fold largely due to the growth of the synthetic resin industry, particularly the production of Bakelite (Walker 1964). A pamphlet published by the Museum of Funeral Customs (Springfield, Illinois) reproduced the recipe for an embalming fluid named "Formula No. 3 (1905)," recommended by the National Funeral Directors Association, which consisted of 11 lbs. formaldehyde, 4 lbs. glycerin, 2.5 lbs. sodium borate, 1 lb. boric acid, 2.5 lbs. sodium nitrate, 1 fl. oz. eosin (1 percent solution), and water to make 10 gal. An advertisement for the Egyptian Chemical Company (Boston, Massachusetts) promoted an embalming fluid containing both alcohol and formaldehyde called "Gold Crown Alcofom," patented in 1904.

A brief description of how formaldehyde was manufactured from methanol ("wood spirit") at this time was provided by Hiscox and Sloan (1939):

> This extremely poisonous preservative is obtained by passing the vapors of wood spirit, in the presence of air, over copper heated to redness. The essential parts of the apparatus employed are a metal chamber into which 4 parallel copper tubes or oxidizers discharge by a common exit tube. This chamber is fitted with inspection apertures, through which the course of the process may be watched and controlled. The wood spirit, stored in a reservoir, falls into a mixer where it is volatilized, and intimately mixed with air from a chamber which is connected with a force pump. The gases after traversing the oxidizer are led into a condensing coil, and the crude formaldehyde is discharged into the receiver beneath.
>
> The small amount of uncondensed gas is then led through a series of two washers. The "formol" thus obtained is a mixture of water, methyl alcohol, and 30 to 40 per cent of formaldehyde. It is rectified in a still, by which the free methyl alcohol is removed and pure formol is obtained, containing 40 per cent of formaldehyde, chiefly in the form of acetal. Rectification must not be pushed too far, otherwise the formaldehyde may become polymerized into trioxmethylene. When once oxidation starts, the heat generated is sufficient to keep the oxidizers red hot, so that the process works practically automatically. (Hiscox and Sloan 1939, 362)

In 1899, one researcher remarked on how widespread the use of formaldehyde was becoming but "how little has been written" about it (Drowne 1899, 179), while noting that he himself had started to use it in 1895, not because he needed a fixative, but because it was much less expensive and easier to carry in the field than alcohol. A pint of formaldehyde at that time cost 55 cents; in a 3 percent solution this made a little more than 4 gal. By comparison, alcohol was $2.00/gal. when diluted to 70 percent. In 1900, the director of the Jersey Biological Station wrote about his five years of experience using formaldehyde, stating that "the glamour of novelty has dimmed, and the enthusiasm with which its advent was hailed in many quarters has been sobered by the passage of time and by the stern teaching of experience. The opinion that Formalin would, in the main, displace Alcohol from its honoured position as the one reliable preservative has disappeared" (Hornell 1900, 86). After discussing the efficacy of formaldehyde preservation on various taxa, Hornell nevertheless concluded that "when we consider this high preserving quality, its greater relative cheapness to alcohol—a gallon of strong solution costs on an average less than one shilling—and its readiness to mix with water, spirit, and other fluids in any proportion, we may account for the introduction of formalin as one of the greatest services ever rendered to the naturalist in remote lands" (Hornell 1900, 89). Thus we find that not only did biologists begin using formaldehyde because it was cheap and easy rather than because they needed a fixative, the mythology of formaldehyde started almost immediately. For example, Drowne wrote that: "it does not make much difference whether formalin is mixed with fresh or salt water" (Drowne 1899, 180). Of course, we now know that buffering formaldehyde solutions correctly is important to both the fixation and the long-term preservation of specimens.

Not everyone immediately accepted the use of formaldehyde. The 1922 edition of the British Museum's *Handbook of Instructions for Collectors* stated that, "The use of formalin as a preservative is not recommended for Reptiles, Batrachians, or Fishes. Batrachians and Fishes, which have a soft and delicate skin, are in particular injuriously affected by this substance" (Anon. 1922, 47). It should be noted that the same publication also recommended field storage and shipping of specimens in zinc containers, apparently without realizing that formaldehyde corrodes metals and the corrosion product would discolor the specimens.

Less than thirty years after Blum's pivotal publication of 1893, formaldehyde had replaced the use of most other fixatives and pseudo-fixatives, including alcohol, for most histological preparations (Mallory and Wright 1918). During this time, formaldehyde was increasingly recommended for field use with little understanding of its chemical actions on the tissues. So let us pause here in the chronology and ask a rather basic question—why fixation? Why was fluid preservation changed from a one-step preservation process to a two-step process of fixation and preservation? At this juncture, a quantum leap was made in how specimens were chemically preserved, and the change was not based on science, but on cost and ease of use of a fixative. Prior to 1893, fixation was something histologists did for purposes of making histological preparations. Biologists began using fixatives not because they wanted to make histological preparations, but because a fixative was cheaper and easier to use than a pseudo-fixative (alcohol). Is an initial fixation step necessary to preserve specimens permanently in fluid preservative? Although we have more than 350 years of experience preserving specimens in ethyl alcohol without fixatives, we have little more than one hundred years of experience fixing specimens in formaldehyde before they are preserved in alcohol (and we now know that formaldehyde degrades the quality of DNA in tissues). Perhaps most significantly, the change from preservation directly in alcohol to a process of formaldehyde fixation followed by alcohol preservation was made despite the fact that no controlled comparative studies were conducted to demonstrate whether or not formaldehyde fixation prolongs the useful life of a preserved specimen.

The immediate purpose of fixation (and preservation) is to stop the physical and chemical changes that occur upon the death of tissue, to preserve its gross form and appearance (Drury and Wallington 1980), and to sterilize the specimen (Stoddard 1989), which means a chemical treatment of the tissue that prevents autolysis (the degradation of proteins into amino acids) and coagulates the cell contents into insoluble substances. The purpose of fixation (as opposed to direct preservation using a chemical that does not form cross-links) is to prepare the tissues for some subsequent histological treatment, yet very few fluid-preserved specimens are used in this way. Furthermore, there are a variety of fixatives available, so is formaldehyde really the best one to use? As Galigher and Kozloff (1971, 64) have cautioned, "A fixing agent which preserves certain structures admirably may injure or destroy others." Furthermore, we really don't know much about how even the most common fixatives work in the tissues. As summed up by Hopwood (1973, 76), "Our poor understanding of fixation is in many ways rather surprising as a huge system of knowledge has been built up which depends on this ill-comprehended step. This applies both to normal and to pathological tissues." Much of the data we do have on the effects of fixatives (and preservatives) on specimens is contradictory—for example, Hendrickx et al. (2003) found that fixation and storage in 4 percent formaldehyde solutions decreased concentrations of copper while increasing the concentrations of cadmium, lead, and zinc in the tissues of isopods, but Gibbs et al. (1974) found increased levels of cadmium and copper and lower concentrations of mercury in fish tissues, with other effects if the specimens were transferred to ethyl or isopropyl alcohol.

Tissue is composed of cells, the extracellular matrix, and tissue fluid (Stoddard 1989). Tissues vary in "the ratio of the volume of their cellular component to that of their matrix" (Stoddard 1989, 2; see table 3, tissue matrix types). The texture, form, and mechanical properties of the tissue depend on cell-to-cell adhesion, cell-to-matrix adhesion, and mutual adhesion of the matrix macromolecules (Stoddard 1989). The connection between the cell and the matrix requires the consumption of energy for its continuous renewal, so at cell death, a disruption in this connection begins. The rate of breakdown varies with the tissue. The purpose of fixation is to prevent this breakdown, hence the fixative must reach cells quickly, which is why perfusion and injection are the superior techniques to reach inaccessible areas of tissue (Stoddard 1989, 5), and why several field workers have cautioned that it is important to keep the interval between death and fixation or preservation as short as possible (e.g., Berger 1955).

After cell death, several things begin to happen—any bacteria present in the specimen begin to multiply, intracellular enzymes are broken into amino acids by autolysis, the resulting amino acids diffuse out of the cells rendering the cell proteins no longer coagulable by chemical reagents (Presnell and Schreibman 1997), and enzymes, bile, and gastrointestinal acids start the lysis of tissues. In addition, there can be damaging osmotic effects if fixation is too slow or is incomplete (osmotic pressure refers to the pressure produced by a solution in a space enclosed by a differentially permeable membrane; when there is a higher concentration of water on one side of a cell membrane than another, water will flow across the cell membrane until balance is achieved). As the fixative inhibits a cell's membrane pumps and alters the membrane permeability, ion-balance is disrupted and the cytoskeleton may be damaged—if a net inflow of water occurs as the cytoskeleton fails, the cell will swell and may rupture. This is an important reason for carrying out fixation in buffered, isotonic salt solutions (the cytoskeleton refers to the protein fibers composing the structural framework of a cell; isotonic means having the same osmotic pressure as the fluid phase of a cell or tissue; osmotic shock means the bursting of cells suspended in a dilute salt solution) (Stoddard 1989). As stated by Parker (1963, 1453), "The problem of gain or loss in weight is one of osmotic exchange of water and the amount of initial change (excluding size effect) may largely depend upon the state of osmoregulation or amount of body water at the time of original measurement." For example, when marine organisms are placed in formaldehyde fixative, they may be subjected to an osmotic pressure that is double that of the seawater they came from (Thibault-Botha and Bowen 2004; Steedman 1976c). Stoddard (1989, 8) offered the further caution that "It is not desirable simply to use a high concentration of fixative, as that simply fixes the outside of the specimen very quickly so reducing its permeability and preventing rapid and uniform fixation of its interior." A summary of the literature describing biomass changes that result from fixation and preservation for invertebrates is provided in tables 13 and 14; for vertebrates in tables 15 and 16. Although the goal of several studies of biomass changes in fluid fixatives and preservatives was to establish a standard for correcting length or weight measurements in order to more accurately compare preserved and fresh specimens, changes of body proportions differ depending on the organism and numerous other variables, making general predictions concerning the long-term effects of these changes impossible.

We are also cautioned that "much work has been done on the osmotic properties of fixatives, particularly formaldehyde fixatives, but little has been done on long-term formaldehyde preserving fluids" (Steedman 1976c, 187) and much of the work that has been done has shown contradictory results. For example, "It has been shown that fixation may be improved by the use of neutral additives such as sodium chloride and sucrose; that hyper-osmotic fixatives may shrink or swell tissues, and that hypo-osmotic fixative fluids may do the same" (Steedman 1976c, 187).

FIXATIVE PH RANGE

Many common histological fixatives are extremely acidic (Freeman et al. 1955); this is particularly a problem with formaldehyde because it is "the simplest member of a homologous series—the saturated aliphatic aldehydes" (Jones 1976, 155) and the most reactive member of the series as well. The acidic properties of fixatives are very important for good fixation:

> The effect of pH on formaldehyde fixation is profound and should be considered during fixation. In an acid environment, the primary amine target groups (–NH_2) attract hydrogen ions (–NH_3^+) and become unreactive to the hydrated formaldehyde (methylene hydrate or methylene glycol). Similarly carboxyl groups $(COO)^-$ will lose their charges (–COOH) which may also affect protein structure. The hydroxyl groups of alcohols (–OH), including serine and threonine, may become less reactive in an acid environment. Because the major methylene cross links are between lysine and the free amino group on side chains, the extent of cross-linking would be expected to be reduced in unbuffered 4 percent formaldehyde, which is slightly acidic. (Eltoum et al. 2001a, 179)

The pH of pure, aqueous formaldehyde is about pH 2.5 to 3.5 due to the presence of "traces of formic acid" (Walker 1964, 106) either as an impurity (remaining from manufacture) or as a result of the oxidation of part of the formaldehyde. When diluted with water, formaldehyde solutions are still acidic, in the range of pH 3.0 to 4.6. Formaldehyde reacts quickly and easily with oxygen in the air to oxidize into formic acid, so formaldehyde fixatives must be neutralized or buffered. This can be tricky, because decalcification of some types of tissues may start at pH 6.4 and below, but the clearing (loss of opacity) of some types of tissues can start at pH 7.0 and above (Lavenberg et al. 1984). Furthermore, too much buffering agent (e.g., raising the pH to 7.0 or above) affects the osmotic strength for the solution, because a strong salt solution shrinks tissues (by extracting water from them), while a weak salt solution may allow too much water to rush into the cells, causing them to burst (Fox and Benton 1987). For this reason, the buffer for formaldehyde must be selected carefully. Most buffers are not able to maintain a stable pH in the long term (Fox and Benton 1987), particularly because of the presence of oxygen—Steedman (1976b) cautions that thin-walled plastic containers may allow enough oxygen to diffuse through to lower the pH of formaldehyde solutions.

One has to use caution in adopting buffers recommended for histology or other applications for the fixation of specimens intended to be maintained permanently in fluid preservatives. For example, Steedman warned that "Much work has been done in the leather industry but, while this advanced general chemical knowledge of fixation, it was concerned principally with white fibrous tissue. As few animals within the invertebrate phyla have any of this tissue in large quantities, much of this work had little practical application to marine zooplankton" (Steedman 1976b, 116). Egginton and Cordiner (1995) compared the effects of six standard fixatives (all gluteraldehyde- or formaldehyde-based with various buffering agents) and found that each different fixative and buffer combination produced a different quality of fixation of fish muscle tissues, including differences in tissue volume and contrast; most of the differences in fixation were attributed to aldehyde concentration and osmolarity of the fixative. A review of the literature on formaldehyde fixation for histochemistry pointed out that "The pH of the fixing solution is a controlling factor in the reaction of formaldehyde with proteins in aqueous media" (Puchtler and Meloan 1985, 203), and that fixation proceeds slowly in unbuffered formaldehyde, with maximum tissue fixation occurring in the range of pH 4 to 5.5. However, it has also been suggested that the optimum pH for a fixative should match the pH of the specimens' blood, body fluids, and muscle tissue in life (usually around pH 7.2); Steedman (1976b) argued that in death the pH of proteins is around pH 5.5 to 6.8, so this is unnecessary.

Some buffers, such as borax, are effective initially but fail in the long term and thus are unacceptable for use in fluid preservation (Hughes and Cosgrove 1990; Taylor 1977). In addition to problems with acidity, the use of unbuffered formalin results in rapid but uneven fixation, and causes osmotic damage (Stoddard 1989). Such traditional buffers as calcium carbonate, magnesium carbonate, and marble chips may form hard crystalline deposits in muscle and other tissues (Quay 1974). Figure 2.1 shows a specimen that was fixed in a formaldehyde solution that contained an excess of sodium borate buffer, and was then transferred to an alcohol preservative, causing the excess buffer to form salts as it exuded from the specimen. Smith (1947) cautioned against using buffered formaldehyde mixtures that were too alkaline and reviewed several recommended buffers, settling on hexamine as the best choice; however, we now know that hexamine tends to clear specimens. Motoda et al. (1976) found that borax, trioxane, magnesium chloride, and hexamine buffers in formaldehyde all caused clearing of marine zooplankton, and that pigments faded with hexamine (the authors recommended the use of formaldehyde and seawater without any additional buffering agents). Markle (1984) compared marble chips to a phosphate buffer in a 3 percent solution for the storage of larval fishes and found that the phosphate buffer better regulated the pH; Markle (1984) included a table of recommended amounts of the sodium phosphate monohydrate and anhydrous sodium phosphate dibasic to regulate the pH within the desired range pH 5.9 to 7.7. Formaldehyde buffered with calcium carbonate or magnesium carbonate "drawn from the storage reservoir and used for fixation very promptly becomes more acid as the tissue is fixed" (Lillie 1954, 32). Carter (1997) addressed the problem of maintaining a stable pH in formaldehyde solutions that had been prepared with seawater; for seawater/formaldehyde mixes, he concluded that the most effective buffering agent was 0.05M sodium-ß-glycerophosphate.

Leonard Stejneger (1851–1943), of the Smithsonian Institution, wrote an influential paper on field preservation in 1891 (Stejneger 1891), adding a supplement in 1911 on the use of formaldehyde (Stejneger 1911). In 1900, Stejneger had used formaldehyde when he stopped in Cuba on his way home from collecting in Puerto Rico. Stejneger had not planned to collect in Cuba and therefore had no preservative solution with him, but changed his mind and noted that "the resident port physician very kindly supplied us with some formalin" (Stejneger 1917, 259). In his 1911 preservation technique supplement, Stejneger recommended diluting formaldehyde with twenty times as much water, and cautioned that "The solution should be tested from time to time for acidity and, if necessary, neutralized by the addition of a small quantity of bicarbonate of soda" (Stejneger 1911, 1). Stejneger's caution to test the pH of formaldehyde solutions was widely ignored in subsequent practice and in published preservation instructions, despite a plethora of recipes for neutralization of formaldehyde appearing in print. As can be seen from table 1, the majority of guides to field fixation ignore the issue entirely, or recommend an inadequate buffer. It is worth noting that pH indicator strips have been available for a long time, dating back to Robert Boyle (1627–1691), who "prepared the first chemical indicator, in the form of a paper impregnated with syrup of violets, which would turn red in acid and green in alkaline fluids" (Pilkington 1959, 156).

Currently, the recommended buffering agent for the best control of pH and long-term stability of formaldehyde solutions is two salts, a monobasic sodium phosphate

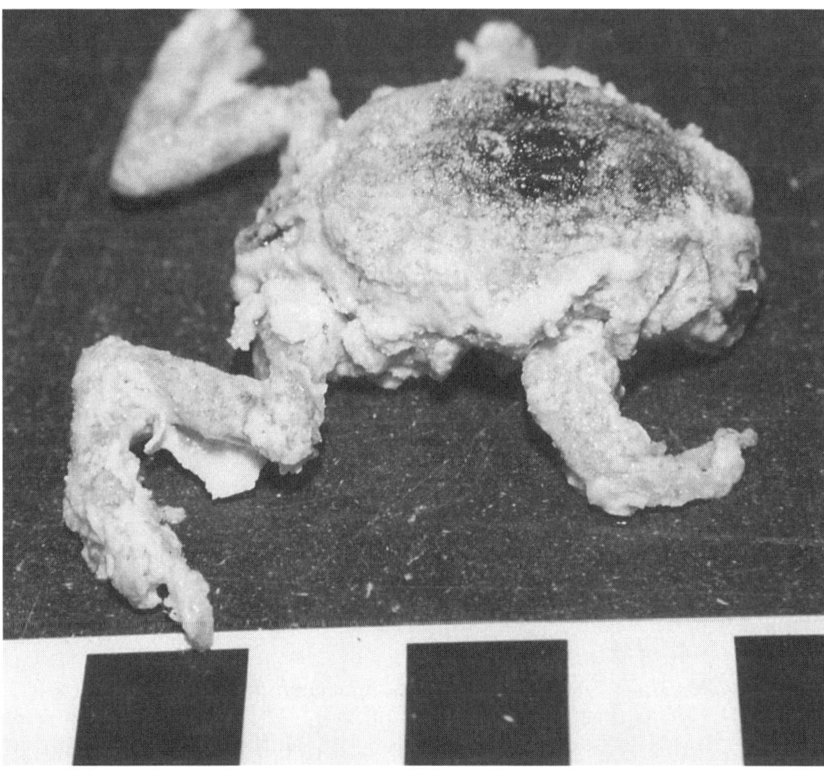

Figure 2.1. Specimen fixed in formaldehyde with excess sodium borate buffer, after transfer to 70 percent alcohol preservative.

monohydrate and a dibasic sodium phosphate anhydrate (Drury and Wallington 1980; Lillie 1954). The use of 4 g monohydrated acid sodium phosphate and 6.5 g anhydrous di-sodium phosphate per liter of one part formaldehyde with nine parts water should result in a pH of approximately 7.0 (Lillie 1954). However, in seawater, phosphate buffers produce insoluble phosphates and thus cannot be used for marine buffers (Markle 1984).

Kotrba and Golbig (2009) have proposed maintaining a neutral pH in formaldehyde fixative and preservative solutions by using a substrate-bound ion exchange material in the form of ampholytes within the specimen container.

NARCOTIZATION

Narcotization is necessary to relax contractile invertebrates so that they may be extracted from their shells or their body parts may be extended. For example, sea anemones are particularly difficult to work with because of the structure of their nervous system, and some invertebrates are more susceptible to narcotization at some points in their circadian cycle than others (Simon Moore 1999). Narcotizing agents may be introduced to the water the specimen is in prior to euthanization, or may be a component of the killing solution or fixative (Wagstaffe and Fidler 1955). In most cases of contractile invertebrates, failure to properly narcotize the specimen will result in a useless preparation. Common narcotizing agents include menthol and chloral hydrate; some narcotizing agents may cause degradation of DNA in tissues. Because of the importance of narcotization in preserving invertebrates and the possibility that some chemicals used in the process may affect fixation and preservation, details of recommended narcotization agents are included in tables 1 and 5.

FIXATIVES FOR BOTANICAL SPECIMENS

As mentioned earlier, shortly after its discovery as a fixative, formaldehyde was also recommended for botanical specimens. In 1924, Conant reported that plant specimens could be fixed in a solution of formaldehyde and glacial acetic acid, then preserved in strong alcohol. Several botanical fluid fixatives and preservatives were popularized by Kew Gardens in *The Herbarium Handbook*. The original composition of the well-known "Kew mixture" of 1989 was 53 percent industrial methylated spirit (IMS, or ethyl alcohol containing 9 percent water and 2–4 percent methanol), 37 percent water, 5 percent formalin ("dilute formaldehyde"), and 5 percent glycerol (Forman and Bridson 1989). A note in the second edition (1992) stated that due to controls on hazardous chemicals, the original formula at Kew had been withdrawn and was replaced by a mixture of "70% alcohol (ethanol, or ethanol + trace of methanol), 29% water and 1% glycerol. If using 90% strength Industrial Methylated Spirit (IMS) the mixture is 78% IMS, 21% water and 1% glycerol. With different strengths of IMS the proportions must be adjusted accordingly" (Forman and Bridson 1989, 210). In the third edition of *The Herbarium Handbook* (Bridson and Forman 1998) the recipe for Kew mixture was given as ten parts IMS, one part formalin, one part glycerol, and eight parts water. Other fixatives and preservatives recommended in various editions of *The Herbarium Handbook* included FAA (formaldehyde, industrial methylated spirit, and acetic acid) with the caution that this solution "makes specimens brittle" (Forman and Bridson 1989, 72); Copenhagen mixture (a combination of ten parts ethyl alcohol with one part glycerol, and eight parts deionized water); and various alcohol-glycerol mixes (typically 5–10 percent glycerin).

To preserve the natural colors of Monotropa and other plants, Nieuwland and Slavin (1928) recommended submerging the plants in alcohol in a test tube; adding 0.5 g of sodium sulfite, and then 0.5 cc of concentrated hydrochloric acid. Additional fixatives for botanical specimens are listed in table 1; see also the discussion of botanical preservatives in chapter 3.

TEMPERATURE, TIME, AND RATES OF PENETRATION OF FIXATIVES

The sooner after death that a specimen is fixed, the better the results will be, both because fixatives and preservatives penetrate faster in fresh tissue, and because the deterioration of the tissues begins immediately after cell death. For these reasons, the penetration rate of a fixative or preservative into the tissues is critical to making a good, well-preserved specimen. The duration of time in fixative is another important factor, so how long should tissues be in a fixative, and at what temperature? A review of recommended preservation practices (table 1) shows a wide range of variation; very few of them are based on experimental evidence or analysis of the long-term effects on the specimens, therefore these recommendations should be used with caution. Specimens should be exposed to the fixative for a sufficient length of time to allow it to penetrate and fix the tissues—among other factors, this length of time will depend on the type of tissue, thickness of the tissue block, interval between death and fixation, and fixation temperature as well.

Fixation and penetration proceed faster at warmer temperatures, but warmer temperatures also make the tissues more vulnerable to bacterial attack and increase the rate of autolysis of proteins that have not yet been reached by the fixative. One study showed that a 10 percent formalin solution fixed tissue samples adequately in 48 hrs. at 20–25°C, or in 24 hrs. at 35°C, but concluded that because autolysis is also speeded up at higher temperatures, it is better to fix at lower temperatures (Lillie 1954). In another study, it was also pointed out that higher temperatures (37°C as opposed to 25°C) fix tissue faster (Eltoum et al. 2001a), so some workers have recommended fixing specimens at higher temperatures (some even place covered trays of specimens in fixative in direct sunlight during field fixation); but as previously noted, autolysis of proteins is also accelerated at 37°C (Drury and Wallington 1980).

The rate of diffusion of the fixative or preservative through the tissues, in general, depends on the temperature and tissue type. In his review of the literature on different proteins and the take-up of formaldehyde, Steedman (1976b) reported that 100 g of gelatin will react with 4.0–4.8 g of formaldehyde, but 100 g of casein will react with only 0.6–2.5 g of formaldehyde. In general, for most fixatives "the time of fixation is approximately equal to the $(\text{distance})^2$ the fixative must penetrate. Most fixatives such as neutral buffered formaldehyde will penetrate tissue to the depth of approximately 1 mm in 1 hr" (Eltoum et al. 2001a, 180). The penetration rate of formaldehyde is described by pathologists as "fairly rapid" while the penetration rate of gluteraldehyde is described as "slow" (Kiernan 1990, 26–27). Gluteraldehyde penetrates at approximately 0.2 mm per hour (Hopwood 1967). Under-fixation and over-fixation both shorten the long-term useful life of specimens. Under-fixation results in tissue that degrades over time; over-fixation may leave tissue more susceptible to deterioration and may produce unwanted side-effects such as discoloration (e.g., Stuart 1995).

Various recipes have been suggested for mixtures that would speed up the penetration of the fixative, but these generally have serious drawbacks. For example, "Substitution of alcohol as the diluent of formaldehyde solution results in faster fixation, greater hardening, loss of fats and lipoids, better preservation of glycogen, poorer preservation of iron bearing pigments, and sometimes partial lysis of red corpuscles" (Lillie 1954, 33).

In any case, caution is needed when interpreting the histological literature on penetration rates and other aspects of fixatives and preservatives, particularly recommendations for fixatives other than formaldehyde, because the vast majority of histological work uses tissue blocks of three cubic centimeters or less, and penetration rates can be vastly different with larger tissue blocks or in whole specimens. Formaldehyde is one of very few fixatives with the ability to penetrate tissue blocks larger than three cubic centimeters.

Many of the multi-component fixatives have limited effectiveness. Dempster (1960) compared the penetration rates of several fixing fluids using 1–2 cm thick tissue blocks of a variety of fresh tissues, and concluded that "the fixing reagents in a mixture do not penetrate the tissues *en masse*, but . . . each reagent penetrates tissue in a characteristic sequence" (Dempster 1960, 68). Dempster's experiments showed that, in general, the rates of penetration slowed dramatically over time, and that colder temperatures, dilution, and inert isotonic media slowed the rates of penetration, while heat and high concentrations of the reagent accelerated penetration. The mucus agglutination of fishes (e.g., eels, sculpins, sturgeons) may significantly slow down the penetration of the fixative and thus the rate of fixation, particularly at colder temperatures (pers. comm., Dirk Neumann). Dempster (1960) also showed that although formaldehyde penetrates slowly initially, it rapidly overtakes the rates of penetration of other reagents with time. Unfortunately, most histological fixatives are not suitable for fixing whole animals because the fixatives do not penetrate rapidly enough or deep enough into the tissues.

Because of the need for the fixative or preservative to penetrate tissues as rapidly as possible, the means of distribution can be critical. For example, Stejneger instructed collectors to "inject a quantity of strong alcohol by means of a hypodermic syringe" into specimens (Stejnener 1891, 10) or to cut open the abdominal cavity of the specimen. In some cases, smaller specimens were injected with fluids through the mouth and anus to avoid making holes with hypodermic needles in the skin. Without doubt, perfusion through the carotid or femoral artery (or the brachial artery in the wings of birds) is the best method for distributing a fixative or preservative rapidly to the tissues of most specimens (Hangay and Dingley 1985a; Jones and Owen 1987; Quay 1974; Schultz 1924). Perfusion only takes a few minutes per specimen for most vertebrates; the drawback is that cuts must be made into the specimen to gain access to a vein and an artery, and the blood is flushed from the specimen as the fixative penetrates. For reasons of convenience and tradition, in field preparation, injection is far more common than perfusion. Perfusion, however, remains the standard in histological preparation of large tissue masses and in embalming. For example, Lillie (1954, 30) stated that, "From the point of view of the histochemist, the practice of hardening an entire human brain without perfusion, by immersion in dilute formaldehyde solution or other fixative before dissection, can only be condemned."

A technique for fixation by saturating tissues with formaldehyde gas in a hyperbaric chamber has been proposed (Clique et al. 2007). Specimens of euthanized fish were placed in a container and a solution of 5 percent formaldehyde (liquid) was decompressed until the formaldehyde was released as a gas, with excess oxygen eliminated by exposure to a sulfite bath. The specimens and formaldehyde gas were then subjected to pressure of 2.9 bar for 24 hr. to fix the tissues. The advantages to hyperbaric fixation are that the human worker is not exposed to formaldehyde and the exposure of the specimen to formaldehyde is decreased; the disadvantages are that the equipment is cumbersome, expensive, and not portable.

THE PENETRATION-FIXATION PARADOX

Histologists have noted what is called the "penetration-fixation paradox" of formaldehyde: formaldehyde is quick to penetrate tissue, but slow to fix tissue (Fox et al. 1985, 846). Not only that, formaldehyde penetrates at a constant rate, despite its concentration. This has been explained (Fox and Benton 1987; Pearse 1980) as follows: formaldehyde in water rapidly becomes hydrated to form methylene glycol. When tissue is immersed in a formaldehyde solution, it is penetrated rapidly by the methylene glycol and the fraction of formaldehyde present. The covalent chemical reaction of the fixative solution with the tissue depends on the formaldehyde present being consumed after forming bonds with the tissue components, and more formaldehyde forming from dissociation of methylene glycol. The equilibrium between formaldehyde as carbonyl formaldehyde and methylene glycol explains most of the mystery of why formaldehyde penetrates rapidly (as methylene glycol) and fixes slowly (as carbonyl formaldehyde). At this time, the molecular mechanism of tissue fixation is not very well understood.

According to Steedman (1976b, 119), the rate of penetration of formaldehyde into a specimen is determined by "(a) concentration of the fixative; (b) temperature of the fixative; (c) porosity of the outer covering of the animals; (d) porosity of the cell membranes; [and] (e) size of the specimen." Penetration can be slowed by the presence of lipids in the tissues (Steedman 1976b). Nevertheless, several other authors have stated that the rate of penetration depends on its rate of diffusion through the tissues, not on the concentration of the formaldehyde (Fox et al. 1985).

It is also important to have a sufficient amount of fixative available for the tissues. Steedman (1976b, 150) cautions that "The lower the specimen to fixative ratio the better the quality of the fixed specimen. A maximum ratio of 1 part specimen to 9 parts fixative is recommended." Lillie (1954) recommends that the volume of the fixative should be fifteen to twenty times that of the tissue to be fixed.

LIPIDS AND FIXATION

Lipids "undergo a considerable number of reactions with aldehydes during fixation" to the point that "it is difficult to demonstrate lipid histologically in tissues which have been stored in formaldehyde for more than four or five years" (Jones 1973, 39). It has also been suggested that either acrolein or glutaraldehyde may be better fixatives than formaldehyde for histological purposes (Jones 1973). The reactions between lipids and aldehydes during fixation include the conversion of unsaturated fatty acids to 1,3-glycols and other irreversible products (Jones 1973, 1976). When lipid preservation or extraction is the goal, Morris (1972) recommended against the use of either formaldehyde or methanol preservation because both substances hydrolyze (or protonize) and degrade polyunsaturated fatty acids in zooplankton.

FORMALDEHYDE AND FIELDWORK

Many collectors worry about the strength of formaldehyde that is purchased in remote areas because it is very easy to dilute formaldehyde with no immediately noticeable differences in the fluid. Steedman (1976a) provided instructions for a sodium sulfite method for formaldehyde assay (using sodium sulfite, a thymolphthalein indicator, and sodium hydroxide). Waller and McAllister (1986) described how to prepare formaldehyde test strips using filter paper containing an acid/base indicator and a mixture of sodium sulfite and sodium metabisulfite—the resulting test strips react rapidly to an aqueous solution of about 1.5 percent formalin or greater by changing from yellow to red, and "will not respond to the low levels of formalin that are present in most alcoholic preservative fluids from prior fixation in formalin, and hence will rapidly distinguish formalin solutions from alcohol solutions" (Waller and McAllister 1986, 93). Commercial formaldehyde test strips are available from Merc Quant and VWR International (in Europe), but because these are designed to detect trace amounts of formaldehyde, the sample must be greatly diluted to assay fixation-strength formaldehyde.

PARAFORMALDEHYDE

The solid form of formaldehyde is called paraformaldehyde. Paraformaldehyde is available as a powder or in larger solid pieces. Without the use of a catalyst, heat, or a chemical to boost pH, it usually takes several days for paraformaldehyde to dissolve in water (Steedman 1976b). The initial pH of paraformaldehyde dissolved in seawater is usually around 8.2. Taub (1962) recommended using sodium hydroxide and hot water to prepare formaldehyde solutions from paraformaldehyde, but this produces aqueous formaldehyde with a pH of 12. One modification of Taub's method calls for adding 60 ml distilled water to a conical flask along with 0.2 g sodium hydroxide; the mixture is then heated to 80–85°C and the formaldehyde prills or flakes added in very small amounts, allowing each addition to dissolve completely before adding the next. Once 40 g of paraformaldehyde is added, allow the mixture to cool and top up with water to make 100 ml (Steedman 1976b). Huheey (1971) suggested using calcium carbonate as a catalyst, which eliminates the need for very hot water, but produces aqueous formaldehyde with a pH of 11.6. Alconox dissolves the paraformaldehyde easily, but Nelson and Sparks (1999) cautioned that the resulting solution will cause clearing and loss of pigments. Steedman (1976b) reported that the use of a borax buffer with paraformaldehyde produces a solution with a pH of 8.2; and sodium glycerophosphate produces a solution with pH 7.2. Ehmann (1989) suggested preparing packets of dry chemicals to take in the field for easy mixing—one plastic bag containing 80 g of paraformaldehyde, 20 g of anhydrous sodium carbonate, and 0.5 g of a wetting agent (a detergent); and a second plastic bag containing 21 g of citric acid. In the field, the contents of the first bag are mixed with 2 liters of water, stirred constantly until dissolved, then agitated periodically over the next 4 hrs. at 20–30°C. The contents of the second bag are then added, stirred for five minutes, and the carbon dioxide gas is allowed to escape for 24 hrs. The resulting mixture will be a 10 percent concentration of formaldehyde with a neutral pH.

POST-FORMALDEHYDE FIXATION WASHING

It has long been recommended that specimens be rinsed or soaked in water following formaldehyde fixation, based on the assumption that formaldehyde bonds are water soluble. Despite this widespread practice, several surveys have shown that trace amounts of formaldehyde are likely to be found in alcohol preservatives containing formaldehyde fixed specimens—for example, Waller and Simmons (2003) found residual traces of formaldehyde in all samples of alcohol preservatives they tested. In their review of the formaldehyde fixation literature, Puchtler and Meloan (1985, 203) found evidence that "only adsorbed and loosely bound formaldehyde is displaced by washing in running water," with bound formaldehyde remaining in the tissues, and that "Bonds linking the remaining formaldehyde to tissue differ in stability. Some formaldehyde can be dislodged by prolonged treatment with water; but residual bound formaldehyde cannot be removed even by weeks of washing at room temperature" (Puchtler and Meloan 1985, 203).

See further discussion of transfer of specimens between fluids in chapter 3.

UNWANTED EFFECTS OF FORMALDEHYDE

When specimens are stored in a preservative that contains formaldehyde, the solution may become cloudy if there is a drop in temperature—this "misting" of the formaldehyde is due to the formation of paraformaldehyde. The cloudiness may sometimes be cleared simply by allowing the fluid to warm back up to regular storage temperature, but if not then 10 ml of methanol (methyl alcohol) may be added for each 2.5 liter of formaldehyde stock solution to reverse the polymerization of the formaldehyde (Simon Moore, pers. comm.). Steedman (1976b) recommend adding a small amount of NaOH to clear the paraformaldehyde. Should the formation of paraformaldehyde become so severe that a specimen becomes coated, the specimen should be soaked in lukewarm water (Dirk Neumann, pers. comm.) or in 80 percent ethyl alcohol for a few days, then carefully brushed clean (Simon Moore, pers. comm.), taking care to brush in the direction of surface structures such as scales, and never across or against them. Great care should be exercised in using either of these methods on delicate specimens that are subject to damage from abrupt osmotic changes.

In his instructions for preserving amphibians and reptiles, Joseph R. Slevin recommended that the collector use "Tin or aluminum pans" (Slevin 1927, 232) for preserving specimens, and that the specimens be tagged before preserving with numbered tags "made of pure tin" (Slevin 1927, 236). Slevin also strongly recommended preservation directly in alcohol, noting that:

> Many collectors prefer to use formalin instead of alcohol. Its use has some advantages and many objectionable features. It is easily transported, both reptiles and amphibians may be preserved in the same solution, and the careless or inexperienced collector who uses it will have less difficulty in preserving his specimens than if he uses alcohol, for specimens placed in formalin never decay. On the other hand, in my judgment, formalin specimens are never as satisfactory as well prepared alcoholic specimens. They usually turn black or a dull leaden gray. Frequently they have a puffy swollen appearance and are very easily broken. Formalin is very irritating to the skin and eyes of those who use it. The tin labels attached to specimens hardened in formalin often are corroded, even years after the specimens have been removed to alcohol. (Slevin 1927, 241–242)

The first undesirable side effect of formaldehyde fixation that Slevin noted was excessive absorption of water, which is an artifact of the procedure employed. The second—discoloration of specimens—was the result of exposing formaldehyde that was not neutral buffered to metal trays and tags (for many years tin or aluminum trays were the standard for field use). Many collectors noted that the trays or tags corroded, but few made the connection to the discoloration of the specimens. The darkening of tissues may occur with formaldehyde even if metal trays and tags are not used because animal pigments may be affected by light, UV, oxidation, structural changes, and the pH of fixatives or preservatives. Specimen darkening may also be caused by pre-fixation freezing or by prolonged storage in formaldehyde. Stuart (1995) fixed lizard specimens in phosphate buffered 10 percent formalin (one part formaldehyde with nine parts water) in plastic trays and found that the specimens began to darken after as little as six hours of exposure, and documented that the specimens became increasingly dark with longer exposure to the fixative by comparing specimens after 24 hours, 7 days, 30 days, 91 days, and 182 days in the fixative.

Formaldehyde fixation can also cause demineralization of bone and tissues, including the etching of otoliths, if the pH of the solution is sufficiently acidic (Kristoffersen and Salvanes 1998). The use of unbuffered formaldehyde may result in otoliths turning opaque within twelve hours of fixation (McMahon and Tash 1979).

GLUTERALDEHYDE

Glutaraldehyde is "a clear, colourless to pale straw-coloured liquid smelling of rotting apples" (Hopwood 1973, 48) that penetrates tissues relatively slowly, forms cross-links slowly (Hopwood 1973), and causes a net shrinkage of fixed tissues (Hopwood 1967), but our knowledge of the reactions of glutaraldehyde with lipids is limited. Glutaraldehyde does have some desirable characteristics as a fixative. For example, "The ability of glutaraldehyde to rapidly denature protein has also been exploited by using it as a disinfectant, where it is more efficient than formaldehyde or various phenol derivatives" (Hopwood 1967, 83). Glutaraldehyde is said to inhibit enzymatic activity better than formaldehyde does (Hopwood 1967). In histological practice, gluteraldehyde is used mostly to fix tissues for electron microscopy (Hopwood 1973), but it is not without problems. Lall et al. (1996) pointed out that to prepare plant specimens for electron microscopy, fixation in gluteraldehyde had to be done in a vacuum to speed the penetration of the fixative to avoid deformation of the intercellular spaces and diffusion of the organelles into the spaces. Furthermore, there is a problem in long-term storage of glutaraldehyde due to the inevitable formation of a precipitated polymer (Hopwood 1973). Steedman (1976b) concluded that gluteraldehyde is not better than formaldehyde, that alkaline gluteraldehyde solutions lose activity rapidly, solutions of gluteraldehyde in seawater tend to have low pH, and gluteraldehyde solutions are more prone to turbidity than formaldehyde solutions.

Glutaraldehyde penetrates tissues much slower than formaldehyde (Drury and Wallington 1980), and Hopwood cautions that, "If fixation in glutaraldehyde is limited to 2–6 h then the blocks or slices or tissue should not be thicker than 1 mm" (Hopwood 1967, 89). Hopwood studied comparative penetration rates for fixatives, using the penetration formula $t = Kd^e$ (*t* is time, *K* is a constant specific for the fixative, *d* is the depth of penetration, and *e* is an exponent relating to rate of diffusion). In Hopwood's studies, the *K* value of gluteraldehyde was shown to be 0.34, but that of formaldehyde was 2.0. Hopwood also found that glutaraldehyde caused a net shrinkage rate of rat liver of about 6 percent in 18 h at 4°C, while by contrast, formaldehyde caused very little shrinkage (Hopwood 1967).

In comparative tests, Steedman did not find glutaraldehyde to be superior to formaldehyde for bulk fixing of whole specimens, found glutaraldehyde to be more irritating to the eyes and more expensive than formaldehyde, and found that gluteraldehyde lost its reactivity faster than formaldehyde (Steedman 1976a).

ALDEHYDE SAFETY

Formaldehyde is very toxic (150 times more toxic than methanol), and is readily absorbed by inhalation, ingestion, or though skin contact (Browning 1965; Pabst 1987). Indeed, a suicide by formaldehyde ingestion was recorded in Toledo, Ohio, shortly after formaldehyde went into commercial production in the United States (Levison 1904). The adverse health effects of formaldehyde exposure were not understood for years (e.g., Pope 1931), and it was nearly sixty years after formaldehyde first went into use before the dangers of exposure were taken seriously. Formaldehyde vapor is particularly irritating to the mucous membranes, and repeated contact will cause dermatitis (Browning 1965). Exposure to formaldehyde vapors may cause conjunctivitis, respiratory disturbances (including asthma), headache, general weakness, changes in body temperature (Anon. 1989; Browning 1965; Gamble 1983; Levine et al. 1984), and cancer (Gamble 1983). There is evidence linking inhalation of acetaldehyde to nasal and laryngeal tumors in lab animals (NIOSH 1976, 1991). Symptoms of overexposure to formaldehyde may include headache, dizziness, eye irritation, burning in the chest, and disturbances of memory, mood, equilibrium, and sleep (Kilburn et al. 1985). Gluteraldehyde is a severe eye irritant (Steedman 1976b).

Formaldehyde fumes are easily absorbed by flexible contact lenses and may become trapped behind lenses (Cohen et al. 1979). Kilburn et al. (1985) reported disturbances of memory, mood, equilibrium, and sleep as well as irritation of the eyes and upper respiratory tract in histology technicians regularly exposed to formaldehyde. The recent literature on the health risks of formaldehyde were reviewed by Arts et al. (2006) and Hawks et al. (2010). Formaldehyde is found in a wide range of products that we encounter outside the laboratory (Preuss et al. 1985), which makes it difficult to assess only laboratory exposure.

ALTERNATIVE AND PROPRIETARY FIXATIVES

Because formaldehyde is a dangerous chemical, there is a lot of interest in finding safer alternatives. For some specialized uses, particularly for thin slices of tissue, nonformaldehyde fixatives have been shown to work very well (e.g., Warmington et al. 2000), but none of these fixatives have been demonstrated to be useful on whole animals or even with large tissue masses. The vast majority of the evaluations of alternative fixatives have been based on histological samples (e.g., Gillespie et al. 2002).

Osmic acid (osmium tetroxide) has been used as a fixative for very small specimens but it is toxic and stains internal organs black (Moore 2009a). Picric acid is a component of several histological fixatives (e.g., Bouin's fluid and Dubosq-Brasil's fluid), but it stains specimens yellow and becomes explosive if it crystallizes (Moore 2009a). Acetic acid has been recommended in combination with alcohol and/or formaldehyde but is not a very satisfactory fixative. In the 1960s and 1970s Dowacil powder was a popular fixing agent, but when combined with protein it produced formaldehyde, and had very poor powers of penetration (Moore 2009a). Moore (2009a) reported that a plant fungicide called Chinosol (8-hydroxyquinoline sulphate) was tested as both a fixative and preservative but was not effective in long-term use.

There are a number of proprietary fixatives on the market (table 4; see also discussion of "Alternative and Proprietary Preservatives" in chapter 3, and table 17). Most commercially available proprietary fixatives contain ethylene glycol or propylene glycol [1,2-Propanediol, $CH_3CH(OH)CH_2OH$], and none work well as a primary fixative although some are reasonably satisfactory in the short term as replacements with post-formaldehyde or post-gluteraldehyde fixed specimens. Glyoxal ($C_2H_2O_2$), a dialdehyde derived from ethylene glycol, has also been suggested as a replacement for formaldehyde in classroom chemical demonstrations but has not been shown to be very satisfactory (Ealy 1991). Limited testing has indicated promising results for some of these proprietary fixatives, including NoToXhisto (Acton et al. 2005). A variety of alternative fixatives were reviewed by Lamph and Bennitt (2009).

A few alternative fixatives use aldehydes as cross-linking agents, but most proprietary fixatives do not form cross-links. Some of the proprietary fixatives are dangerous in their own right, and none of them can penetrate whole animal tissue as rapidly and as deeply as formaldehyde does. Some proprietary formulas use carbonyl-bis (L-methionine p-nitrophenyl ester) as a cross-linking agent.

In many cases, the exact composition of proprietary fixative solutions is unknown because chemical companies keep their formulas secret, making it difficult to determine what other deleterious effects these replacement fixatives might have on the specimens or their use in future research.

3

Preservation

As discussed in chapter 2, preservation is the prevention of postmortem changes in tissues. The oldest method of intentional, long-term preservation of organic material is dehydration, in the form of mummification, which has been practiced in many areas of the world. Intentional mummification dates back at least eight thousand years in the region of present-day Chile (Arriaza 1995), and to around five thousand years ago in Egypt (El Mahdy 1989); in both locations, humans and other animals were mummified. Purpose-made mummies are known from the Torres Straits Islands, Melanesia, several areas of South America, the Near East, the Canary Islands, and other diverse locations (Chamberlain and Pearson 2001). Mummification practices often developed in areas that had a naturally dry climate and alkaline soils that helped dehydrate tissues. Over the course of thousands of years, the processes of mummification usually became more complex as techniques were developed to remove internal organs, accelerate the dehydration of tissues, apply resins and oils as pest repellants, and improve the wrapping and sealing of the mummies. The two most important factors in successful mummification—and thus the long-term maintenance of mummies—are dehydration (to stop autolysis and other inherent processes of decay) and protection of the mummified tissues from biological agents of deterioration (primarily through the application of oils, resins, and wrappings). Fluid preservation shares a common basis with mummification, in that alcohols preserve tissues mainly by removing water, and all successful fluid preservatives protect the tissues from agents of biological deterioration.

Good fluid preservatives protect the tissue from decay or deterioration over the long term, give the specimen as normal an appearance as possible, and "afford it mechanical protection when handled" (Stoddard 1989, 23). Protecting specimens from decay and deterioration means that most preservatives must act as disinfectants (table 6).

PRESERVATION WITHOUT FIXATION

Although in modern practice fluid preservation is a two-step process of fixation and preservation, the oldest extant fluid-preserved specimens were not prepared with a fixative (sensu stricto) but rather by direct preservation without using a cross-linking fixative agent (see discussion of fixation in chapter 2). The drawbacks to direct preservation in alcohol include (1) the poor penetrative properties of alcohol; (2) the dehydration of tissues by alcohol; and (3) the dilution of the alcohol preservative as water is extracted from the specimens during preservation. For most specimens, the alcohol preservative must be injected or perfused into the tissues due to its poor penetrative properties, or the specimens must be cut open to allow the fluid to penetrate. Because alcohol preserves, in part, by rapidly extracting water from tissues, most historic instructions for direct preservation in alcohol call for the initial preservative solution to be changed after twenty-four hours because the alcohol becomes diluted (the amount of dilution is relative to the volume of the specimen, the type of tissues, and the type and strength of the alcohol preservative).

Because fluid preservation is a trade-off between dehydration and preservation, the best results for direct alcohol preservation will be obtained by using 70 percent ethyl alcohol (at which strength it is a good biocide and preservative but does overly dehydrate tissues), with a change of alcohol after twenty-four hours. The use of higher concentrations of alcohol for initial preservation (as suggested by some modern authors) is not recommended because concentrations higher than 70 percent dehydrate specimens excessively and, as the outer tissues dehydrate, will impede the deeper penetration of the alcohol preservative. However, there are some exceptions—in a study on post-preservation manipulation, dry mounting, and DNA extraction of ant specimens, King and Porter (2004) concluded that direct preservation for at least twenty-four hours in 95 percent ethyl alcohol or isopropyl alcohol was preferred over other alcohol concentrations because 95 percent did not produce excess flexibility or brittleness (so the specimens could be mounted as dry preparations) but allowed for reasonable DNA extraction—it is important to note that this study did not address the best alcohol concentration for specimens that would be kept long-term in a fluid preservative. In addition, King and Porter (2004) did not evaluate shrinkage from dehydration (shrinkage is less problematic for a specimen with a hard exoskeleton than for soft-bodied invertebrates or vertebrates).

FREEZING

Specimens intended for fluid preservation should not be frozen because freezing and thawing are destructive to tissues (Quay 1974). Freezing causes loss of structural integrity in tissues, shrinkage, and changes in color (see tables 13 and 15). Furthermore, the formation of ice crystals disrupts the arrangement of myofibers and disrupts and distorts sarcolemmas (the connective tissue sheaths that envelope muscle fibers) (Mills 1975; Scott and Aquino-Shuster 1989). Higgins and Karlsson (2013) reviewed literature relating to the fact that freezing and thawing single suspended cells is more successful than freezing and thawing tissue due to intracellular ice formation. In addition, freezing delays the time interval between death and preservation, allowing autolysis and biological deterioration to begin (see table 7). Fixative and preservative fluids do not penetrate frozen tissues, which results in further deterioration of the tissues as the specimen must be thawed prior to fluid preservation.

When it is necessary to make fluid preparations from frozen specimens, it is recommended that the specimens be submerged in the fixative or preservative fluid to thaw, and progressively injected with the fluid as thawing permits.

TRANSFER BETWEEN FLUIDS

Since the early 1900s, the traditional process of preserving specimens in fluid has been to collect, euthanize, and fix the specimens (usually in a formaldehyde-based fixative) in the field, maintain them in the fixative for a few days to a few weeks (sometimes for months), then wash the specimens for twenty-four to forty-eight hours in water and transfer them directly to a storage strength alcohol preservative fluid. There are a number of problems with these practices. It has long been presumed that washing formaldehyde-fixed specimens for twenty-four hours was sufficient to remove the formaldehyde, based on the assumption that formaldehyde forms water-soluble bonds in the tissues. However, Puchtler and Meloan (1985) reported that short-term washing or soaking only removes loosely bound formaldehyde from tissues, and that formaldehyde continues to be removed from the tissues in storage for a number of years, which explains why most alcohol preservatives used with formaldehyde-fixed specimens contain small traces of formaldehyde (Waller and Simmons 2003; see discussion in chapter 2). After the washed specimens are transferred to an alcohol preservative, a translucent cloudiness may appear in the preservative, or white formaldehyde "needles" may form in the bottom of the container—these are both indications that the mixture contains sufficient formaldehyde to allow the precipitation of paraformaldehyde. When paraformaldehyde forms, the alcohol mixture may become sufficiently acidic to damage the specimens, so the specimens should be removed, rinsed in storage strength alcohol, and placed in fresh storage-strength preservative (see discussion of paraformaldehyde formation in chapter 2). Another problem with washing formaldehyde-fixed specimens in water is that while the specimens are in water, enzymatic or bacterial activity may initiate deterioration of the specimens (Taylor 1981). Furthermore, specimens absorb water during the washing process, which will then be extracted from the tissues in an alcohol-based preservative, diluting it below the desired strength.

Several factors influence the quality of preserved specimens during the transfer from fixative to preservative, including the interval between fixation and preservation, fluid quality, specimen volume, and extent of dehydration and/or hydration of tissues (table 7). An often overlooked problem with the traditional process is that a rapid change between differing concentrations of water can cause tissue distortion and damage. Fixatives and preservatives vary widely in osmotic pressure—it is a very abrupt change to transfer a specimen directly from 96 percent water (the 4 percent formaldehyde fixative solution) to 100 percent water (to wash the specimen) and then to 30 percent water (the 70 percent ethanol preservative). The necessity to stage tissues through progressive concentrations of chemicals has long been recognized in the histological literature (e.g., Mallory and Wright 1918), but staging whole animals between concentrations of fluids during fixation, washing, and preservation has rarely been understood to be important. As Waller and Strang (1996, 80) noted, "The requirement for stepping specimens through increasing concentrations has been noted by several authors . . . although this requirement has not always been linked with a need to limit osmotic pressures within specimens."

Osmotic pressure rises steadily with ethanol concentrations below about 75 percent, and rises much more rapidly at concentrations above about 80 percent (Waller and Strang 1996). This suggests that solutions with approximately equal concentration increments are most appropriate for stepping specimens up to higher ethanol concentrations. Steps of about 20 percent are generally recommended for transferring specimens from a formaldehyde fixative to a 70 percent ethyl alcohol preservative; to reach concentrations above 75 percent, progressively smaller concentration steps should be used (Waller and Strang 1996). Moore (2001) recommends washing fixed specimens in water and then staging them through ethanol concentrations of 20 percent, 40 percent, 60 percent, and 80 percent, leaving the specimens for twenty-four to forty-eight hours at each concentration step; stepped specimens are then placed fresh 80 percent ETOH for storage (Moore prefers the term concentration ladder rather than concentration steps to indicate the progressive nature of the transfer). If air becomes trapped in the specimens during staging, they should be stepped back down to water (alcohol vapors will damage the seals on vacuum pumps), placed in a vacuum to release the air, then stepped back to 80 percent ethanol for storage (Moore 2001). The concentration of alcohol at each step should be carefully monitored as the mixture will become dilute with water and other fluids from the specimens; larger specimens will need more time at each stage of concentration than smaller specimens (e.g., small invertebrates may require only an hour or two, whereas large vertebrates may need twenty-four hours plus per stage).

Studies of osmotic effects during transfer between fluids, including shrinkage and swelling, support the staging of specimens through differing concentrations of alcohol (e.g., Jones and Owen 1987; Lafromboise et al. 1993; Steedman 1976). Although Cushing and Slowik (2007) found that transfer directly from a low concentration of alcohol (ca. 45 percent) to a high concentration (75 percent) did not seem to be detrimental to specimens of spiders, the specimens in their tests had been previously preserved in a high concentration of alcohol (75 percent) so were in all likelihood dehydrated before their experiment began.

PRESERVATIVE QUALITY

Ethyl alcohol (also called ethanol, ETOH, grain alcohol, spirit of wine, or Cologne spirit) is a colorless, hygroscopic liquid, miscible with water and ether, a good solvent for many substances, that burns with a flame that is almost nonluminous (Browning 1965). In commercial production ethyl alcohol is made from the natural fermentation and distillation of molasses, grains (primarily corn, malt, wheat, rye, barley, rice, or bran), pineapple juice, whey, cellulose pulp, or potatoes, or synthetically by the hydrolysis of ethyl sulfuric acid and diethyl sulfate, or from cracked petroleum gases (Browning 1965). Ethyl alcohol can be ingested, inhaled, or absorbed. It primarily affects the brain, but chronic exposure has been shown to be detrimental to other organs in the body as well (Browning 1965).

Preservative solutions—whether prepared with alcohol, formaldehyde, or other chemicals—should be mixed with deionized, distilled, or purified water to avoid the formation of precipitates such as calcium chloride, oxidation products from water lines, or high amounts of chloride that may damage specimens (see figure 3.1). Measure carefully when mixing preservative solutions, and check the final concentration before using the mixture. When water and ethanol are mixed, there is a contraction in volume of the solution of about 2 percent; for example, at 20°C, mixing 70 mL of ethanol and 30 mL of water will result in 96.84 mL of solution, with an alcohol concentration of 72.28 percent (Waller and Strang 1996, 73).

Denatured alcohol is not recommended for the preservation of scientific specimens because of the possibility that the additives used to denature it might interfere with good preservation or later use of the specimen (e.g., chemical analyses). In the United

Figure 3.1. Calcium carbonate precipitates in alcohol formed after diluting with unfiltered tap water.

States there are more than one hundred approved denaturants, including acetone, ammonia, benzene, boric acid, coal tar, formaldehyde, gasoline, hydrochloric acid, methanol, phenol, red mercuric iodide, salicylic acid, thymol, and toluene (Waller and Strang 1996; CFR 27); many denaturant chemicals will react with the specimens or specimen storage containers. A common denaturant suite in over-the-counter ethanol in the United States is a mixture of acetone, denetonium benzoate, and methyl isobutyl ketone (acetone is a ketone; denetonium benzoate is an extraordinarily bitter chemical which makes the alcohol nearly impossible to swallow; methyl isobutyl ketone is a solvent probably used to keep the other denaturants in solution). In several European countries, non-denatured ethyl alcohol is not readily available. For example, in England, museums must use industrial methylated spirits (IMS) (ethyl alcohol which contains 5 percent methyl alcohol). In some countries common denaturants include butanone, also known as methyl ethyl ketone (MEK), or $CH_3C(O)CH_2CH_3$, camphor, and naptha. MEK degrades plastic and acrylic materials; camphor may affect evaporation rates and pose human health issues (Dirk Neumann, pers. comm.). Camphor, a turpentine derivative ($C_{10}H_{16}O$), has been used for thousands of years as a pest repellant and medicinally, despite the fact that, in high doses, it may cause neuromuscular problems such as seizures and convulsions. Naptha (also known as petroleum ether or petroleum spirits) is a volatile solvent.

OLD FLUID PRESERVATIVES

Old fluid preservatives may contain chemicals that were added in an attempt to improve preservative strength (see table 1). Table 8 provides a list of chemicals used in fluid preservation with the date of their first mention in the literature. Common additives to fluid preservatives include chromic acid, picric acid, glycerin, arsenic, and mercuric chloride ($HgCL_2$, commonly called corrosive sublimate)—the latter two are surprisingly common additives in medical and anatomy collections (see the section Anatomical and Histological Fluid Preparations in this chapter). In a survey of fluid-preserved anatomical specimens in the Mütter Museum in Philadelphia, the fluid in 21 percent of the containers tested positive for arsenic, and 38 percent tested positive for lead—these contaminants could have been additives to the preservative fluid, or could be the result of leeching from the containers or the container seals (Thede 1996). Table 9 lists fixation and preservation techniques frequently encountered in anatomical collections.

Old fluid preservatives, or fluid preservatives containing old specimens, should always be treated as if they might be toxic. For example, droplets of elemental mercury were found in two lizard specimens preserved prior to 1901 (Simmons et al. 2007). The specimens showed signs of darkening, indicating that mercuric chloride (corrosive sublimate) was probably added to the alcohol they were preserved in. The mercuric chloride could have been converted to elemental mercury by heterotrophic bacteria, interaction with proteins, reduction and oxidation reactions, exposure to ultraviolet light, reaction with the alcohol preservative (an organic solvent), or some combination of these processes. For further discussion of safety measures when working with old preservatives, see the section on Health and Safety in chapter 5.

BOTANICAL USE OF FLUID PRESERVATION

Botanical specimens, particularly those intended to be used for anatomical or morphological studies, have been preserved in fluids for many years, although the dominant method for preserving plants is to make dry specimens attached to herbarium sheets (Pinzl, IN PREP-a). The fluid-preserved collection at the herbarium of the Kew Royal Botanical Gardens in England was not started until 1930, although the collection contains specimens prepared as early as the 1830s (Tredwell 2006). The Marie Selby Botanical Gardens claims the largest collection of fluid-preserved flowers and plants in the Western hemisphere with more than twenty-six thousand vials, preserved in a mixture of denatured alcohol, water, and glycerin (Holst 2003).

The development of fluid preservation for botanical use has paralleled developments in the fluid preservation of animals. Nevertheless, much of the literature on fluid preservation in plants is directed at preserving colors, or preserving plants for studies of anatomy or morphology. In 1899 Stone recommended fixing plant specimens in 4 percent formaldehyde and the use of nitric acid to preserve colors; Nieuwland and Slavin (1928) recommended preserving specimens of Monotropa in a fluid preservative with a reducing agent, reporting that alcohol with sodium sulfite and hydrochloric acid worked best (see table 1). Not all color preservation methods were satisfactory—Moore (2010) noted that sometimes fluids with copper salts were used to preserve chlorophyll, giving the specimens a blue-green tinge. In their masterful two volume work, Wagstaffe and Fidler (1968) recommended fixation in formaldehyde in various strengths and the use of sulfur dioxide, copper acetate, copper sulfate, sodium citrate, tertiary butyl alcohol, zinc, and other chemicals to preserve colors in plants (reviewed in table 1). Bridson and Forman (1998) recommended a variety of fluids for various taxa (reviewed in table 1). Despite these recommendations, fluid preservation has remained an ancillary form of botanical preservation. DeWolf (1968, 69) stated that "the best (but far from the most convenient) way to preserve plant material for study is to place the material in jars of preserving fluid. Such a collection, however, is not an herbarium." DeWolf (1968) noted that fluid-preserved specimens are bulky, prone to evaporative loss, it is necessary to remove the specimens from their jars for comparison with other specimens, and that glass containers are difficult to transport. DeWolf stated that "the standard botanical preserving fluid" was formol-acetic-alcohol, popularly known as FAA (see table 1). Condon and Whalen (1983) recommended that specimens of fruit, flowers, and other organs of plants that were attacked by pests and might still contain live pests should be preserved in alcohol.

Often, the botanical use of fluid preservatives is in an intermediary stage between collecting and later pressing and drying plants. Stackhouse (1800) recommended immersing flowers in a saturated mixture of powered alum in water prior to pressing them. Condit (1948) reported that on the Perry expedition to Japan in the 1850s, the expedition botanist painted dried plant specimens with a mixture of corrosive sublimate (mercuric chloride), strong spirits of wine, and camphor to combat both mold and insect pests. Swingle (1930) and Moore (1950) both suggested oxy-quinoline sulfate (hydroxyquinoline sulfate), a disinfectant, rather than alcohol or formaldehyde for this purpose. Hodge (1947) recommended moistening the paper layered between pressed plants with alcohol (using locally available beverage alcohol when no stronger alcohol was available) and sealing the plants and papers inside a metal container for transport to a laboratory for drying. After a conversation with Paul H. Allen in Panama about protecting dried plant specimens from mold, Schultes (1947) and Fosberg (1947) recommended using formaldehyde mixtures for this purpose. Schultes (1947) used a solution of one quart of formaldehyde mixed with a quart and a half of water to moisten the drying sheets, and recommended the use of a rubberized bag for transport. Fosberg (1947) added that although Allen had applied a mixture of 70 percent alcohol and sufficient formaldehyde "to give it a strong odor" with a spray gun, he obtained better results using a mixture of one part formaldehyde, one part 95 percent ethyl alcohol, and one part water (substituting two parts 70 percent ethyl alcohol when 95 percent was not available), applied with a paint brush. Johnson (1948), following a suggestion in Fosberg (1947), tried dipping cones of spruce (*Picea*) and hemlock (*Tsuga*) in Fosberg's mixture, but found the needles fell off the twigs when drying. Huxley and Bryant (2007) recommended a mixture of one part formaldehyde to ten parts water for temporary storage of seaweeds during field collecting when the plants are to be later dried and attached to herbarium sheets. Balick et al. (1982) tested several field preparation methods on specimens of fresh coca leaves (*Erythroxylum* spp.) and cautioned that the alkaloid components of the specimens were extracted when the plant specimens were soaked in water, 95 percent ethanol, "white rum," FAA (ninety parts ethanol, five parts glacial acetic acid, and five parts formalin), a 1:1 formaldehyde and water mixture, or undiluted formaldehyde. Ceska and Ceska (1986) stated that alcohol, glycerol, and other chemicals should not be used to treat plant specimens that are to be later pressed, as these chemicals can dissolve the pigments in the specimens. They also cautioned against the practice of dipping herbarium specimens in mercuric chloride as this will destroy the flavonoids necessary for chromatographic research. However, Ceska and Ceska (1986) did endorse preserving duplicate specimens of *Wolffia* in 70–90 percent ethanol to preserve the fronds (but not formaldehyde, as this made the fronds brittle). Kendrick (1965) cautioned that fleshy fungi should not be preserved in alcohol, formaldehyde, FAA, or lactic acid as all of these preservatives cause loss of color, extreme fragility, and produce heavy, bulky specimens that are less accessible than dry specimens. Harrington (1947) reported on ten years of experience using formaldehyde fumes to preserve flowers and inflorescences by enclosing them in jars containing a 2–3 cm layer of cotton soaked in full-strength formaldehyde, claiming that colors and fine features of the plants are better preserved in this way.

GLYCERIN

Glycerin, also known as glycerine and glycerol, is a viscous polyol (sugar alcohol), sometimes used as an additive to other preservative solutions (see table 1) or used by itself as a preservative. Due to its viscosity, glycerin may be used as a preservative for delicate anatomical specimens, cleared and stained specimens (see the following discussion), or for moisture-sensitive rocks, minerals, or fossils. For example, some fossils from the Messel Shale of Germany are stored in glycerin to prevent desiccation (Manchester et al. 1994). Some Messel Shale fossils include feathers that are sufficiently well preserved that their ultrastucture can be studied with a scanning electron microscope (Vinther et al. 2010).

Glycerin was sometimes added to alcohol with the idea that if the alcohol evaporated, the glycerin would protect the specimen from dehydration (Beirne 1955; Browne 1884); some authors also claim that the addition of glycerin keeps specimens flexible. No studies have demonstrated that a small amount of glycerin will provide protection from dehydration or keep the specimens flexible (despite anecdotal support for these ideas), and the treatment may put the specimen at risk. Glycerin is hygroscopic and thus absorbs water vapor from the air in high humidity situations if containers are not well sealed, or if containers are left open while specimens are being examined. As the glycerin absorbs moisture, it dilutes the preservative while introducing airborne contaminates (including mold spores and airborne bacteria) into the container (Jones and Owen 1987). Figure 3.2 shows a glycerin storage medium

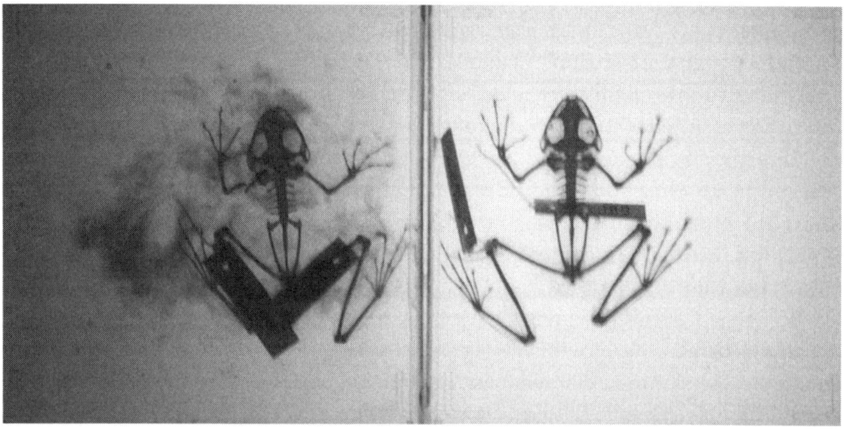

Figure 3.2. Bacterial growth in glycerin preservative (left) and glycerin preservative without bacteria (right).

supporting bacterial growth. In addition, glycerin added to fluid preservatives makes specimens slippery and hard to handle; if enough alcohol evaporates, the glycerin concentration may increase to the point of clearing the specimens (Boase and Waller 1994). Evaporated alcohol-glycerin mixtures leave behind a medium that encourages the growth of mold and bacteria (Levi 1966).

Because glycerin supports the growth of bacteria and is hygroscopic (thus pulling contaminants into containers), a few crystals of thymol (2-isopropyl-5-methylphenol) or a few drops of camphor are often added to glycerin preservatives as biocides, or glycerin may be used in a 1:1 mixture with ethanol (although this substantially lowers the viscosity of the preserving fluid). Glycerin may clear tissues if the pH of the solution becomes too basic.

ISOPROPYL ALCOHOL

Isopropyl alcohol is also known as isopropanol, propanol-2, dimethyl carbinol, and rubbing alcohol (Browning 1965). Previously used trade names for isopropyl that may still be encountered in the literature include Perspirit, Propol, Hartosol, Optal, and Avantine. Isopropyl is a clear, colorless liquid with a distinct odor, miscible with water, alcohol, and ether. It is manufactured from propylene, as a by-product of the cracking of petroleum hydrocarbons, and from the hydrogenation of acetone with catalysts (Browning 1965).

Isopropyl alcohol was first made in 1855, but it did not go into commercial production in the United States until 1920 (Hatch 1961). It was recommended as an alternative preservative to ethanol, without being tested, as early as 1922 (Griffin 1922). A notice in *Turtox News* in 1934 complained of the high taxes on ethanol, the red tape necessary to buy it tax-free, and the additives used to denature ethanol, and recommended isopropyl as a replacement for ethanol despite the fact that "Sufficient experimentation has not been carried on in connection with its use in biological work to warrant a dogmatic statement from us that it can *always* be used as a substitute, but our preliminary work justifies this *belief*" (Windsor 1934, 216). There was a conflict of interest in this recommendation as *Turtox News* was the newsletter of a company that sold isopropyl alcohol. The article went on to say, "Where can you buy isopropanol? Turtox will gladly accept your orders for it" (Windsor 1934, 217). At that time, anhydrous isopropanol was $3.75/gal. (for an 88 percent solution), but "the grade used for general preservation" was just $2.50/gal. (Windsor 1934, 217), much less expensive than ethanol.

Different alcohols penetrate the epidermis at different rates. The permeability of the epidermis (measured as the solute absorbed per mL of tissue) for methanol is 0.60, for water is 0.88, for ethanol is 0.90, and for isopropanol is 1.3, thus isopropyl penetrates the epidermis faster than ethanol (Schaefer et al. 1982). The low penetration rate of methanol is a factor in its poor performance as a preservative. It is important to note that in part due to this more rapid penetration ability, isopropyl is twice as toxic as ethanol (Monick 1968). As a secondary alcohol, isopropyl is also better at dissolving lipids than ethanol, and is more reactive with oxygen in the air—isopropyl reacts with oxygen to form ketones and unstable peroxides, which may damage preserved specimens (van Dam 2003). Gerrick (1968) noted that animal pigments were highly soluble in isopropyl compared to formalin, but did not make a comparison between isopropyl and ethanol.

In 1922, Philip Pope (1928) preserved one newt each in solutions of 40 percent, 50 percent, 60 percent, 70 percent and 80 percent isopropanol, as well as one specimen in 70 percent ethanol for a control. Pope examined the newt specimens after six years in the preservatives, and concluded that the 40–50 percent isopropyl preserved specimens were fine, even though he could not find the control to compare them to. The advantages that Pope listed for isopropyl were that unlike ethanol it had no legal restrictions on its purchase, it was inexpensive, it could be used in low concentrations, it was not irritating (like formaldehyde), and it did not soften bones or teeth (as did formaldehyde) (Pope 1928). However, subsequent workers have reported that 45–50 percent isopropanol has been shown to cause considerable shrinkage of specimens, it can be difficult to mix thoroughly, it is prone to layering in specimen containers, it may soften bone, and it may cause clearing of tissues if the concentration falls below 30–40 percent (Fink et al. 1979). DiStefano et al. (1994) found that 50 percent isopropyl failed to prevent tissue decay in fish and crayfish, and concentrations above 75 percent caused crayfish to become brittle (similar concentrations of ethanol did not cause embrittlement). Howmiller (1972) measured greater weight loss in specimens of macrobenthic invertebrates preserved in 70 percent isopropyl (with 5 percent glycerin) than specimens preserved in 70 percent ethanol (with 5 percent glycerin) or 10 percent formalin. Moku et al. (2004) recorded shrinkage to 91 percent of initial length in myctophid fish larvae preserved in 70 percent isopropyl, compared to shrinkage to 97.6 percent for 70 percent ethyl alcohol and 94.9 percent in 90 percent ethyl alcohol. By contrast, Bininda-Edmonds and Russell (1992, 1993) found no significant difference in wing area or mass between specimens preserved in 70 percent ethyl alcohol and 45 percent isopropyl alcohol.

King and Porter (2004) reported observational data that long-term storage in isopropyl alcohol results in greater fading of ant specimens, probably due to the greater propensity of isopropyl and other long-chain alcohols to extract lipids, proteins, and pigments from cellular membranes at high concentrations (Goates and Knutson 1994).

Some workers, particularly ichthyologists, have opined that they prefer isopropyl-preserved specimens because they are more flexible than alcohol-preserved specimens, and several ichthyologists have anecdotally reported no problems with fish collections long preserved in isopropyl, and no greater fading of patterns than in ethanol. Walker et al. (1995) claimed that bony fishes preserved in alcohol are more flexible and "seem to shrink less" than those stored in ethanol, incorrectly citing Laframboise et al. (1992) as having "found noticeable dehydration when transferring specimens from 45 percent ISO to 70 percent ETH and a graded series of solutions was needed to minimize this shrinkage" and the figures in Lai (1963) as supporting their claim (Walker et al. 1995, 3). In fact, Laframboise et al. (1992) reported on specimens that had been initially preserved in ethanol, changed to 45 percent isopropyl in the 1960s, and then changed back to ethanol because "by the 1980's specimen deterioration and some pattern loss was increasingly evident" after two decades in an isopropanol preservative and housed in an inadequate storage environment. Their study of the transfer of the specimens from isopropanol to ethanol was prompted by the discovery in 1986 that the isopropyl concentration in the collection ranged "from 6% to 40% with an average near 20%" (Laframboise et al. 1992, 28). They reported that the specimens were transferred from this average 20 percent isopropyl (not 45 percent as Walker et al. [1995] stated) to 75 percent ethanol through six different transfer regimes, including direct transfer and staging through graded series, thus their project did not provide a means of evaluating shrinkage in isopropyl compared to ethanol, contrary to the statement of Walker et al. (1995). Lai (1963) conducted one of the few significant controlled studies comparing the effects of preservatives on proportional measurements of fishes (*Pimephales promelas*, in the class Actinopterygii, which are bony fishes). Lai took six standard measurements per individual of nearly five hundred specimens in twelve different test and control groups. The means, standard deviations, and standard errors were calculated and reported for each group in table format as well as represented in the figures referred to by Walker et al. (1995). Lai (1963, 18) found that "different concentrations of [unbuffered] formalin (5% to 15%) used for fixation do not significantly affect proportional measurements subsequently obtained," but that "different concentrations and kinds of alcohol used as preservatives significantly affect the proportional measurements subsequently obtained," such that "more cases of significant difference occurred when groups preserved in isopropanol were compared with groups preserved in ethanol than occurred when comparisons involved groups preserved in different concentrations of ethanol (65 to 75 percent)." Contrary to what Walker et al. (1995) claimed, Lai also (1963, 18) reported that "transfer of specimens from 40% isopropanol to 70% ethanol caused no change in their proportional measurements" but "the reciprocal transfer significantly changed proportional measurements."

According to Stoddard (1989), isopropyl-preserved specimens are flexible because the tissue matrix undergoes more breakdown than tissues do in ethanol; some ichthyologists have anecdotally reported greater fading in isopropyl than ethanol for some fish taxa. In a controlled accelerated aging study, Von Endt (2000) found that isopropyl causes more deterioration of collagen and keratin than does ethanol, which would contribute to the greater flexibility of isopropyl-preserved fish.

When isopropyl is mixed with water, the resulting solution undergoes contraction, so solutions should be mixed on a weight basis rather than a volume basis (Hatch 1961).

FLUID PRESERVATION FOR DNA EXTRACTION

For purposes of later DNA extraction from animal tissues, there is general agreement that the most successful procedure is to remove a sample of fresh tissue from the specimen immediately after it is euthanized and rapidly freeze the sample to –80°C and store it at that temperature (the specimen itself should be preserved separately as a fluid specimen or dry specimen). The next best procedure is to preserve the tissue sample in 95 percent ethyl alcohol, preferably storing it at a temperature below 0°C. The third choice is to preserve the tissue sample in a fresh, stable DNA buffer solution (see reviews in Carter [2004] and Dessauer et al. [1996]). Absolute alcohol (100 percent ethanol) and other ETOH concentrations above 96 percent are not recommended for DNA preservation because the alcohol is chemically dehydrated (e.g., with benzenes) and traces of the dehydrants may remain that will degrade DNA (Dirk Neumann, pers. comm.).

It is possible to extract DNA from fluid-preserved museum specimens, even those that were fixed in formaldehyde (e.g., Chatigny 2000; Fang et al. 2002; Klanten et al. 2003), despite the fact that the fixative action of formaldehyde degrades the DNA (e.g., Dubeau et al. 1986; Haselkorn and Doty 1961), as do certain other chemicals and procedures that may be employed in specimen preparation and storage (such as exposure to ultraviolet radiation during tissue extraction or storage). Graves and Braun (1992) raised a caution about the use of museum specimens as DNA sources due to limited sample size for many taxa and the damage caused by sampling techniques.

Ben-Ezra et al. (1991) compared the effects of formaldehyde, ethanol, Bouin's solution, Zenker's solution, B-5 (a mercuric chloride–based fixative) and Omnifix (an ethanol-based proprietary product) on DNA extraction and amplification after fixation times of six hours, twenty-four hours, forty-eight hours, seventy-two hours, and one week; they found that the best results came from tissues fixed in ethanol or Omnifix, that consistent results from formalin-fixed tissues were only obtainable for fixation times of forty-eight hours or less, and that Bouin's solution, Zenker's solution, and B-5 yielded poor to negative results. There are indications that some buffers, such as DMSO, are unstable for DNA preservation due to destabilization of salts (Dirk Neumann, pers. comm.). Several studies have attempted to find alternative fixatives or storage fluids for DNA samples. Greer et al. (1991) compared eleven fixatives and three fixation times, concluding that acetone and 10 percent neutral buffered formalin gave the best results; Kuch et al. (1999) tested a commercial laundry detergent, 90 percent ethanol, and a conventional DNA buffer, concluding that the laundry detergent worked best; and Leal-Klevezas et al. (2000) tested preservation of blood in ethylene glycol and propylene glycol, concluding that a 20 percent mixture of the two was most successful.

Early attempts to extract usable DNA from formaldehyde-fixed tissues showed some promise (e.g., Crisan et al. 1990; Dubeau et al. 1988; Goelz et al. 1985; Impraim et al. 1987; Jackson 1978; Nuovo and Richart 1989; Nuovo and Silverstein 1988; Rogers et al. 1990; Warford et al. 1988), but a standard, reliable technique has not yet been developed—specimen tissues and preservation regimes are probably too variable for development of a single, universal standard technique.

In a comparative study of preservatives and temperatures on arachnid DNA (Vink et al. 2005), the best results were obtained by preserving specimens in a commercial product named RNA*later* (the ingredients of which are a trade secret; see table 4) or propylene glycol; the authors recommended that tissues should be stored at –80°C. The best DNA extractions from tissues preserved in 95 percent ETOH were from specimens stored at –80°C. The study found no significant difference in DNA from specimens preserved in 95 percent ETOH and stored in 70 percent ETOH and from tissues maintained in 95 percent ETOH. By contrast, Mandrioli et al. (2006) reported significantly better DNA quality and extraction yield from adult cabbage moth (*Mamestra brassicae*) specimens stored in 100 percent ethanol at 4°C compared to 75 percent ethanol, which gave poor results (however, they found the best preservation was in acetone, not ethanol). Dean and Ballard (2001) reported lower yields from insects euthanized in alcohol, and Houde and Braun (1988) were unable to extract usable DNA from alcohol-preserved bird specimens.

Carter (2004) demonstrated that DNA preservation is better in ethyl alcohol than in IMS (probably due to the presence of methanol in IMS). Chakraborty et al. (2006) concluded that good DNA could be extracted from fish specimens frozen at –20°C or preserved in 95 percent ethanol for up to five years, but not for specimens fixed and preserved in 10 percent formalin; the authors were able to obtain DNA from specimens fixed for just seven days in 10 percent formalin that was buffered with phosphate and subsequently stored in ethyl alcohol. Some workers have now returned to preserving entire specimens in alcohol, without fixatives or other chemicals (Chakraborty et al. 2006).

It should be noted that attempts to extract DNA from tissues in 70 percent ETOH have produced variable results. For example, Criscuolo (1992, 1994) was successful even with specimens preserved one hundred years previously in brandy, as was Persing et al. (1990) using specimens preserved between 1945 and 1951, but Seutin et al. (1991) were much less successful using tissues recently preserved for just six to eleven weeks. In some studies, the authors have made the assumption that they were using specimens that were fixed in formaldehyde without knowing whether or not the specimens had been, or what buffer (if any) had been added to the presumed formaldehyde fixative (e.g., Hughey et al. 2001), or whether or not the preserving fluid had been contaminated by trace amounts of formaldehyde or other chemicals. Stuart et al. (2006) reported successful extraction of DNA from bone of both skeletons and fluid-preserved specimens, including specimens that were probably fixed in formaldehyde.

Williams (2007) investigated the possibility of shipping specimens in 24 percent ethanol after preservation in 100 percent ethanol to see if the DNA was damaged, and concluded that the yield and quality of the DNA were affected by the preservative used, the length of time the samples were in the preservative prior to exposure to 24 percent ethanol, and the length of time the samples were in 24 percent alcohol (shorter time periods yielded better quality DNA).

It has been noted that most specimens preserved directly in ethyl alcohol (without formaldehyde fixation) have white eyes (caused by the lens turning opaque), while most specimens that are formaldehyde-fixed do not have white eyes. When selecting fluid-preserved specimens for DNA extraction, those with white eyes will probably yield better DNA than those without white eyes (Russ et al. 1997).

EVAPORATION

The loss of preservative fluid due to evaporation has been a problem in fluid collections since the first specimens were preserved in alcohol 350 years ago. Loss of specimens due to evaporation of alcohol so vexed Chamberlin (1925) that he proposed preserving specimens in mineral oil instead of ethanol (see table 1). As Rob Waller (pers. comm.) has pointed out, the entire circumference of a container closure has evaporation potential. In a collection of twenty thousand jars with an average lid circumference of 30 cm (about the size of a standard quart or liter jar lid) there would be a 6,000 m (or 6 km) long line of evaporation potential.

Evaporation from jars can be significantly reduced by wrapping the jar/lid junction with polypropylene/acrylic-adhesive (PPA) tape (Gillette 2004; Steigerwald and Laframboise 1996), by using Teflon plumber's tape around the jar threads, or by inserting a sheet of Teflon film or parafilm between the container and the closure (Simmons 2002).

The complicating factor with evaporation is that ethanol evaporates from a mixture faster than water because alcohol and water form a binary azeotrope rather than a true solution (see discussion in chapter 1). Nevertheless, in most collections the practice has long been to top-up containers with storage strength alcohol, which results in a steady decline in preservative concentration over time, despite a few published recommendations to use a stronger solution for topping up (e.g., a replacement solution of alcohol "80 percent or stronger" was recommended by Anon. [1938b]). In the binary azeotrope formed by ethyl alcohol and water, each compound retains its own properties, thus the vapor pressure of alcohol and water are different. The vapor phase concentration "approximately quantifies the tendency for solutions to become more dilute in ethanol as they evaporate" (Waller and Strang 1996, 77–78), because "at 50% relative humidity, the solution concentration lost by evaporation would be equivalent to about 95% v/v ethanol" (Waller and Strang 1996, 77–78). The lower concentrations of alcohol may allow bacterial growth in containers (see figure 3.3). At room temperature, the vapor pressure of ethanol is about 55 percent greater than that of water. There is some data on this phenomenon from a few studies of collections that were topped up with storage strength 70 percent ethyl alcohol (see table 18). In a small, unpublished study conducted in 1987, I compared alcohol evaporation from containers with no lids and loose lids and found that in open containers, after eighteen days, there was a total fluid loss due to evaporation of 66–81 percent with a loss of ethanol concentration of 16 percent. In the containers with loose lids, over the same time period there was a total fluid loss of 3–12 percent with a loss of ethanol concentration of 2–3 percent.

When bacterial or mold growth does occur in a container, the specimens should be removed, cleaned, staged back to storage strength preservative, and placed in a clean

Figure 3.3. Bacterial growth resulting from low concentration of alcohol preservative.

container. The contaminated preservative should be discarded, the container should be thoroughly washed, and the container mouth and closure carefully inspected for flaws before being returned to service. If the specimen is not fragile, bacteria or mold growth can be carefully removed with storage-strength preservative and cotton swabs. Moore (2005a) described the removal of a gelatinous fungal mycelium from a fragile fluid-preserved specimen in a container with a low preservative concentration by cleaning the specimen in 30 percent ethyl alcohol, repairing the specimen with celloidin, and restoring it to a 70 percent alcohol preservative.

CLEARING AND STAINING

Clearing and staining is a technique in which the cartilage or bones of a specimen are stained, other tissues are rendered transparent, and the specimen is retained as a fluid preparation. The technique employed in clearing and staining may be as simple as the supposedly reversible procedure of dipping a frog toe in cedar oil (Noble 1917), but more commonly the process results in a permanent alteration of the specimen (see table 10 for a review of clearing and staining techniques). The cartilage stains originally employed were methylene blue or toluidine blue, which were later replaced with a more stable stain (alcian blue). Bone has traditionally been stained with alizarin, which was used as early as 1559 by Levinius Lemnius (1505–1568), who extracted it from madder root, *Rubia tinctorum* (Harris 1960; Hollister 1934); alizarin was first prepared as a reagent in 1868 (Hollister 1934).

During the clearing and staining process, the specimen is preserved, skinned, and gutted, and large muscle masses are removed. The specimen is stained for cartilage, bone, or both, cleared with a strong base (usually potassium hydroxide, as first published by Schultze in 1897), and then subjected to treatment with a digestive enzyme such as pancreatin or trypsin to reduce the quantity of tissue remaining on the bones. Because cleared and stained specimens are so fragile, they are usually stored in glycerin or dilute glycerin (see previous discussion). Moore (2006) notes that most cleared and stained specimens require little conservation other than treatment for fungus, and cautions to always stage specimens between solutions to avoid problems with osmotic pressure. In a related study of skeletal preparation, von Endt et al. (1999) found that trypsin maceration at 35–38°C at neutral pH did not result in changes to amino-acid composition of the bone. Mabee et al. (1998) found that clearing and staining fish specimens resulted in an overall shrinkage of 3–6 percent from live length (see table 15).

ANATOMICAL AND HISTOLOGICAL FLUID PREPARATIONS

A bewildering array of chemical mixtures have been used in the preparation of anatomical and histological fluid-preserved specimens (see table 9). For example, a widely used 1831 anatomical preparation reference by Usher Parsons (1788–1868) stated that:

> Many preparations, and almost all pathological ones, are best preserved in some antiseptic liquid. Those generally employed are, 1. Alcohol, either pure or diluted in the form of proof spirit; 2. Alcoholic solutions of a metallic, an alkaline, or an earthy salt; 3. Aqueous solutions of the same salts; 4. Oils, particularly oil of turpentine; 5. Acids; 6. Acids and alcohol; 7. Alcohol and oils; 8 Ammonia.
>
> The alcohol used for preserving specimens may vary in strength, according to their size and thickness. All those that are thick and massy, should be put into pure rectified spirits; smaller ones may require only half the quantity of alcohol with water, and such as are thin, as membranes, require only common proof spirit.
>
> The salts usually added to proof spirit, to increase its antiseptic quality, are, muriate of mercury, nitrate of potash, muriate of soda, muriate of ammonia and allum. When muriate of mercury is used, its weight should be equal to one fourth that of the preparation. The other salts may be considerably short of the strength of saturated solutions, especially when dissolved in spirit, as the unavoidable evaporation of spirit will cause the salt to crystallize.
>
> The essential or volatile oils, and especially that of turpentine, are recommended for cartilages, fibrocartilages, tendons, and fibrous membranes to give them a beautiful transparency. They also preserve heads better. . . . All those parts that are injected with matter that is soluble in alcohol, or which have bones in them, which acids would destroy, may be preserved in volatile oils, provided they are not bulky.

> The acids used, are sulphuric, nitric, muriatic, acetic and pyroligneous, diluted very much with distilled water. . . .
>
> Proof spirit with a small portion of sulphuric, nitric, or muriatic acid is valuable for preserving the brain and nerves. (Parsons 1831, 113–115)

Parsons further recommended that preservative fluids be clear and free of color, and cautioned that all fluid preservatives will "destroy the color of raw surfaces, and thus change their natural appearance, and destroy the distinctive hue of the different textures" (Parsons 1831, 115–116), but admitted that "They are the only means however of preserving morbid specimens although they materially alter their natural appearance" (Parsons 1831, 116). Parsons offered a solution to this problem for preserving natural colors, based on a mixture recommended by "Mr. Gaskoin, in the London Medical Gazette 1828" (Parsons 1831, 116)—wash the specimen carefully with a solution of muriate of ammonia using a soft sponge, then wrap the specimen in linen and immerse it in a solution of one part saturated ammonia solution with two parts rectified spirit of wine for three days. Unwrap the specimen and allow it to macerate for 10–15 days, then rinse it again and preserve it in a solution of one part saturated ammonia, with salt, and one part distilled water and rectified spirit of wine. More salt may be added so the specimen appears less "corrugated."

In general, for preparing fluid-preserved specimens, Parsons (1831) recommended maceration in water, removal of unnecessary tissue, then suspension in a glass container of alcohol. Hollow structures could be filled with "curled hair, wool, cotton, or the like, and small ducts and vessels are exhibited by the introduction of bristles, quills, or bougies" [a bougie is a thin, flexible surgical instrument that can be inserted in a canal or other passage] (Parsons 1831, 120) which were removed after the tissues had hardened for ten days or so. Parsons recommended suspending specimens in fluid using glass floats or corks rather than the then standard method of using threads pulled over the rim of the container because the threads act as capillary siphons. He cautioned against the practice of adding a layer of oil to the surface of the alcohol to reduce evaporation, as the oil could discolor the fluid and the specimen.

Jean Nicolas Gannal (1791–1852), in his *History of Embalming and of Preparations in Anatomy, Pathology and Natural History* (Gannal and Harlan 1840), published in French in 1838 and in English translation in 1840, reviewed the fluids recommended for preserving anatomical specimens. Gannal stated that dilute acids could be used to preserve the suppleness of the tissues, but cautioned that muriatic acid could turn the tissue surface gelatinous or transparent; nitric acid tended to tarnish and contract tissue; and sulfuric acid bleached the tissues. Gannal stated that alkaline liquors were rarely used, but that "Those salts derived from the combination of acids and earths, the alkalies or metals, may be employed like the acids diluted with water" (Gannal and Harlan 1840, 165). Concerning alcohol, Gannal wrote that:

> Alcoholic liquors are most generally used for the preservation of animal substances, if they are more costly they are liable to fewer objections. Brandy, rum, tafea, are colored by a resinous substance which clouds their transparency, and which is liable to be deposited. The alcohol of cherries, of grain, of cider, or of wine, is preferred at present, which

> can be procured well rectified and transparent, and which may be afterwards weakened with distilled water, so as to obtain alcohol very limpid, marking from 22° to 30° of Baume's areometre. (Gannal and Harlan 1840, 167) [Tafea is an alternate spelling for tafia, an alcoholic beverage made from sugar cane, similar to rum but rarely aged; Baumé 22–30° is about 55–74% ethyl alcohol.]

In 1893, Pierre Augustine Fish (1865–1931) summarized "old and new methods" of brain preservation. The fluid-preservation techniques included hardening the brain in alcohol prior to placing it in a carbonated or pure alkali solution for two days, and preserving the brain in a solution of creosote water. Fish argued that:

> Alcohol is the oldest and most universal preservative employed. It has good "fixing" properties but needs considerable attention in order to produce the best results. For fixing, it is frequently used in conjunction with some of the various salts, or in case some non-alcoholic fixer is used, it supplements or completes the hardening thus begun. As a preservative, it is generally used at the ordinary commercial strength—ninety to ninety-five percent, although for most tissues eighty or even seventy-five percent seems to suffice. (Fish 1893, 397)

For "general utility, economy and certainty of result" (Fish 1893, 398) hardening brain specimens in 2–5 percent potassium bichromate was recommended, followed by immersion in absolute alcohol to remove the "abnormal color" of the fixative. Other brain fixatives discussed included corrosive sublimate (mercuric chloride, either aqueous or alcoholic). Various chemicals could be added to reduce the caustic action of the mercuric chloride, including camphor or tincture of iodine. In the late 1890s, W. Lloyd Andriezen (1868–1906), a British pathologist, reported that he preserved human brains by using "sublimate and alcohol fixation" (Andriezen 1894, 574). (Note: sublimate refers to mercuric chloride.)

Very few analyses have been conducted of anatomical, histological, or medical museum specimens, despite their prevalence in museums. In 1994, Christine Thede analyzed 193 samples of fluid-preserved specimens from the Mütter Museum in Philadelphia, some dating back to the mid-1800s (Thede 1996), checking container and specimen condition, pH, formaldehyde concentration (by titration), and the presence of arsenic and lead. Of the samples tested, 21 percent contained arsenic and 38 percent contained lead (see previous discussion); the average pH was 4.2; formaldehyde concentration averaged above 10 percent.

In addition to the chemicals mentioned earlier, methyl benzoate or turpentine were often used in conjunction with anatomical dyes or with mercury injection to render tissues transparent (Simon Moore, pers. comm.).

In their wonderful volume, *Medical Museum Technology*, J. J. Edwards and M. J. Edwards wrote that "The use of formalin did not merely add something new to existing methods but it completely changed the whole process of fixation and preservation" (Edwards and Edwards 1959, 73).

The major pre- and post-formaldehyde fixatives and preservatives used in the preparation of most anatomical and histological preparations are summarized in table 9. The history of anatomical and histological fluid preparation reflects a particular emphasis on the preservation of colors in specimens, as reviewed by Silvester (1922). The best known of the post-formaldehyde techniques for color preservation is the one developed by Johann Carl Kaiserling (1869–1942) and its numerous variations, first published in 1896 (Kaislerling 1896, with revisions in 1897 and 1900) and many other variations since (e.g., Judah 1922; see table 9 for others). An interesting series of publications came out of the need during World War I to preserve postmortem tissues with war wounds to develop better treatment methods, including a proposed universal preservative used by the British that allowed for easy preparation and shipment of specimens (Abbott 1918) and a technique developed by Judah (1918) that did not use glycerin.

MOUNTING SPECIMENS INSIDE CONTAINERS

Mounted fluid-preserved specimens are most often attached to glass or acrylic plates or frames inside a glass or acrylic container. Historically, many other kinds of backing plates and frames have been used, including Bakelite, plaster, and metal. Specimens were often suspended by threads or monofilament from fixed mounts or attached to glass or cork floats. Some older jars were manufactured with loops on the underside of the closure for the attachment of threads for suspending specimens; wooden, metal, or plastic mounts may be also be found wedged or glued to the inside of container or closure or across the mouth of the container (see figure 3.4). Metals, wire, most plastics, and colored thread should not be used to mount specimens, as metal will corrode in a

Figure 3.4. Metal bar across mouth of glass jar for suspension of specimen.

preservative, plastic will deteriorate, and colors will leech from plastic and thread and discolor the preservative and the specimen. Some specimens require special mounts—for example, some marine invertebrates are supported by acetate discs beneath the mantle attached by a thread to a suspension point above (Simon Moore, pers. comm.).

Moore (2009b) recommended celloidin (pyroxylin, or nitrocellulose) adhesive for mounting specimens preserved in alcohol, and gelatin for specimens preserved in formaldehyde solutions. According to Moore (2009b), celloidin is known by at least three other names, necoloidine, pyroxylin, and collodion. Necoloidine is an 8 percent solution of pyroxylin (dry nitrocellulose flakes moistened in 70–80 percent alcohol to form a gel that produces a hard, semi-opaque solid with an adhesion strength of about 100 g per 0.5 mm layer up to about 5 kg for a 2 cm thick layer). Pyroxylin is nitrocellulose flakes dissolved in a 50:50 mixture of absolute ethyl alcohol and di-ethyl ether. Pro-celloidin or collodion are other names for pyroxylin powder. To use celloidin as an adhesive, remove the specimen from the preservative, allow it to dry slightly, and then pipette the solution on the surfaces to be joined. Allow the adhered specimen to set for about ten minutes before returning it to the preservative solution (see figure 3.5). Celloidin adhesives are reversible by using the solvent they were prepared with.

Gelatin-based adhesives may also be used as jar sealants and are reversible in warm water. Moore (2009b) recommended using only leaf gelatin in order to obtain sufficiently long molecular strands of gelatin. The gelatin leaves should be hydrated in cold water until limp, then carefully heated using a hot water bath until they are molten (overheating will break the strands, making the adhesive ineffective). To the warm molten gelatin solution add a mixture of 2.5 mL glycerol (to prevent dehydra-

Figure 3.5. Alcohol-preserved snail specimen adhered to glass with celloidin.

tion) and 1.5 mL of glacial acetic acid (as a fungicide and glass-binding agent), then pour the solution onto a glass sheet and allow to form a soft, rubbery, and sticky mass which should be stored in an air-tight container. To use a gelatin adhesive, remove the specimen from the preservative, allow to dry slightly, and then apply a melted gelatin solution to the surfaces to be joined. Return the specimen to the preservative when the gelatin has cooled.

GLYCOL, PHENOL, AND PHENOXETOL AS PRESERVATIVES

Various ethylene glycol and propylene glycol compounds and their derivatives as well as phenol have been proposed as alternative preservatives (see table 17), particularly for specimens that are on exhibit. The limitations of the antibacterial action of ethylene glycol have been known for some time. In addition, glycol-based preservatives may cause paper labels to become transparent and then deteriorate (Simon Moore, pers. comm.).

The use of a modified phenol-based embalming fluid for fluid preservation of birds was suggested, but with the caveat that "phenol slowly decalcifies bone; toluene might prove to be an adequate substitute" (Berger 1955, 300). The same author later noted that many of the components of other preservatives, including embalming fluids, "may seriously interfere with microtechnic staining methods" (Berger 1956, 452).

I have worked with specimens on exhibit in two different proprietary glycol-based compounds, and noted that the manufacturers of these two solutions make no claims for the effectiveness of their products for long-term storage; rather, the chemical mixtures are sold as "holding fluids" that are safer alternatives to formaldehyde and alcohol for classroom use of fluid-preserved specimens. After keeping specimens in both fluids for a little over a year, both Carosafe and Ward-Safe turned unacceptably cloudy and the specimens had to be returned to ethanol preservative. The cloudiness is probably due to a chemical change involving the glycol compound, but no further studies have been done to investigate this phenomenon.

The disinfectant actions of some chemicals used as or proposed as preservatives are summarized in table 6.

Phenoxetol (also called 2-phenoxyethanol or ethylene glycol monophenyl ether) was suggested as a preservative for biological organisms in 1956 (Owen and Steedman 1956) based on a short (six-month) trial period. The preservative properties of phenoxetol for whole animals were discussed in the context of evaluating how well it preserved proteins, and based on its antibacterial and antifungal properties (Nakanishi et al. 1969). At that time, it had already been in use for an unknown period at the British Museum for postfixation preservation of tissues. A test of phenoxyethanol as a preservative for human cadavers initially obtained good results, but over a very short time frame of just three years (Wineski and English 1989). When specimens preserved for approximately ten years in phenoxetol began to deteriorate (Crimmen 1989), it was determined that the phenoxetol was unable to penetrate into dense tissues (Moore 1997, 1999). An analysis of the antibacterial properties of phenoxetol was published in 1944 that noted that "the substance could not be described as a powerful bactericide" (in fact it was less powerful than phenol) but that it was highly active against a few specific bacterial strains (Berry 1944, 175).

NOVEC FLUID

A "designer fluid" produced by 3M corporation (Novec HFE-7100) has been proposed for use as holding solution for specimens on exhibit (Drahl 2008). Novec is a methoxy-nonafluorobutane compound ($C_4F_9OCH_3$) that was developed to clean, degrease, rinse, and dry electronics. It is clear, colorless, and odorless, has a high boiling point, low surface tension, and a density of 1.52 g/cm^3.

The Smithsonian Institution began using Novec in 2006 in an exhibit of marine life as an experiment. While Novec has wonderful optical characteristics and specimens can be seen very clearly in it, it has several serious drawbacks, including the fact that it is not a preservative—Novec forms an envelope around the specimen that is supposed to keep certain preservatives (such as formaldehyde) trapped in the specimen, but remove others (e.g., alcohol). As a consequence of Novec's designed cleaning action, as the fluid removes alcohol from specimens, they begin to deteriorate. In order to keep specimens from deteriorating while in the Novec, the specimens on exhibit at the Smithsonian were transferred from alcohol to a formaldehyde solution. Over time, however, Novec also began to extract formaldehyde from the specimens. Because the density of Novec is approximately 1.5 times that of water, fluid-preserved specimens that are placed in it become so buoyant that they must be restrained to remain below the surface of the fluid, which can cause serious damage to the specimens.

Unlike most other preservatives, Novec can be reused, but it is extremely expensive.

MINERAL OIL

One promising alternative fluid that has been suggested for exhibition purposes but rarely used is mineral oil (Chamberlin 1925; Waller 1992). Mineral oil is a petroleum distillate that is a clear, colorless fluid composed mostly of alkanes and cyclic paraffins, with a density of approximately 0.8 g/cm^3, available in varying grades. Mineral oil is an excellent barrier against water vapor, so it would also likely form a good barrier against alcohol and formaldehyde as well (Waller 1992). Chamberlin (1925) recommended killing and dehydrating arachnids and small insects in 95 percent ethanol; making one change of alcohol; transferring the specimens to a solution of 25 percent carbolic acid crystals and 75 percent xylene for some time; transferring the specimens to pure xylene (to remove the carbolic acid, which is immiscible in oil); and storing the specimens in "one of the highly refined, colorless and highly viscous mineral oils" (Chamberlin 1925, 634). Haly (1892) suggested displaying preserved specimens in "carbolized oil," a mixture of coconut oil and carbolic acid mixed to achieve "specific gravity of 10° and 20° below proof-spirit" (Haly 1892, 212).

UNIVERSAL FIXATIVES

There have been several attempts to concoct what one author referred to as "the universal fixative" or universal preservative, a name that was first applied to FAA, a mixture of 6 1/2 cc of formaldehyde, 2 1/2 cc of glacial acetic acid, and 100 cc of 50 percent alcohol (Conant 1924, 1926). FAA was first promoted as "a universal killing and fixing agent" for both plant and animal tissues and credited to T. E. Rawlins of the University of Wisconsin and University of California, who used it in his work on tobacco plants (Conant 1924, 25). One caution about such mixes, aside from the lack of comparative data and the lack of understanding of how they work chemically in the tissues, is that "the fixing reagents in a mixture do not penetrate the tissues *en masse*, but that each reagent penetrates tissue in a characteristic sequence" (Dempster 1960, 68).

CRITERIA FOR EVALUATING ALTERNATIVE FIXATIVE AND PRESERVATIVE FLUIDS

Andries van Dam (2003) proposed a set of criteria for evaluating alternative fluid preservatives, including antiseptic action (the fluid should be bacteriostatic and fungistatic); a water/fat solubility coefficient greater than one (to minimize migration of the biocide into fatty tissue, which causes a loss in antiseptic strength); pH range (the antimicrobial activity should be pH dependent); stability and reactivity of the preservative (the fluid should be resistant to deterioration from exposure to light, oxidation, buffers, and container and closure materials); toxicity (the fluid should have low toxicity); and flammability (the fluid should have low flammability). These criteria and others are presented in table 12; the disinfectant mechanisms of some preservatives in table 6.

Analyzing a variety of fluid preservatives, including some used in food and cosmetics, van Dam rejected glycols (they are not true biocides), corrosive biocides (e.g., halogens, anilides, peroxygens), benzoates, bis-phenols, phenols, quaternary ammonium compounds, and other chemicals, suggesting instead that DMDM-hydantoin is worth investigating. DMDM-hydantoin (also known as dimethylodimethyl hydantoin, or 1,3-Bis(hydroxymethyl)-5,5-dimethylimidazolidine-2,4-dione) is a formaldehyde-releasing compound widely used in cosmetics. It is also known by the trade name Glydant. There are contradictory studies concerning how dangerous exposure to it is. In addition, there is evidence that with long-term use DMDM-hydantoin causes extra-molecular protein cross-linking, resulting in gradual denaturation of the proteins and thus damage to the specimen (Simon Moore, pers. comm.).

Carter (2012) compared tissue preservation by DMDM-hydantoin (10 percent), formaldehyde (4 percent, buffered and unbuffered), and ethanol solutions (absolute alcohol and 80 percent industrial denatured alcohol) in small cubes of lean pork muscle tissue, using Fourier transformation infrared (FTIR) spectroscopy to monitor chemical changes brought about by the preservatives, which are detected by differences in molecular-evel vibrations. By closely analyzing the Amide I and Amide II peaks produced by FTIR, Carter was able to compare the effects of denaturation on the proteins in the samples—the alcohol-preserved samples showed little variation from fresh samples, whereas the formaldehyde and DMDM-hydantoin–prepared samples showed much greater denaturation; the chemical effects on tissues of the DMDM-hydantoin were very similar to those of formaldehyde.

4

Effects of Fixatives and Preservatives on Specimens

As noted in chapter 1, many of the early descriptions of fluid-preserved specimens proclaimed them to be perfectly preserved, but often the writer was describing the preserved animal without having seen a living specimen of the species, or was making a subjective comparison to other preservation methods available. Fixation and preservation may affect specimens through dehydration, rehydration, shrinkage, swelling, alteration of colors, changes in skin texture, or extraction of chemical components. Despite the fact that some of these changes have been observed since the very beginning of fluid preservation, few have been investigated in detail. Most fixation- and preservation-induced changes are ignored by the workers who use specimens for scientific work, other than noting colors in preservative compared to colors in life. Nevertheless, numerous chemical mixtures have been proposed to reduce or eliminate fixative- and preservative-induced changes, but their efficacy is mostly anecdotal. There are few controlled studies investigating changes to specimens reported in the literature, with the notable exceptions of Lai (1963) and Lee (1982), and several papers addressing biomass changes in invertebrates and fish. A summary of reported changes in invertebrate specimens is provided in tables 13 and 14, and for vertebrates in tables 15 and 16.

In general, specimens tend to swell in fixatives and then shrink when transferred to preservatives. In formaldehyde-based fixatives, most swelling (including weight gain) occurs within the first few hours or days of exposure (depending on the fixative and the type of specimen). Transferring the specimen to an alcohol preservative usually induces shrinkage and weight loss, with most of the change occurring within the first few days of exposure (depending on the fixative and the type of specimen). These changes are due to the fact that formaldehyde-based fixatives are mostly water (the standard 1:9 dilution of formaldehyde is about 95 percent water), which causes specimens to hydrate, while most alcohol-based preservatives are around 50 percent to 30 percent water, which causes specimens to dehydrate. The use of more complex fixatives and preservatives tends to produce more complicated effects. In addition, some workers have recorded handling-induced changes, usually in the form of dehydration or interference with the penetration of fixatives or preservatives into tissues.

For invertebrates, the most commonly reported changes include specimen swelling, shrinkage, weight loss, color changes, embrittlement, and loss of structural integrity (some specimens become flaccid). For example, the pattern of swelling in fixatives followed by shrinkage in preservatives was reported in a number of taxonomic groups by Donald and Paterson (1977) and Mills et al. (1982). Changes in the structural integrity of specimens have been reported in microplankton (Stoecker et al. 1994), actinopods (Beers 1976a), crayfish (DiStefano et al. 1994), and octopuses (Voight 2001). The loss of minerals or organic components of specimens have been reported in crustaceans (Hopkins 1968) and isopods (Hendrickz et al. 2003). Variable preservation was reported in ciliates by Modigh and Castaldo (2005). See table 13 for more details gleaned from published reports.

For vertebrates, most of the work on fixative- and preservative-induced changes has been done with fish, particularly investigations of shrinkage to compare measurements of live or fresh specimens with those of preserved specimens. Some authors (e.g., Shetter 1936; Sigler 1949) have calculated conversion factors for comparing fresh and preserved specimens. Lai (1963) and Lee (1982) studied the effects of formaldehyde fixation and alcohol preservation on morphometric characters in fish and anura, respectively.

The loss of mineral content in fishes was reported by Gibbs et al. (1974) and damage to otoliths by McMahon and Tash (1979). Loss of muscle definition (Winokur and Hillyard 1992) and changes in skin pustularity (Nelson 1971) have been reported in anurans. Stuart (1995) investigated the darkening of color patterns with increasing time in formaldehyde in lizards. Scott and Aquino-Shuster (1989) reported changes induced by freezing frog and snake specimens prior to preservation. See table 15 for more information and further examples of induced changes in vertebrate specimens.

CHANGES IN BODY DIMENSIONS AND BIOMASS

The effects of fixatives and preservatives on specimens that has drawn the most attention of researchers are dimensional and other biomass changes, as summarized in tables 13 and 14 for invertebrates, and tables 15 and 16 for vertebrates. Nevertheless, little research has been done on the specific effects of fixatives and preservatives on the particular tissues involved. We do know, for example, that elastin (a fibrillar protein in connective tissue and blood vessels) swells with exposure to formaldehyde and alcohol, but does not go into solution (Mukherjee and Hoffman 1971), but such studies are rare.

The literature on changes in dimensions and biomass during fixation and preservation has been reviewed by a few authors (e.g., Ellis 1987). These changes seem to be due primarily to dehydration as water in the tissues is replaced by alcohol. All preserving alcohols cause tissue shrinkage, but the amount of shrinkage is variable, with less shrinkage reported to occur in methyl alcohol and more shrinkage reported to occur in isopropyl alcohol compared to ethyl alcohol (Ciferri 1971). Preservation techniques can diminish or exaggerate some of the biomass changes—for example, preservation directly in alcohol (without a fixative) may result in shrinkage of cellular contents (syneresis) thereby causing distortion because alcohol penetrates tissues rapidly and has low specific gravity (Moore 1999). In general, the effects of biomass changes are much more pronounced in invertebrates than in vertebrates.

Many researchers have tried to establish a percentage of expected change that can be used to make better comparisons between preserved and nonpreserved specimens. Unfortunately, a review of the results of various studies shows too wide a range in variation to be useful, probably because of the variety of factors causing dimensional changes. For example, for anurans, Lee (1982) found dimensional changes of as much as two orders of magnitude in postpreservation in one toad species, with snout-vent shrinkage of 6.19 percent, while Deichmann et al. (2009) found snout-urostyle shrinkage of 0.31–5.6 percent in fourteen species. Fish studies have reported shrinkage of anywhere from 2–12 percent and more in body length (see table 15). Klauber (1943) and Reed (2001) reported shrinkage in length of snakes but no significant changes in body mass. Thibault-Botha and Bowen (2004) reviewed a large amount of the literature on body mass changes and concluded that gelatinous zooplankton are much more prone to shrinkage than other zooplankton due to their high water content and lack of strong structural features or hard coverings, and that body mass loss is generally higher in species with a higher surface to volume ratio. Some studies showed increases in length or weight following initial shrinkage of tissues. Radtke (1989) found significant shrinkage occurred at death in larval cod, but no shrinkage occurred when live larvae were placed in 95 percent ethanol, leading him to conclude that the shrinkages reported in the literature "were probably due to the handling associated with the preservation of specimens in these studies" (Radke 1989, 1893) rather than the preservatives. Theilacker (1980) found that handling of anchovy specimens may cause significant shrinkage prior to fixation or preservation. As summarized by one researcher who reviewed the literature, "The effects of preservatives on fish morphometrics are difficult to predict because of variance related to type of preservative, duration of preservative, origin of species (marine or freshwater fish), species, life stage, and others" (Sagnes 1997, 910).

Changes have also been noted in skin features due to differential preservation such as dermal pustularity of microhylids (Nelson 1971) and changes in the patterns of snakes (Smith 1955) and lizards (Smith 1975).

CHANGES IN COLOR

Color changes may result from chemical alterations of the tissues, physical alterations of the tissues, or both, and may involve the loss, acquisition, or alteration of color. Color is conferred chemically by pigments, and structurally conferred by interference, light scattering, or refraction. Fluid fixatives and preservatives may both extract pigments and cause structural changes. For example, lipochromes (responsible for yellows, oranges, and reds) are alcohol-soluble carotenoids (Pettingill 1970). Fry (1985) described how the loss of color (yellow) due to submersion in alcohol in some bird specimens led to an erroneous description of a taxon as a new subspecies; Fry confirmed the color loss by subjecting dry specimens to submersion in alcohol, finding that green feather colors became blue-green, pure yellow became ivory, scarlet became pale buff, bright pink became buffy white, carmine became gingery brown, yellows and reds were suppressed, and that alcohol preservative had no effect on blue, orange-buff, or black. Another example of a color change that affected the usefulness of the specimen was in a note by Bliss (1872) concerning the unexpected appearance of a diagnostic vermillion spot on the abdomen of an alcohol-preserved cyprinoid fish that was not present when the fish was alive. Dissection demonstrated that the color was a true pigmentary color, and was probably visible only when the fish was in reproductive mode or preserved in alcohol.

Green-to-blue color changes in many fluid-preserved vertebrates (particularly amphibians and reptiles) occur when xanthophores (responsible for yellows) are leached out by preservatives and the remaining iridophores are altered by dehydration, which affects the interference of light. In at least one instance, an artist, apparently working with a freshly preserved specimen, captured the green-to-blue color change in mid-process in a snake specimen (Simmons and Snider 2012).

Prum et al. (1994), investigating how colors are produced in bird facial skin, noted that in one genus of birds, a bright green structural color produced by the reflection of light on ordered collagen fibers was better preserved in a 2.5 percent gluteraldehyde fixative than in standard 10 percent formalin. In specimens fixed in formaldehyde and preserved 70 percent alcohol, the bright green changed to violet or blue, due to shrinkage of the collagen fiber arrays.

Color changes in specimens may also result from reactions with copper (in the form of copper wire or a copper container), oxidation of metal containers or lids, or metal needles, pins, or probes. Leonhard Stejnener's instructions for field preservation (of amphibians and reptiles) recommended the use of a color standard due to the loss of colors in preservative solutions (Stejnener 1891), a suggestion which has been ignored by most fieldworkers since (Simmons 2002).

Specimens are sometimes darkened by exposure to formaldehyde, a condition commonly referred to as formaldehyde brown or formaldehyde gray, which may be caused by several factors including the interaction of acidic formaldehyde with metal trays or tags, or the formation of what is commonly referred to as formalin pigment which may form as the formaldehyde reacts with the hematin in hemoglobin that escapes from the red blood corpuscles at cell death. Studies have also shown that specimens become darker the longer they are left in formaldehyde (Stuart 1995; see also the section, Unwanted Effects of Formaldehyde in chapter 2).

As a consequence of the color changes during preservation, there have been many attempts to concoct a preservative mixture that will retain colors. Probably the best known of these is the Kaiserling's method (see table 9), originally proposed for anatomical specimens. It is important to note that there are many variations of the recipes for Kaiserling's solutions in the literature; Kaiserling himself published several variations (e.g., Kaiserling 1896, 1897, and 1900), as did other workers (e.g., Craig 1914; Edwards and Edwards 1959).

Various published methods proposed to preserve color in fishes were reviewed by Borodin (1930), who found all of them to be unsatisfactory, at the same time complained about the damage that formaldehyde does to scales and bones (most likely due to the use of inappropriately buffered or unbuffered formaldehyde). Based on the ingredients then found in British-made IMS that were used for a few jars of fishes from the Red Sea that had retained much of their color, Borodin recommended a formula consisting of thirty parts alcohol, two parts formaldehyde, one and one-half parts wood tar, and sixty-six and one-half parts water saturated with common salt for the preservation of color in fishes. Gerrick (1968) used fifteen commercial antioxidants dissolved in either 10 percent formalin or 40 percent isopropyl to preserve colors in four species of fish for a two-year period. The best preservation of color was 1 percent erythorbic acid in 10 percent formalin. Ionol CP-40 preserved reds, but the other antioxidants failed. He found that "isopropyl alcohol, while an excellent solvent for antioxidants, was ineffective as a color-preserving fluid because animal pigments were highly soluble in it" (Gerrick 1968, 240).

One of the more creative attempts to preserve color was based on experiments conducted at the Colombo Museum (Sri Lanka), using coconut oil and carbolic acid with glycerin (Haly 1892). More recently, an antioxidant (BHT, or butylated hydroxytoluene, sold under the trade name Ionol or Ionol-40), was advocated to preserve colors (Smith 1995; Waller and Eschmeyer 1965; White and Peters 1969). Windsor (1971) recommended the use of a 50 percent solution of ammonium sulfate to preserve color in frogs, although he noted that the preservative caused excess dehydration of the specimens. Hildebrand (1968) recommended two procedures for color preservation. The first was a variation of a method published by Sheim (1951), which called for fixing specimens in a solution of 50 mL of formaldehyde buffered with 5.9 g dibasic sodium phosphate and 4.7 g monobasic sodium phosphate with sufficient tap water to make 1000 mL for about a week; then preserving the specimens in a fresh solution of the same formula plus 5 g sodium hydrosulfite that was aged for a week and filtered (the author cautions that it is not suitable for fatty tissues and should not be used in metal containers). Hildebrand's second recommendation was a variation of the Owen and Steedman (1956) method in which specimens are fixed for a minimal amount of time in formalin or alcohol to which 10 g of sodium acetate per liter has been added, then washed twelve to twenty-four hours in running water, and stored in either a 1 percent mixture of ethylene glycol monophenyl ether in water or a 0.2 percent solution of para-hydroxybenzoic acid in water. Zenke (1991) describes a method using ethylene glycol to preserve the silvery luster in some fish species; a review of several recipes for color preservation can be found in Kessler (1989). Similarly, botanical researchers have proposed a number of chemical preservatives or additives, most targeted toward preserving particular colors in plants (table 1; see discussion under Botanical Use of Fluid Preservation in chapter 3).

SOLVENT EXTRACTION IN FIXATIVES AND PRESERVATIVES

The preservative fluid around a specimen includes the components extracted from the specimen. Alcohol is a good lipid solvent, and has been used for years to extract lipids from tissues (Bloor 1943; Johnson 1971). A study by von Endt (1994) showed that in preservative fluids, fats are hydrolized and migrate within the specimens, and are ultimately removed by the alcohol. The amino acid profiles he found in preservative solutions indicate that the specimens undergo a general protein loss, as well as some structural protein loss (von Endt 1994).

Extracted lipids in preservatives tend to float to the top (lipids are insoluble in water) and oxidize into fatty acids, which can lower the pH of the fluid and cause tissue breakdown (Dingerkus 1982; Moore 2002a; Moore 2005b). Figure 4.1 shows

Figure 4.1. Dissolved lipids in preservative with specimens.

lipids dissolved in an alcohol preservative; figure 4.2 shows lipids dissolved in a formaldehyde preservative (floating near the surface of the solution). Moore (2002a) reports on an instance of a stoat in a preservative so contaminated by emulsifying lipids that the pH of the alcohol preservative was 3.6, low enough to decalcify the animal's skeleton.

A relative estimate of the amount of lipids in the preservative can be made by pipetting a small amount of fluid into a Petri dish of distilled water on a black background and evaluating the turbidity—the lipids will appear as a white cloudiness around the pipette (Moore 2005b).

A study by Edwards et al. (2002) determined that formaldehyde fixation caused a decrease in stable isotopes in fish specimens.

Preservative fluids may become discolored due to solvent action, and this discoloration may affect the specimens. Figure 4.3 shows ink bleeding into an alcohol preservative (the label was added to the jar before the ink was sufficiently dry). Preserving fluids that have become discolored due to the presence of dissolved lipids may be cleaned by dropping charcoal particles in the fluid and shaking it, then filtering the charcoal particles out (Simon Moore, pers. comm.).

A few studies have concentrated on specific changes that can affect the research use of collections. Hendrickx et al. (2003) determined that using a preservative fluid in a pitfall trap (4 percent formaldehyde with a small amount of detergent to decrease surface tension) could cause significant alterations in metal concentrations in woodlice. Hopkins (1968) found that some crustaceans underwent a carbon loss of

Figure 4.2. Dissolved lipids floating at surface of formaldehyde preservative.

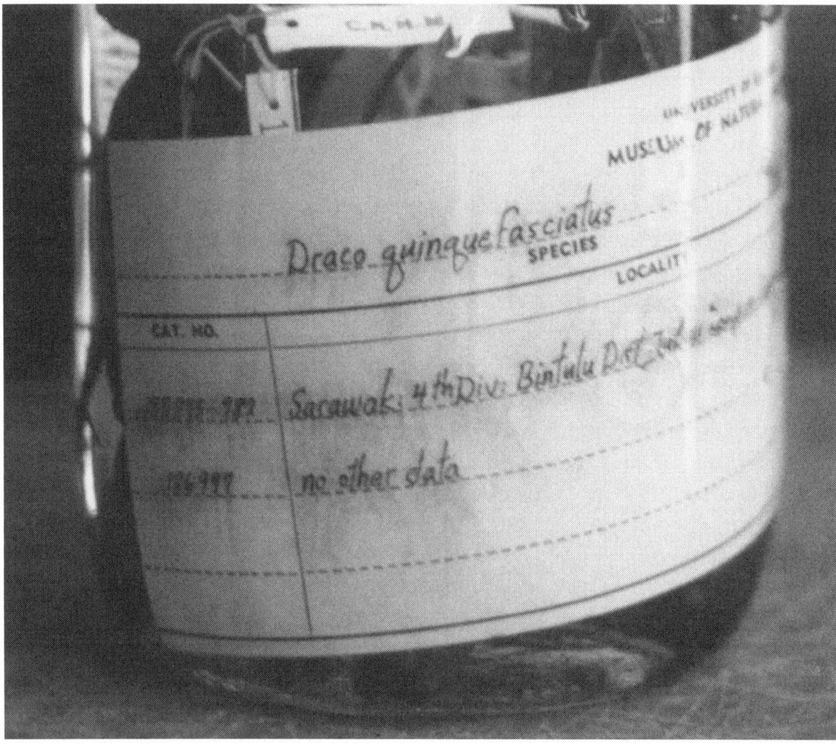

Figure 4.3. Ink bleeding from label in alcohol preservative.

17–23 percent and nitrogen loss of 19.2–21.0 percent in fixatives and preservatives. Morris (1972) found that marine zooplankton lost up to a third of their total lipids, suffered degradation of polyunsaturated fatty acids in the lipids, and some hydrolysis of lipids when preserved in 10 percent formaldehyde or 100 percent methanol.

Much work remains to be done on the effects of fixatives and preservatives on specimens, particularly the changes that affect the comparison of live and preserved specimens for scientific studies. Based on the evidence available to date, it is unlikely that a simple correction factor can be calculated for most taxa, due to the many variables that contribute to changes induced by fixation and preservation, including how specimens are handled, how fixatives and preservatives are mixed, which buffers are used, and how specimens are stored. A further complicating factor is that for most preserved specimens now in collections, there are no reliable records of how specimens were handled, fixed, and preserved.

5

Managing Fluid-Preserved Collections

Collections management includes everything that is done to care for and document collections and to make collections available for use (Simmons 2014). Good collections management makes a significant difference in prolonging the useful life of fluid-preserved specimens, particularly by (1) establishing links between the specimens and their associated data; (2) maintaining specimens and data in optimum condition; and (3) making the specimens and data available for appropriate use (Simmons 2002). Collections management should be based on the principles and practices of preventive conservation (which means caring for the specimens in ways that ensure their long-term usefulness), with particular emphasis on a stable collection storage environment and a well-ordered collection storage array.

Managing a collection, particularly a large collection of fluid-preserved specimens, can seem like an endless task—between checking fluid levels, monitoring the storage environment, processing incoming specimens, and making the specimens available for use, it is easy to get bogged down in details (Simmons 2002). To complicate matters further, the trend in collections over the last twenty years has been larger, more heavily used collections cared for by fewer and fewer collections care workers (Simmons 1993, 2013; Simmons and Muñoz 2006). Good organization and forward planning are keys to successful collections management.

In North America, the roles of curators, collections managers, technical assistants, and other collections care workers have been reviewed and analyzed by Cato (1991) and Simmons (1993) to find better ways to manage collections. Beginning in the 1970s, as natural history collections became larger and more complex and increasing demands were placed on curators to produce scientific research, the old system of managing collections with a scientist-curator and a minimally trained preparator proved to be inadequate. As a result, a new profession of collections managers evolved to fill the need. The duties of collections managers typically include tasks formerly assigned to preparators, and the collections management responsibilities formerly assigned to curators, resulting in a system that enables curators to focus on research while still providing the intellectual leadership necessary for research collections. In Canada, Waddington (1989) and Yamamoto (1985) described the collections management system in use at the Royal Ontario Museum, which allows each of the ten science departments to control their own acquisition, documentation, and access to specimens, while overseen by a coordinator of collections management who ensures that institutional standards are maintained. Cato et al. (1996) developed guidelines for providing mid-career training in collections management for museum staff members. In the United Kingdom, the National Museum of Wales collections management system includes all aspects of collections care, including registration, documentation, and conservation (Howlett and Horák 2006); this comprehensive form of collections management was implemented by developing a suite of collections management policies with a multidisciplinary group to identify collections management–related issues, combining conservation staff and collections care staff in the same curatorial departments rather than in separate units, and adding a director of collections and research to the administration. Using this system, the museum "significantly raised its standards of collections care" (Howlett and Horák 2006, 63).

The standardization of equipment, storage furniture, and supplies in an institution can result in significant savings. For example, when all fluid-preserved collections within the same institution adopt the same standard sizes of containers and closures, it results in a significant economy of scale cost savings. The use of standardized procedures for collections care tasks can greatly simplify the care of a large collection while ensuring continuity of quality. Standardized procedures should be based on recognized best practices (Simmons 2002). With standard containers and closures and standardized procedures, a single lab can service all of the fluid-preserved collections housed in the institution.

Many traditional practices, formerly accepted without question, should be reconsidered (Simmons 2013). For example, shelving for fluid-preserved collections is usually selected to maximize the use of space without consideration of other resource use, such as the amount of staff time required to select and return specimens and monitor specimen condition. Although it is possible to pack many more containers of specimens on large shelves, a single row of containers of fluid-preserved specimens on a shelf allows for fast and efficient monitoring of fluid levels, visual inspection of fluid quality, and makes it easier to maintain order in the storage array. When the shelves contain two or more rows of containers it becomes necessary to move the containers in the front one-by-one to check the containers behind them, which is time-consuming, increases the risk of disrupting container seals, and increases the likelihood of misplacing containers. A summary of traditional treatments and practices that are not recommended in the management of fluid-preserved collections is provided in table 23.

McGinley (1993) proposed the first useful quantitative method to assess the overall status of curation and conservation of natural history collections using a system called the Collection Health Index (CHI), based on a unit-by-unit analysis of the collection. Williams et al. (1996) adapted the CHI for use in vertebrate collections.

The CHI remains a reasonable quantitative measure that provides very useful comparative data across a variety of museum collections for determining where collections care resources can best be expended.

Although directed specifically at collections of amphibians and reptiles, the majority of the collections care and management issues discussed in Simmons (2002) can be applied to any fluid-preserved collection. Further information on managing collections in general can be found in Buck and Gilmore (2010) and Simmons (2014).

IDENTIFICATION OF FLUID PRESERVATIVES

Specimen containers should be labeled to indicate what preservative they contain. This is best accomplished by using a label format that indicates what fluid the specimen is housed in (e.g., 70 percent ethanol, 10 percent formalin). In some collections, different styles of containers or closures are used for specific preservatives, but it is more efficient to indicate the preservative on the label.

It is rare to have reliable, detailed preservation histories of fluid-preserved specimens, particularly in older collections, which makes identification of fluid preservatives (and diagnosis of preservation problems) difficult. In systematic collections, the most common preservative is ethyl alcohol, but most preservatives also contain small amounts of formaldehyde or other fixatives, depending on how the specimen was initially prepared (Waller and Simmons 2003; see also discussion in chapter 2). Older fluid-preserved specimens may have been exposed to any number of other fixatives or preservatives, which are probably present in at least residual amounts in the specimens or in the preservative surrounding the specimen. For example, Simmons et al. (2007) reported on elemental mercury found inside old specimens, probably as a result of their prior preservation in a mercuric chloride solution (see discussion in chapter 1).

Determining the composition of unknown preservative fluids is a challenging task, particularly in a large collection or if collections management funds are limited. Although with practice it is easy to distinguish formaldehyde, ethyl alcohol, and isopropyl alcohol by their odor, this practice is not recommended due to the health risks of inhaling formaldehyde fumes (see the section in this chapter, Health and Safety). The most accurate method to determine the composition of a fluid is to have a chemistry lab analyze a sample, but this is both costly and time-consuming. As an alternative, Moore (1999) recommended three methods for distinguishing preservative fluids:

1. The use of leuco-fuchsin impregnated indicator strips that react to the presence of formaldehyde (as described by Waller and McAllister [1986]).
2. The use of a digital density meter—these devices are expensive, but well worth the cost for a large collection. Moore (1999) cites data from Carter (1994) for specific gravity at 20°C of the following common preservatives:
    a. Ethanol, 80 percent—0.859 g/mL

   b. Formaldehyde, 4 percent—1.009 g/mL
   c. Formaldehyde, 10 percent—1.014 g/mL
   d. Formaldehyde, 40 percent—1.079 g/mL
3. The use of a simple, homemade gravimetric device (Moore 1983, 1994, 1999) using weighted plastic heads from map pins.

Unfortunately, since Moore published the instructions for making this device, most manufacturers of map pins have changed the composition of the plastic they use to one of greater density that does not float in alcohol and thus cannot be used to make the gravimetric device—other small plastic disks, beads, or spheres that are lightweight enough to float in alcohol can be substituted. Moore's gravimetric device is the basis for the Alcomon Indicator System disks (see the following discussion).

In addition to Moore's suggestions, Nicol (1994) noted that when a strip of Goatskin Parchment paper (available in Great Britain) is dropped into alcohol (> 30 percent) it quickly becomes saturated and sinks, but when dropped into formaldehyde solutions it floats for a long time (sometimes curling along the edges), and floats flat on the surface of phenoxetol solutions.

It is not uncommon in fluid collections to find containers of unknown preservatives that might contain formaldehyde, or to find fairly high concentrations of formaldehyde (left over from fixation) in preservative solutions. There are three methods that can be used to detect formaldehyde in even very small amounts: (1) commercial formaldehyde test strips; (2) lab-made formaldehyde test strips; and (3) Schift paper indicators. Commercial test strips are usually sold in kits with sufficient supplies for about one hundred tests and are designed to detect formaldehyde at very low levels; for use in determining formaldehyde concentration in fluid-preserved collections, it is necessary to substantially dilute the sample to be tested if you are checking for 10 percent formaldehyde. Waller and McAllister (1986) published instructions for making test strips designed to detect levels of formaldehyde commonly found in museum collections. The Schift test uses an indicator that turns magenta in the presence of formaldehyde (Schift papers have a shelf life of just six to eight weeks).

CHECKING FLUID CONCENTRATION

The original test of alcohol concentration, developed by Robert Boyle (1627–1691), was to soak a strip of paper or cloth in the alcohol mixture and then try to light it—alcohol of concentration greater than 50 percent will burn. Since 1705, the standard measure has been proof (a mixture of alcohol and water with a specific gravity of 0.91984, containing 0.495 by weight or 0.5727 by volume absolute alcohol), as described in chapter 1. Alcohol concentration today is measured as density (mass per unit volume; table 11).

There are other means of assessing alcohol concentration. With practice, a rough estimate of alcohol concentration can be made by dropping an ice cube into a container of alcohol and watching how quickly it sinks—the ice will sink faster the higher the concentration of alcohol. Moonshiners in the Appalachian mountains of the southeastern United States used a technique known as proofing whiskey "by the bead" (Dabney 1974). The technique involves shaking a half-full jar of moonshine (or another alcohol mixture) vigorously to make it foam up. In higher alcohol concentrations larger bubbles (beads) form and subsequently disappear faster than in lower concentrations (beads will not form in alcohol concentrations below 46 percent). Experienced moonshiners could determine proof with great accuracy by using this method, but a "false bead" could be created by adding a little oil, glycerin, or lye to the whiskey to make the bubbles behave as if the proof were higher. The basis of the technique is that in liquids of higher specific gravity (or greater viscosity) smaller bubbles form and take longer to disappear. The higher the ethanol concentration, the lower the specific gravity of the liquid. Another moonshiner technique (not recommended!) was to toss a glass of moonshine into a fire and estimate proof by how much the flames blazed up (Dabney 1974). A related method of estimating alcohol concentration, long known to wine connoisseurs, makes use of a phenomenon variously called tears of wine, curtains, or church windows. In a clean, clear glass, when an ethanol mixture such as wine or whisky is swirled gently or the glass is tilted, a ring of liquid forms on the side of the glass just above the surface, and droplets can be observed sliding down the glass into this ring of liquid—with practice, the proof of the mixture can be estimated by observing the size and movement of the droplets. The phenomenon, technically known as the Marangoni effect or the Gibbs-Marangoni effect, was first described by James Thomson (1822–1892) in 1855, and named for the Italian chemist Carlo Marangoni (1840–1935), who studied the effect for his dissertation, and J. Willard Gibbs (1839–1903), who published on the effect in his monumental paper, *On the Equilibrium of Heterogeneous Substances*, an important work on thermodynamics published in two parts between 1876 and 1878. The basis of the phenomenon is that a liquid with higher surface tension pulls more strongly than a liquid with lower surface tension, creating a surface tension gradient. As the water and ethanol mixture is drawn up the side of the glass by capillary action, the alcohol in the thin film evaporates faster than the water (due to its higher vapor pressure), which lowers the surface tension, causing more liquid to be drawn up; as the liquid moves up the side of the glass, gravity causes it to form droplets that fall back down the side of the glass. The higher the concentration of ethanol in the wine or whiskey, the more "tears" form (at 20°C, the surface tension of water is 72.9 dynes/cm, and the surface tension of ethyl alcohol is 22.3 dynes/cm). The Marangoni effect also explains the whorls that are observed when adding water to a mixture of alcohol with a concentration greater than 46 percent.

The first accurate instrument developed to test fluid density was the hydrometer. Hydrometer readings are commonly expressed as specific gravity (the ratio of fluid density to the density of pure water at a given temperature). Fluids with a specific gravity of less than 1.000 are lighter than water; fluids with a specific gravity of greater than 1.000 are heavier than water. A hydrometer consists of an elongated tube with a weight at one end, a scale inside the tube, and sometimes a thermometer inside the tube as well. When the tube is placed in a fluid, it floats in an upright position (see figure 5.1). The scale should be read at the meniscus of the fluid. Because fluid density fluctuates with temperature, a correction must be made to obtain an accurate reading.

There are a variety of scales used for hydrometers. The scale most widely used for measuring alcohol concentration is the specific gravity scale. Another commonly used scale is the Baumé scale, developed by the French chemist Antoine Baumé (1728–1804) in 1768 for reading dense fluids such as syrups. Other hydrometer scales include the Tralle scale (which is percent ethyl alcohol by weight), the API (American Petroleum Institute) scale; the Brix scale (measured in degrees that indicate the percent sucrose by weight at a standard temperature), the Balling scale (used in the brewing industry), and the Twaddle scale (used for fluids with a specific gravity greater than 1.000).

Figure 5.1. Hydrometer in alcohol preservative showing meniscus.

For an accurate reading, a hydrometer should be clean and dry before it is inserted into the fluid, and it should be at the same temperature as the container of fluid to be tested. The container of fluid to be tested must be tall enough to allow the hydrometer to float freely, and the sample to be tested should be thoroughly mixed before inserting the hydrometer. After placing the hydrometer in the fluid to be tested, allow it to stabilize for a few moments as the air bubbles clinging to it dissipate. Read the scale on the hydrometer by focusing on a point just below the plane of the fluid surface, then slowly raising your line of sight until the point where the hydrometer intersects the surface of the fluid appears to change from an ellipse to a straight line.

By far the most accurate and most rapid way to check alcohol concentration is with a digital density meter. Hydrometers have a fairly wide margin of error, even when corrected for temperature; a good digital density meter is much more precise, quicker and easier to use, and can be used with small samples of alcohol (a hydrometer requires a sufficient volume of fluid to float the tube; a digital density meter works with just a few cubic centimeters of fluid).

The Alcomon Indicator System (Alcomon Company, Netherlands), developed by Andries van Dam of the University of Leiden (based on Moore [1984]), can be used to gauge the approximate concentration of ethanol solutions. The system uses two small polypropylene disks with iron filler (for proper buoyancy), one red and one orange, that are placed in a container with the specimens. Both disks sink if the ethyl alcohol concentration is 60 percent or above; the red disk sinks and the orange disk floats if the concentration is 50–60 percent; both disks float if the alcohol concentration falls below 50 percent. Using this system, alcohol replacement can be based on standard strength solutions (e.g., if the red disk sinks and the orange disk floats, the volume in the containers should be reduced to 50 percent and then topped back up to the standard fill level with 95 percent ethyl alcohol). The disks can be left in the containers over the long term. Alcomon Indicator System disks are not reliable if additives are used in alcohol solutions (other than methanol), and there must be room in the container for the disks to float and sink without touching the specimens. Although the Alcomon system has not yet undergone independent testing, materials evaluation by the manufacturer predicts a minimum twenty-year life for the disks in alcohol preservatives.

A pair of Alcomon Indicator System disks can be inserted into a glass tube and used with a suction bulb to perform rapid checks of alcohol density in the lab or in the field as well (Andrew Bentley, pers. comm.).

REUSE OF OLD ALCOHOL

Alcohol evaporates from specimen containers faster than water because alcohol and water form a binary azeotrope, not a true solution, causing the alcohol concentration of mixtures to drop over time (see chapter 1). In addition, alcohol becomes more dilute as water is extracted from specimens and picks up contaminates from the specimens themselves (e.g., lipids and proteins). For these reasons, many collections managers opt not to reuse alcohol once it has been in contact with specimens, despite the expense of purchasing new stock solutions. Traditionally, some budget-minded collections managers have filtered used alcohol through cheesecloth or paper filters to attempt to rid it of some of the contaminates. A better procedure is to place a container of used alcohol in a freezer or very cold refrigerator to congeal the contaminates for easier removal when passing it through filter paper (Hawks 2003). The Smithsonian Natural History Museum tested a 19 L capacity factional distillation recycler for recovery of ethyl alcohol and isopropyl alcohol from waste fluids (Keel et al. 2011). The study determined that approximately five to nine hours were needed to distill 13–17 L of fluid. The resulting alcohol had a concentration of 89–95 percent but retained a significant odor, probably due to amines that were not removed by the distillation (however, DNA was removed by the process). Gas chromatography-mass spectrometry of the alcohol before and after distillation showed that other contaminates (very small amounts of ethyl ethers, ethyl esters, and aldehydes) were present in the used ethanol both before and after processing through the recycler, and toluene and xylenes in isopropyl alcohol survived the distillation. The study concluded that the recycler failed to produce alcohol sufficiently clean to use for specimens without the risk of contamination, but that the recycled alcohol might be used for staging specimens between formaldehyde fixatives and alcohol preservatives.

PH OF PRESERVATIVE SOLUTIONS

Theoretically, a preservative solution should be slightly acidic to provide protection against bacteria and mold, and because proteinaceous materials react badly in an alkaline environment (Hawks and Williams 1986). A container of fluid-preserved specimens is essentially a closed system containing a variety of materials that can affect pH, such as the specimens, the chemicals used in fixation, the container and closure, tags and labels, the preservative, and the source of the water used to dilute the preservative. Accurate pH measurements of preservative solutions are difficult to obtain for a number of reasons, including the fact that dissolved lipids clog most electrodes. Waller and Simmons (2003) determined that standard electrodes have a short life when used in preservative fluids—a dispensing electrolyte electrode is needed for more accurate measurements—and recommended that pending further study, pH measurements of fluid preservatives be considered no more reliable than ± 1 pH unit. Commercial pH indicator strips can be used for fluid preservatives, but the indicator strips read about one step lower than actual pH in alcohol-based preservative solutions. In their study, Waller and Simmons (2003) also determined pH by titration to a thymolphthalein endpoint, however, they reported that the concentration of ethanol and formaldehyde affects the behavior of the titration indicators (e.g., the color transition for thymolphthalein shifts from pH 8.2 at 0 percent v/v ethanol to pH 10.7 at 70 percent v/v ethanol, but shifts to a significantly lower pH in the presence of formaldehyde). It has been suggested that a more accurate measurement of pH of preservative solutions might be obtained by diluting the fluid, but this will not work because the pH scale behaves differently in water miscible nonaqueous solvents such as ethanol (Carter 2009). Carter (2009) cautioned that the behavior of pH in water miscible nonaqueous solvents (including ethanol) is difficult to predict, thus the reliability of pH measurements of preservatives is often suspect and it is problematic to compare the results of different studies.

There are a few studies of pH in collections worth comparing, keeping in mind the difficulties in pH measurement discussed earlier—these studies are summarized in table 18. Waller and Simmons (2003) found the average pH of ethanol preservatives (with specimens) to be 6.38 with a range of 5.19 to 7.55; Cato (1990) found an average pH of 6.2 with a range of 5.0 to 7.0; von Endt found an average of 6.5 with a range of 5.7 to 7.3, which he considered to be "moderately acid for a 70% v/v solution of ethyl alcohol and water" (von Endt 1994, 13). Von Endt (1994) also found that a variety of fatty acids and amino acids were in the preservative solution. The specimen itself can affect the pH, as well as trace amounts of formaldehyde, which continues to oxidize to formic acid. Waller and Simmons (2003) found trace amounts (0.07 percent) of formaldehyde in the ethanol they analyzed. Other sources of acid include the labeling materials, attachments, and other inclusions. Studies have demonstrated that Resistall paper, a once-popular preservative-resistant label substrate, can drive the acidification of water, ethanol, and formalin solutions, sometimes in less than four days (Andrei and Genoways 1999; Hargrave et al. 2005; Van Guelpen 1999). Resistall's waterproof coating is made with melamine ($C_3H_6N_6$), a formaldehyde polymer; Resistall paper itself has a pH of 4.5 to 5.2. Despite its proven durability in preserving fluids, Resistall paper is no longer recommended for labels inside containers of fluid-preserved specimens because of its potential for causing acidification problems. Acidic, deteriorating, or suspicious labels should be replaced and, if they contain original data, archived as described by Kishinami (1989, 1992).

Carter (2009) studied the effect of pH on preserved muscle tissue in 80 percent ethanol to follow the behavior of pH in the preservation environment, using Fourier Transformed Infra Red (FTIR) spectroscopy to assess changes in protein structure in the specimens. Carter found that the pH of the preservative solutions changed significantly with the addition of the specimens, moving to the region of pH 5.3 to 7.2. Kotrba and Golbig (2011) compared the pH of ethanol in soda-lime glass and borosilicate glass, and concluded that the soda-lime glass caused the ethanol preservative to become more alkaline over time, suggesting that borosilicate is a superior glass for fluid-preserved collections.

Kotrba and Golbig (2009) have proposed using a solid ion-exchange material such as a substrate-bound ampholyte to buffer pH to the desired range, preferably in combination with a color-change pH indicator inside the container to monitor the capacity of the buffer.

Waller and Simmons (2003) measured ORP (oxidation-reduction potential, or redox potential) in formaldehyde and ethanol preservative solutions and found that serial measurements were distorted, probably due to contamination of the platinum wire in the ORP electrode. They concluded that ORP measurements do not provide useful data and are expensive and time-consuming to obtain. Measurements of conductivity (total dissolved ionic chemical species) by Waller and Simmons (2003) revealed much higher and more variable levels of conductivity in formaldehyde preservatives than in ethanol preservatives.

PREPARING FIXATIVES AND PRESERVATIVES

Fixatives and preservatives should be mixed with distilled or deionized water whenever possible to avoid contaminants. The pH of tap water has a significant effect on the pH of fixative and preservative solutions (Hargrave et al. 2005); most tap water is alkaline due to water treatment with alkaline salts (Fisher et al. 1995).

The final volume of the solution that results from mixing ethyl alcohol and water is not equivalent to the sum of the volumes of the two components due to the contraction in volume that occurs upon mixing (Waller and Strang 1996). Solutions in the range of 40 percent to 90 percent ethyl alcohol contract approximately 2 percent. This contraction error must be accounted for during mixing of alcohol solutions.

Solutions of fixatives and preservatives should be allowed to rest for twenty-four hours before use. This ensures a complete mixture of chemicals, reduces layering, and allows the heat of mixing to subside. When ethyl alcohol and water are mixed, the resulting binary azeotrope may release tiny air bubbles that can adhere to specimens and cause damage to delicate structures (if necessary, air bubbles can be removed with a vacuum pump).

CONTAINERS AND SEALS

A container is a receptacle that holds a specimen and the preservative fluid—usually a jar, a bottle, or a tank. A closure is the means of sealing the mouth of the container, such as a lid and gasket. Closures are either lids or stoppers. A lid fits on top of, or over the top of, a container. A stopper fits inside the mouth of a container. Closures may be simple (no gasket) or complex (with a gasket or a liner). Most fluid specimen containers are glass jars, although many other types of vessels have been used (e.g., snakes preserved fully stretched out in glass tubes closed with cork or rubber stoppers, described in Anon. [1938a] and Strayer [1965]). Available container materials are summarized in table 19.

Sealing containers of fluid-preserved specimens has been a problem as long as fluid preservation has been practiced—a good seal is necessary to keep the fluid in and the contaminants out. For many years, the ideal was considered to be a hermetically sealed container—indeed, the first published account of fluid preservation refers to "a glass-vial sealed hermetically" (Birch 1968; Cole 1944). The phrase "sealed hermetically" harkens back to the alchemical origin of modern chemistry:

> A modern chemist who talks of a flask being "hermetically" sealed is using the very name of the mythical fount of alchemical wisdom, Hermes Trismequistus, a being derived from the Egyptian god Thoth who was said to have communicated on the "facts of life" of the birth of metals by inscribing them on a gigantic emerald slab for the benefit of initiates. (Pilkington 1959, 80)

The first containers for fluid-preserved specimens were simple glass cylinders that were sealed with a pig or sheep bladder, often in combination with a cork, lead, or glass plate, and coated with resins, waxes, and varnish (Parsons 1831). In the Netherlands, containers were frequently sealed with a plate of schist, covered with beeswax, carnauba wax, or colophonium and chalk powder (van Dam 1997). In 1771, Nicola reviewed the recommendations of Réaumur (1748) for sealing containers, and added some innovations of his own, including an oil coating over the seal (Nicola 1771). In 1790, Thomas Pole described a good seal on a container as:

> [A] wet bladder drawn smooth and tight over it, and bound down by a fine packthread, wound six or eight times round the neck of the glass; this being suffered to dry, is to be lightly rubbed over with mucilage, and covered with a fine piece of tin foil, cut so as to extend but just over the edge of the glass, and rubbed down to it as close as possible over this a second bladder is drawn tight as before, and carefully bound down by as many regular turns of packthread as will extend from the rim to the bulge of the glass. (Pole 1790, 261)

The portrait of Albertus Seba that appears in his *Locupletissimi rerum naturalium* (published in four volumes between 1734 and 1765) shows Seba holding in his right hand a jar containing a snake preserved in alcohol and sealed with red wax (Müsch 2001). As a testimony to how long such a seal could endure, if well prepared, Clark (1993) reported that:

> In the collections of the [Natural History Museum] are some preserved mammals which have been stored in glass jars topped with pig's bladder and sealed by red wax. Thomas (1892) established that these specimens were figured by Seba. . . . The mammals have remained perfectly preserved for approximately 300 years because the pig's bladder and wax seal has prevented any evaporation of the preserving fluid over this long period. The major disadvantage of this style of storage jar concerns the practicality of examining the specimens, which would necessitate replacement of the pig's bladder seal. (Clark 1993, 348)

There were many variations on the bladder-and-wax seal. Describing a container in the collection of John Hunter (1728–1793), Harris (1979) wrote that:

> This was a round glass jar 30 cm in height and 23 cm in diameter, sealed at the top with bladder which had been painted with asphalt varnish. The seal had broken resulting in a massive evaporation of the preserving fluid. The upper layer of bladder was removed to reveal a layer of lead beneath which was a further layer of brighter, shinier nature. Qualitative emission spectroscopy showed that this layer was tin with a trace of lead and iron. This was not surprising considering the close contact of the lead layer for over two

> hundred years. Beneath the tin layer was a further bladder layer. Thus the sequence from the top was: bladder-lead-tin-bladder. Pig and sheeps bladder were used since the early seventeenth century for sealing museum jars and drug containers, etc. (Harris 1979, 71)

The use of bladders continued for many years. Burton reported that "Stretched bladders, sealed with paint, were at this time [ca. 1836] often used for covering bottles containing specimens in spirit" (Burton 1969, 389). In 1895, Bean was still recommending that "When cork bottles are used, tie a piece of bladder securely over the cork" (Bean 1895, 238). Reg Harris reported that he personally:

> had the task of "bladdering" museum jars at a medical museum in the late 1930s and the process is still used in restoration work. These bladders were usually obtained from an abattoir in the dry state; they were well-extended and resembled football bladders in their transparent state. After wetting for a few hours the bladder was soft enough to stretch over a jar top and then allowed to dry completely. This formed a good seal provided that the mount was kept in one place and not moved about, in which case leakage was more or less inevitable. As glass jars were improved, glass lids were introduced as a further protection against evaporation under the bladder sequence, the lids in most cases being sealed with a paste made of stockholm tar and red lead (litharge). The paste took several days to harden and the seal was absolute. (Harris 1979, 71)

Jars that were not sealed with bladders were usually sealed with stoppers made of wood, cork, wax, or glass. Stoppers had to be luted (sealed into the mouth of the jar; from the Latin word *lutum*, meaning potter's clay). In 1884, Browne reviewed the problems with common luting agents, noting that wax was susceptible to damage from the alcohol and recommending "lithocolle," the recipe for which he had obtained from a French naturalist. Lithocolle consisted of a mixture of common resin, yellow beeswax (paraffin wax), red ochre powder, and oil of turpentine (turps), with the proportions adjusted to make the resulting lute more or less brittle (Browne 1884). The lithocolle was applied hot, either with a brush or by dipping the jar and stopper into the liquid several times. The files of the Oxford University Museum contain formulae for making several types of jar seals, including (1) five parts gutta-percha and two parts asphalt, heated together; (2) three parts gutta-percha and two parts asphalt; and (3) 1 oz. "Best glue," 3/4 oz. finely ground white lead, and 1 teaspoon glycerin jelly.

A small 1829 publication distributed to collectors (Providence Franklin Society 1829) recommended sealing glass-stopper jars with a compound composed of "Rozin, 2 parts—Yellow Wax, 1 part—Red Ochre, or pounded brick, sufficient quantity to colour. It is useful, again to cover the bottles thus closed, with a piece of cloth, firmly tied, and covered with liquid pitch; and for large bottles, to support the corks with a string, which, by being fastened to the circumference of the bottle, forms a cross above the corks." A 1907 publication (Ouwens 1907) describes how to seal a glass container with a piece of 1–1.5 cm thick cork. The cork is submerged in boiling paraffin wax, then inserted into the container top, leaving a small space at the top that is filled with plaster of Paris and covered with "iron lac" (probably iron lacquer).

In 1899, it was reported that the Naples Zoological Station was using cylindrical glass jars with ground glass stoppers for both exhibition and storage, but due to the expense of glass, also using earthenware jars and basins for working with specimens and thin glass tubes sealed with corks for small specimens (Lo Bianco 1899). The report notes that "with large tubes, it is desirable to put a plug of cotton inside the tube next to the cork, since the alcohol extracts the tannic acid from the cork and is stained brown thereby" (Lo Bianco 1899, 7). The use of glass tubes, plugged with cotton, and placed inside larger containers of alcohol was also reported. Large, flat specimens were preserved in rectangular jars with a glass cover cemented on with gutta-percha cement. Very large specimens were maintained in zinc boxes with zinc closures, although it was noted that with time the zinc became corroded.

The next big innovation after bladders and glass stoppers were glass jars with glass, metal, or ceramic closures, made possible by the development of various bottle- and jar-making machines (Derry and Williams 1960) in which molten glass was dropped into a mold and compressed air used to form the vessel within the mold, and by improved furnace technology. For exhibition, rectangular glass containers with glass or metal lids were long the standard. These containers were usually polished on one face, or one face and both sides, for better visibility of the contents. An advertisement from a 1932 issue of *Science* magazine for jars made by the Empire Laboratory Supply Company in New York features a typical product, described as:

> Jars Rectangular Museum. These jars are specially manufactured to our most rigid specifications. They have flat ground-on lids for permanent sealing and are so constructed that an even and neat seal can be easily accomplished. They are of clear white glass and are triple annealed, which feature reduces the breakage when stored. Each Jar is numbered serially, both body and cover corresponding. Each cover is especially made to fit each jar, which is an unusual performance in the manufacture of the ordinary jar. Each lid is provided with a small vent-hole to allow the escape of gases and refilling. These Jars are also provided with the largest side ground plane and highly polished at the prices enumerated below.

The advertisement also offered glass rods for mounting specimens and "Empire Green Label Cement—for sealing Museum Jars, manufactured from the formula as used by E. J. Judah, of McGill University."

Postbladder jar sealants included a variety of jar cements (most of which were black or dark brown), gelatin compounds (which are usually yellow or white), and greases. When resealing jars, if the historical appearance of old seals is important, they may be replaced with silicon (Horie 1983) or a paraffin sealant that has been colored to match the original sealant. Some workers believe that paraffin soft white is a more stable and reliable seal than silicon (Waterhouse and Graner 2009). The drawbacks to silicon are that the seal is permeable and must be removed from the glass with a commercial silicon release agent. I have had success using Spectrem brand silicon made by Tremco (available in black, white, and clear formulations), which is a medium-modulus, one-part, high-performance neutral-core silicon. Dirk

Neumann (pers. comm.) has reported good success sealing glass-topped jars with a hydrocarbon-based sealant called Alsirol made by J. P. Pöllath (it is sold as stopcock grease). A silicone rubber sealant made by Dow Corning (RTV 731) has been recommended for glass and plastic museum jars (Joram 2005).

The 1959 classic *Medical Museum Technology* (Edwards and Edwards 1959) provides instructions for three methods to seal glass containers: (1) apply putty between the glass closure and the container (not recommended, but quick and inexpensive); (2) use a cement made by melting 100 g of gutta-percha in 100 g of Stockholm pitch and then adding 25 g of paraffin wax (at 54°C), applied warm and with a warmed lid, and placed under a 2 lb weight and left overnight to harden; or (3) a cement made by melting 50 g of gum Arabic and mixing in 50 g sugar, 2 g water glass, and 30 mL formalin, allowing three to four days to set after applying the seal.

To make an effective seal between two glass surfaces, such as the container and a glass plate closure, both surfaces should be finely ground to create a rough surface on which the adhesive can adhere. A grinding surface for glass may be made by spreading a small amount of water and glycerin on a sheet of scrap glass and adding to it a medium grit carborundum powder. The jar surface and closure can then be roughened by rubbing them on the water-glycerin-carborundum mixture on the scrap glass sheet (see figure 5.2).

Natural rubber has been used to make stoppers and gaskets since the early 1800s, although not very successfully at first. Natural rubber is very elastic, but has a short useful life when exposed to fluid preservatives. Once the chemical composition of natural rubber was established by Michael Faraday (1791–1867) in 1826 (C_5H_8), chemists around the world began trying to make rubber in the laboratory or to improve natural rubber. The first successful substitute was butadiene (C_4H_6), developed in Germany, from which "Buna" rubber was developed (the name Buna is derived from *bu* for butadiene and *na* from the chemical symbol for the sodium catalyst used to polymerize it). During World War II, a butadiene and styrene rubber was developed in the United States, which led to the development of other synthetic rubber compounds such as butyl (polyisobutylene-polyisoprene), neoprene (polychloroprene), nitrile (a copolymer of butadiene and acrylonitrile), EPR (ethylene-propylene), and EPDM (ethylene-propylene terpolymer), all of which have been used at one time or another to make jar gaskets. Although the newer synthetics are more durable than natural rubber, they all have relatively short lifespans, particularly when in contact with fluid preservatives.

The first affordable mass-produced glassware suitable for the storage of fluid-preserved specimens was introduced in the mid-1800s (e.g., the Kilner and Mason jar became available in 1858) with a variety of metal clips, screw-on tops, and rubber rings (Shephard 2000). The original jar patented by John Landis Mason in 1858 had a zinc screw cap and threaded neck so that the closure could be screwed down tight against the shoulder of the jar. In 1869, a further innovation came about with the addition of a rubber seal around the rim of the jar under a glass lid. By around 1900, probably the most popular jar in museum use was the glass-top, wire-bail jar

Figure 5.2. Simon Moore demonstrating how to grind a glass jar top using carborundum.

with a rubber-ring gasket. At one time, these jars were considered the best containers available for specimens, but the gaskets either embrittle or dissolve with age and exposure to fixatives and preservatives, leaching products into the preservative fluid and ultimately failing as container seals (see figures 5.3 and 5.4). A similar process occurs with all compressible stoppers currently on the market (see figure 5.5).

For many years, the only screw-top jars available had metal lids that easily oxidized, or rigid Bakelite (polyoxybenzylmethylenglycolanhydride) lids that become brittle over time and unscrewed themselves with minor changes in temperature in the collection storage area. Bakelite, the first true synthetic plastic, was developed in

Figure 5.3. Embrittled jar gasket caused by passage of preservative through gasket material.

1910 by Leo Hendrick Baekland (1863–1944), a native of Belgium who worked in the United States as an independent researcher.

What is currently recommended for fluid-preserved specimens are good quality screw-top glass jars with flexible polypropylene lids, with a polyethylene or Teflon liner or insert (Simmons 2002)—polypropylene is permeable to oxygen, hence the need for a liner (van Dam et al. 2000). It is important that screw-top containers and their closures have complete threads rather than partial threads and seating tabs in order to obtain a good seal (figure 5.6). Many glass containers now in commercial use have partial threads because the closures are applied by machine and are not intended for long-term reuse. Warén et al. (2010) have made an argument for the use of metal lids when polypropylene lids are not available, providing the closures are checked regularly for oxidation and replaced when they show the first indications of damage.

The quality of the seal for lids that do seat well on the threads can sometimes be improved by inserting a layer of Parafilm (a plastic paraffin film) or a polyethylene film between the lid and the jar (Schmid 1981), by wrapping the jar threads with Teflon plumbers tape, covering the jar opening with a sheet of wide Teflon tape (usually sold as artifact wrapping material) before the lid is applied; or by wrapping the jar/closure junction with a quality jar-sealing tape (Gillette 2004; Steigerwald and Laframboise 1996), such as PPA (polypropylene/acrylic) tape.

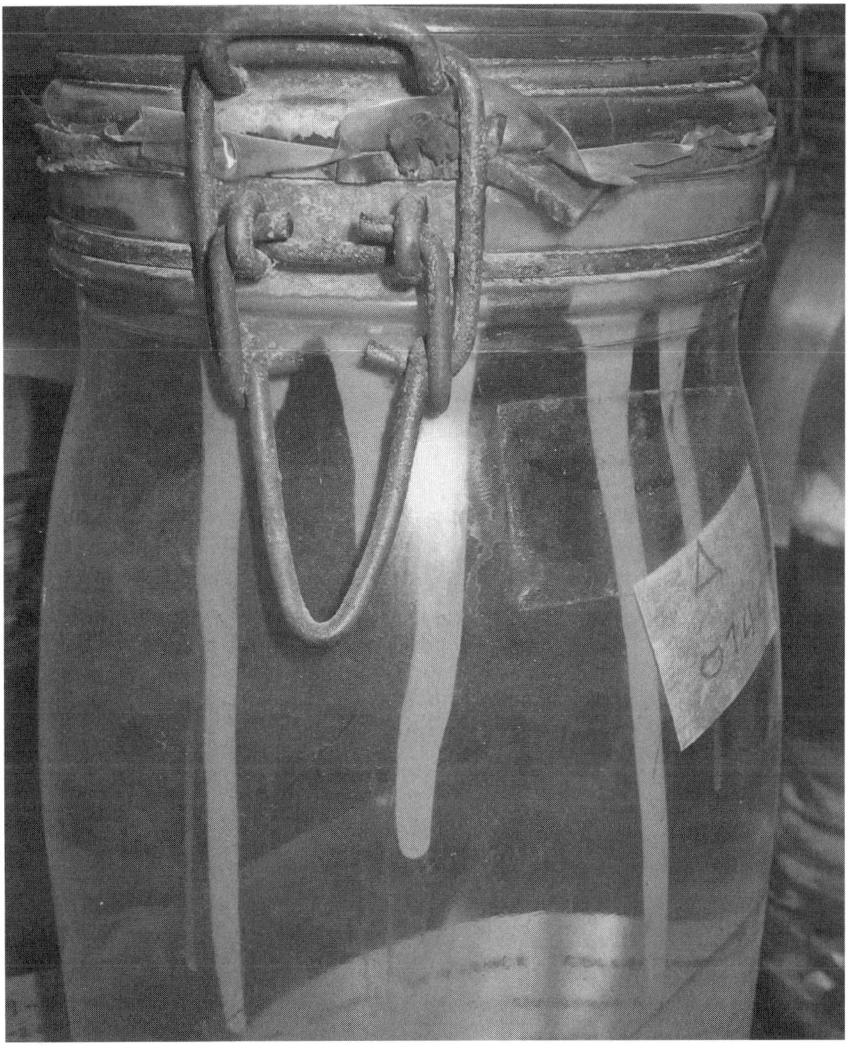

Figure 5.4. Softened jar gasket due to passage of preservative through gasket material.

The most common type of glass used for jars—soda-lime glass—is susceptible to damage from high relative humidity (Owens and Emanuel 1942), as well as various chemicals that may be found in fixatives and preservatives (Anderson et al. 1975). The resistance of soda-lime glass to chemical exposure depends on the conditions under which the glass was formed and annealed (Anderson 1975). Relative humidity above 80 percent causes deterioration of dry glass, as does repeated wetting and drying (Matson 1949; Stockdale and Tooley 1950). Container glass shows permanent weight gain of up to 0.65 percent in weight under conditions of 98 percent

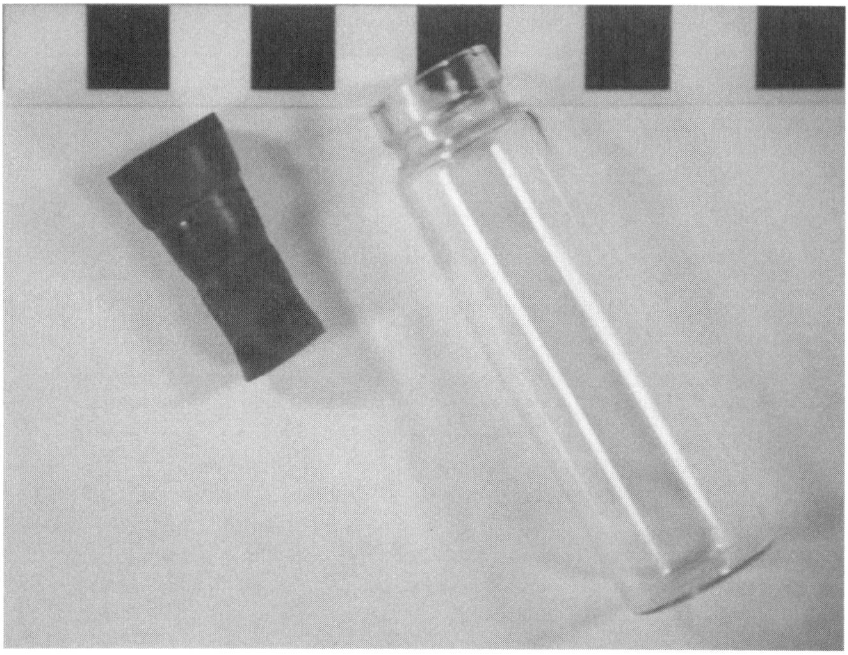

Figure 5.5. Distorted and discolored compressible stopper as a result of contact with preservative.

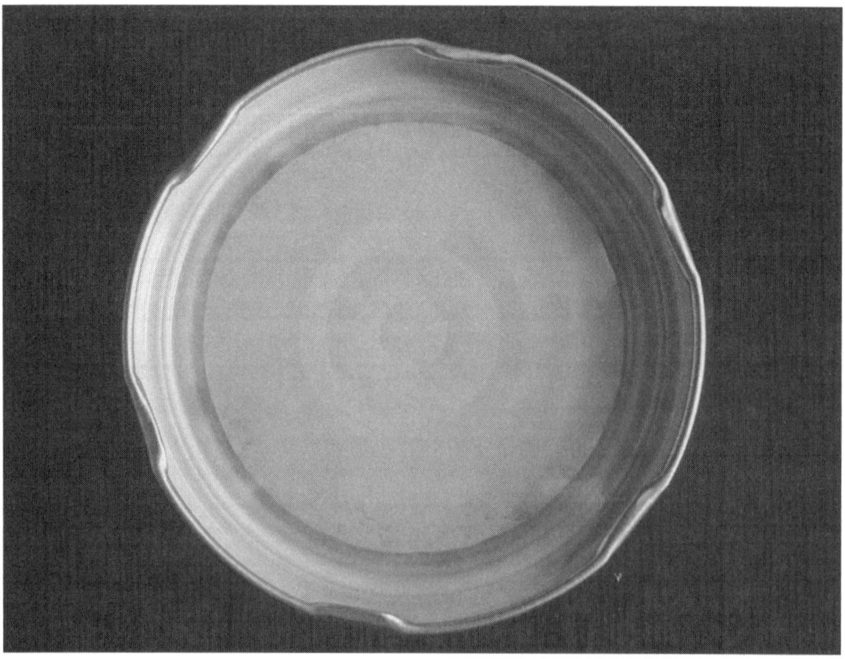

Figure 5.6. Tabs on metal jar lid.

relative humidity at 100°F; with this weight gain, there is also an acceleration of "the formation of reaction products on glass" that can be seen under a microscope (Matson 1949, 128). It is not uncommon in fluid-preserved collections to find glass containers that look dirty (see figure 5.7) or exhibit iridescence when dry—these are indications that the glass is undergoing deterioration. Glass deterioration "begins with many small points of attack and progresses by the coalescence of these spots to form large patches of reaction products" (Stockdale and Tooley 1950, 15). These reaction products may gain enough weight that "spontaneous surface cleaning" of the glass occurs (Stockdale and Tooley 1950, 15). Fortunately, glass deterioration is a slow process, but it does mean that the preservative fluid will also contain the chemical reaction products from the glass. Eventually, some glass containers will become fragile from the loss of components and may easily break when handled.

Borosilicate glass is not susceptible to deterioration by fluid preservatives, but it is more expensive than ordinary glass. Because of the resistance of borosilicate glass to preservatives, a new type of the old-fashioned museum storage jars with glass stoppers is currently on the market. These jars were developed for the Natural History Museum

Figure 5.7. Glass deterioration that causes dirty appearance on surface of dry glass.

(London) due to the shortcomings that had been encountered with other storage containers, including glass, metal, and plastic (Clark 1992, 1993). Unlike the original glass-stoppered jars, which had lids individually ground to fit each jar, the new jars are machined to such a high level of precision that they have interchangeable closures.

Rectangular borosilicate containers were developed by the Museum of Science and Industry in Chicago for the exhibition of a collection of human body slices (Simmons *forthcoming*). The containers are made of Borofloat, a type of boroslicate glass available in sheet form that has an optically flat surface. The glass was water cut to make smooth surfaces for the joints and the plates attached using Hxtal Nyl-l epoxy, which consists of 4,4-isopropylidenedicyclohexaanol-epichrohydrin with poly(oxy) (methyl-1,2-ethanediyl), alpha-hydro-omega-(2-aminomethylathoxy)-ether with 2-ethyl-2-(hydroxymethyl)-1,3-propanediol(3:1) + imidizoles.

ALTERNATIVES TO GLASS CONTAINERS

For some time, collections care workers have been searching for inexpensive, lightweight alternatives to glass containers, particularly because of the rising cost of glass containers and because so many standard jar sizes are no longer available. Most plastic containers are not alcohol resistant, or allow too much oxygen to permeate to be good containers for the long-term storage of specimens. One type of plastic that has been fairly widely used in collections is high density polyethylene (HDPE). HDPE containers have somewhat reasonable resistance to oxygen permeance, but are susceptible to ultraviolet radiation and over time, preservative concentration may drop due to the perfusion of alcohol through the plastic. Furthermore, HDPE is not transparent, so the containers must be opened to check on specimen condition and fluid levels. Some collections have opted for inexpensive transparent polycarbonate containers (Estep 2004). One very promising material is polyethylene terephthalate, or PET (Walker et al. 1999), which has a very good permeability constant for oxygen (see table 20) and is more resistant to alcohol permeation than most other plastics. There is now a fairly good selection of PET containers on the market. PET containers should be used with a flexible, lined polypropylene closure (many of the closures sold with PET containers do not provide a good enough seal for fluid-preserved specimens).

Large containers are a particular problem due to their expense. The cost of a good quality stainless steel 18 gal.-specimen tank is about five to six times the cost of a 30-gal. HDPE barrel. Until they can be replaced, older large specimen containers can sometimes be improved by using a better material for the seal, such as a closed-cell foam weather-strip gasket or extruded polyethylene rope (Suzumoto 1992).

Because rectangular glass containers are very expensive and unavailable in most sizes, many museums have replaced them with acrylic (e.g., Perspex or Plexiglas) containers. However, acrylic containers warp, crack, and discolor over time. Research has demonstrated that this is caused by the fluid preservatives diffusing slowly through the acrylic, creating negative pressure inside the containers (van Dam 2000). With temperature fluctuations, the acrylic then flexes, causing internal stress cracks (van Dam et al. 2000). As the preservative diffuses through the acrylic, it carries with it some of the dissolved lipids and proteins, which discolors the acrylic. The diffusion of the preservative through the acrylic cannot be stopped, but the flexing of the container walls can be regulated with the installation of a valve to equalize pressure; unfortunately, the use of the valve admits air into the container, which will oxidize the preservative (van Dam et al. 2000). Acrylic should not be used with denatured alcohol that contains methyl ethyl ketone.

LABELING

The list of materials that have been used for specimen tags and labels over the course of the history of fluid collections is long, and includes parchment, paper, plastic, metals, textiles, and wood, and a wide variety of inks (see table 21). For a good part of the history of collections, it was relatively easy to find inks made with carbon black pigment (e.g., India ink) that could be used to write on a good quality parchment or 100 percent cotton paper stock. There were some minor deviations, such as the suggestion to waterproof paper labels by coating them with wax (Doolittle 1925), or to use a rubberized laundry tag stock, and various forms of unknown commercially produced waterproof fiber tags were also recommended (Anon. 1938b). Following the introduction of the self-correcting electric typewriter, and then dot-matrix, laser printer, and inkjet technologies, labeling has become increasingly problematic, as the easiest and most rapid means to produce a label or a tag are no longer the most permanent.

Self-correcting typewriters used a form of black lettering (a nonarchival toner) designed to be lifted cleanly off the substrate with an adhesive, hence the lettering formed a very weak bond with the paper substrate. Inkjet printers used a wet ink usually composed of carbon black along with a variety of surfactants, detergents, and solvents that was variably resistant to alcohol, depending on the components. Desktop inkjet technology does not produce as fine a line as desktop laser printers because the lines of ink are not applied continuously to the substrate, but rather are formed as a series of tiny dots of ink that emerge from a sprayhead as bubbles under pressure that break when they strike the substrate.

When submerged in preservatives, the lines and letters produced by laser printers frequently come free from the label substrate, making the labels illegible and posing a contamination risk to the specimens. Laser printer technology is similar to that used in photocopiers. A toner (usually a dry mixture of 5–10 percent carbon black powder with other chemicals such as acrylic or styrene polymers) is electrostatically attached to a metal drum, then transferred to the paper substrate. The toner particles are attached to the substrate surface by a combination of heat and pressure. Desktop machines usually operate with a drum temperature of about 150°C and fairly low pressure; industrial laser printers may use temperatures as high as 390°C with a pressure of 300 psi. Machines that produce higher heat and greater pressure produce more durable labels, but even high heat and pressure do not form strong bonds with the fibers in a paper substrate. Unlike an impact printer or typewriter, if the toner particles separate from the substrate there is no impression that can be read in raking light. The quality of the printed image is also dependent on the quality of the toner and substrate, which initially made it difficult to determine why the lettering on laser-produced labels usually failed to adhere to the substrate. A number of remedies were suggested in the literature, including postprinting enhancement of the adhesion of the toner to the substrate using a hot iron or spraying a chemical fixative over the surface of the label. Zala et al. (2005) tested the adhesion of laser printer–produced labels on a number of paper substrates by subjecting the labels to clean alcohol without specimens and found no loss of letters; it was later demonstrated that the culprit in laser printer labels was dissolved lipids in the preserving fluid.

It is now difficult to find a good paper stock that can withstand long-term immersion in alcohol or formaldehyde solutions (see previous comments on Resistall paper). In the United Kingdom and Europe, goatskin parchment that is resistant to alcohol and formaldehyde is still available.

Currently, the recommended tags and labels for fluid-preserved specimens are those produced by a thermal printer on white, spun-bonded polyethylene stock (Bentley 2004). Thermal printers transfer a dry toner to the label substrate using an electrostatic process, but the dry toner is then melted into the label substrate, making a permanent abrasion and preservative resistant label.

Waller and Simmons (2003) found that attaching labels with synthetic thread caused more damage to specimens than cellulose thread.

Faded ink may sometimes be read with an infrared lamp; tears in old labels can often be repaired using Japanese kozo tissue (Moore 1999).

Because tags and labels immersed in fluids are susceptible to damage, before subjecting specimens with labels or attached tags to any treatments (such as staging the specimens through preservative concentrations) it is recommended that all information from the tags and labels be recorded or photographed.

THE STORAGE ENVIRONMENT

The container and the fluid preservative surrounding a specimen constitute the specimen's microenvironment. The fluid includes not only the preservative chemicals, but also trace amounts of fixatives, previously used preservatives, and the chemical components extracted from the container, labeling materials, and specimen (primarily proteins and lipids). For example, figure 4.3 shows dissolved label ink contaminating the alcohol preservative. Bisulca et al. (2006) have shown that corroding metal tags in fluid preservatives can react with lipids in the specimens to form fatty acid soaps, and that the metal corrosion products can be deposited in the specimen.

Fluid-preserved collections need a stable, cool storage environment. Higher temperatures accelerate deterioration and evaporation, lower temperatures cause preservative problems (e.g., formaldehyde will polymerize at temperatures below 57°F, according to Fisher et al. [1995]). Steedman (1976b) recommended storage temperature for formaldehyde in the range of 10–25°C to avoid the formation of paraformaldehyde haze and deposits.

Some authors (e.g., Horie 1994) have suggested that the preferred storage temperature for fluid-preserved collections should be below 18°C. I find this impractical, particularly for a collection that is heavily used because it would mean that containers would be regularly going in and out of an 18°C storage facility into a warmer laboratory area for use. All things considered, I believe that the best storage temperature for fluid-preserved collections is 18–21°C (65–70°F) with relative humidity around 50 percent. The temperature in collection storage should be a few degrees lower than the temperature in the area where the containers are sealed, as moving the newly sealed containers to a cooler environment will enhance the ability of the seal to prevent evaporation. While planning a fluid storage facility at the Natural History Museum and Biodiversity Research Center at the University of Kansas in the mid-1990s, it was recommended that the storage temperature be lowered to 55°F to reduce the flashpoint of 70 percent ethanol. We were concerned about the temperature differential between storage and laboratory areas, so we selected a few jars from the collection and placed them in a laboratory refrigerator (see figure 5.8), demonstrating that 70 percent ethanol with specimens turned cloudy below about

Figure 5.8. Cloudiness in cold alcohol preservative due to formation of paraformaldehyde and congealing of extracted lipids.

60°F. The cloudiness may have been caused by the formation of paraformaldehyde from the trace amounts of formaldehyde in the preservative, the congealing of lipids extracted by the alcohol (Moore 2002b; Moore 2005b), or other causes, but in any case, it was an unacceptable change in the preservative, and we argued successfully for a temperature of 18°C (65°F) in the new fluid-preserved specimen storage facility.

Fluid-preserved specimens should be kept in the dark when they are not being used. All light is damaging to specimens, but ultraviolet radiation is particularly damaging. Contrary to popular belief, ordinary glass does not protect specimens from UV radiation. The most damaging parts of the UV spectrum are wavelengths of 300 to 400 nanometers, which will pass through glass (Harris 1968; Macleod 1975).

The ratio of specimen volume to preservative volume in the storage container can have a significant effect on the long-term quality of the specimen (Fisher et al. 1995; Simmons 2002). Although a fluid volume to specimen volume ratio of 2:1 was long recommended (Zweifel 1966), a ratio of 7:3 is now recommended based on an analysis of specimen quality and the correlations between low specimen to fluid volume ratios and decreased ethanol concentration (probably because higher specimen volumes cause more dilution of the preservative), and the presence of more unfixed formaldehyde at lower alcohol concentrations (Fisher et al. 1995). As has been known since the dawn of fluid preservation, water extracted from the specimens dilutes the preservative (Simmons 2002; Taylor 1981). Taylor (1981) calculated that adding a volume of 25 percent specimens may dilute 75 percent ethyl alcohol to a range of 58–62 percent, and that adding a volume of 75 percent specimens may dilute the 75 percent alcohol to 20–26 percent.

Accelerated aging tests are needed on fluids, labels, and containers, particularly in regard to preservative quality and its effect on specimens. The only such tests that have been published were done on keratin (von Endt et al. 2000) and collagen (von Endt 2000). Keratin and collagen (in the form of hair and feathers) were subjected to accelerated aging by heating samples in 70 percent ethyl alcohol to 180°C for 1–2 days, with and without the addition of 1 percent formaldehyde, and in 50–55 percent 2-propanol (isopropyl alcohol). The controls were dry samples. The study concluded that collagen dissolves more quickly in fluid preservatives than keratin; isopropyl alcohol causes more damage to both collagen and keratin than ethyl alcohol; keratins are unstable at high temperature; the keratin in feathers is less stable than keratin in hair; and that dry keratin is more stable than fluid-preserved keratin. More importantly, it was demonstrated that even small amounts of formaldehyde in the alcohol promote deterioration reactions in keratin, so formaldehyde-fixed specimens should be washed thoroughly (through steps in concentration, as described in chapter 3) to remove formaldehyde to prolong the useful life of the specimens.

TOPPING UP AND REPLACING PRESERVATIVES

Changing the preservative fluid when it became discolored was long a traditional practice in managing fluid-preserved collections. Preservatives become discolored as they extract lipids, oils, and pigments from specimens and due to chemical interactions with containers, closures, and sealants. Although some collections care workers believe that the color or apparent turbidity of the preservative is an indication of preservative quality, this has not yet been shown to be true (e.g., larger volumes of discolored preservative may appear darker than smaller volumes equally discolored). Waller and Simmons (2003) compared preservative samples taken from a large collection to three color standards (iodine, potassium chromate, and potassium dichromate) and a turbidity standard (suspended deagglomerated alpha alumina), and concluded that comparison to the standards produced no useful information for assessing preservative quality. More work needs to be done on the causes and the significance of the discoloration of fluid preservatives.

Because the fluid around the specimens contains chemicals extracted from the specimens, it is, in effect, part of the specimens, and therefore it is recommended that discolored alcohol solutions not be replaced unless the fluid threatens the useful life of the specimen (e.g., if the alcohol has become so saturated with lipids or oils that it is becoming acidic or staining the specimen).

Traditional thinking in fluid collections management was that containers should be filled as high as possible with preservative to reduce the amount of air available for oxidation (Jones and Owen 1987; Stoddard 1989), but we now know that it is better to fill containers to 90 percent of their volume (for alcohol-based preservatives) or 95 percent of their volume (for formaldehyde-based preservatives) to obtain better container seals against evaporation (van Dam 2000; see the next section, Why Do Closures Fail?).

Over the years, a wide range of recommendations have been proposed in the literature for preservative strength, all without experimental justification (see table 1). For example, the ideal storage strength of ethyl alcohol recommended by various authors has ranged from 56 percent to 95 percent. The recommendations for preservative strength were apparently based on the presumed thickness or permeability of the skin of the specimen or its propensity to dehydrate (e.g., higher concentrations of alcohol were sometimes recommended for reptiles than for amphibians). Given this wide variation in collections, what strength alcohol should be used for storage? Most collections use 70 percent because at that concentration, ethyl alcohol is a biocide, and alcohol preservation is always a trade-off between dehydration and protection of the specimen.

When topping up, the concentration of alcohol added to the jar should be calculated to bring the jar to proper storage strength (Sendall and Hughes 1997). Checking fluid levels in small collections and bringing the preservatives back to storage strength is relatively easy, but is much more time-consuming in large collections that have thousands of containers. It is known that repeated topping up with storage strength alcohol will, over the long run, result in lowered preservative concentrations. Nevertheless, it is very time-consuming to check containers individually and adjust the preservative concentration in each one. One method of managing fluid levels in a large, heavily used collection is to use standard sizes of containers, filling each container to a set level (e.g., 90 percent full for an alcohol-based preservative). As a collections management rule of thumb, if the level of fluid in the container has dropped between inspections by 10 percent or less below the standard fill level, add sufficient storage strength preservative to bring the container back to 90 percent full. However, if the fluid level has dropped by more than 10 percent below the standard level, check the preservative concentration with a digital density meter or hydrometer and refill to 90 percent full with an alcohol solution calculated to return the container to storage strength preservative. For this method to be effective, each container must be inspected before it is returned to the collection, and occasional inspections of fluid levels in collection storage must be made. It is also incumbent on collections managers to track any containers which must be regularly topped up and fix the seal and adjust the preservative concentration before it falls too far. Fluid levels can be marked on the outside of glass containers using colored grease pencils, then easily cleaned off when no longer valid.

An alternative recommendation offered by Fisher et al. (1995) is that the fluid should be replaced if the concentration in the container is below 50 percent or above 80 percent ethyl alcohol, or at a pH below 5.5 or above 8.0 (however, it is very difficult to get an accurate measure of the pH of ethanol solutions; see previous discussion of pH in this chapter). For concentrations in the 50–80 percent range, alcohol should be added to bring the total up to desired storage strength. When topping up, the low fluid level on the containers should be marked with a colored grease pencil, using a different color for each topping up cycle; containers and closures that have lost fluid after two consecutive topping up periods should be inspected or replaced.

A more comprehensive method for managing fluid replacement in a large collection was proposed by Notton (2010), based on the same principles as the regulation of temperature and relative humidity within an acceptable range of variation by emphasizing fluid volume in standard sizes of containers and using collection profiling, regular inspections, accurate determinations of alcohol concentration made with a digital density meter, and the creation of a "topping up table." The system requires the user to determine an alcohol concentration set point and range for the collection (e.g., 70 percent +/– 5 percent at 20°C) and a topping up volume (e.g., containers to 90 percent full). Next, an inspection schedule is established and an estimate is made of evaporation rates within the parameters of the inspection schedule. Using a digital density meter, a "topping up table" is developed based on container volume and measured alcohol strength after evaporative loss. By using the levels of fluid in standard sizes of containers and the "topping up table," lost fluid can be replaced with standard concentrations of premixed preservative.

WHY DO CLOSURES FAIL?

An increase in ambient temperature in the collection storage area increases the air pressure in the headspace of a sealed jar, which results in increased stress on the closure (Horie 1994; van Dam et al. 2000). The internal jar pressure depends on the vapor pressure of the fluid in the container, as well as the thermal expansion rate of the jar, the fluid, and the air above the fluid. The volumetric expansion coefficient is an expression of the thermal expansion rate, or the ratio of change in volume per degree C to volume at 0°C. This is a temperature-dependent variable. For example, at 20°C, ethyl alcohol has an expansion coefficient forty times higher than glass, and water has an expansion coefficient eight times that of glass. For water or ethyl alcohol, an increase in temperature means a rise in the fluid level in the jar, which causes the compression of the air in the headspace, which puts stress on the closure and seal (van Dam et al. 2000). Containers of fluid-preserved specimens should be stored at a constant temperature, preferably a few degrees lower than the place where they were sealed to create a slight negative pressure in the container, which will improve the seal (Horie 1994; van Dam et al. 2000). Horie (1994) calculated that if a one liter container of specimens preserved in 70 percent ethanol at 15°C was sealed and heated to 25°C, the internal vapor pressure would rise by 24 mm/Hg (3,200 Pa); but factoring in the expansion of the glass and liquid, the actual pressure on the closure is 4,000 Pa. If the container has 2 mL of air space, the pressure goes as high as 13,000 Pa.

The ratio of fluid volume to headspace is important. One reason to fill containers full is to reduce the amount of air available for oxidation, but the higher the jar is filled with fluid, the greater the compression of the air volume, and the greater the internal pressure (van Dam et al. 2000), which can cause jar lids to leak or even to break. A decrease results in the opposite effect. For example, if a 100 mL capacity glass jar is heated from 15°C to 25°C, the glass and liquid expansion increase the internal air pressure by 4 percent; if the jar has 2 mL of air space, the air pressure increases by 12 percent (Horie 1994). Figure 5.9 shows a glass jar lid that cracked, and figure 5.10 shows the bottom of a glass vial that was detached as a result of fairly minor variations in storage temperature causing changes in internal pressure. The current recommendation is to fill a container to 90 percent of its volume for ethanol-based preservatives, and to 95 percent of volume for water-based preservatives to reduce stress on the closure from temperature fluctuations (van Dam et al. 2000).

Small glass vials with compressible stoppers have long been favored in invertebrate collections (Riley 1892), despite their high failure rate. Although small glass vials have the advantage of being relatively easy to sort through when looking for specimens, compressible stoppers ultimately fail due to chemical deterioration caused by aging, exposure to ultraviolet light, exposure to the fluid preservative, or all of these causes in combination. The consequences of failure of compressible stoppers ranges from loss of preservative to the chemical contamination of the specimen with the deterioration by-products of the failing stopper—for example, while advocating the use of stoppers, Riley nevertheless reported that "American rubber stoppers are all made of vulcanized India rubber and have the disadvantage of forming small crystals of sulphur about the stopper, which become loosened and attach themselves to the specimens" (Riley 1892, 90). Most published recommendations for small glass vials with compressible stoppers are based on short-term evaluation periods (e.g., Evans 1958). Despite the fact that most compressible stoppers available today are made of a synthetic such as neoprene, they will still deteriorate due to contact with the preservative and with age (see figure 5.5). A far better alternative to small vials with compressible stoppers is to use glass shell vials (straight-sided vials with no neck constriction or threads) that are plugged with inert polyethylene fiber and submerged in

Figure 5.9. Glass jar lid cracked as a result of minor variations in storage temperature.

preservative inside a jar that has a good seal (Brooks 1932; Emerson and Ross 1965; Simmons 1999b). Although it may not be quite as convenient to retrieve a vial from a jar of fluid as it is to pick up a vial in a storage rack, a combination of good labeling and use of only a few vials per jar ameliorates the problems and the specimens are better protected in the long term.

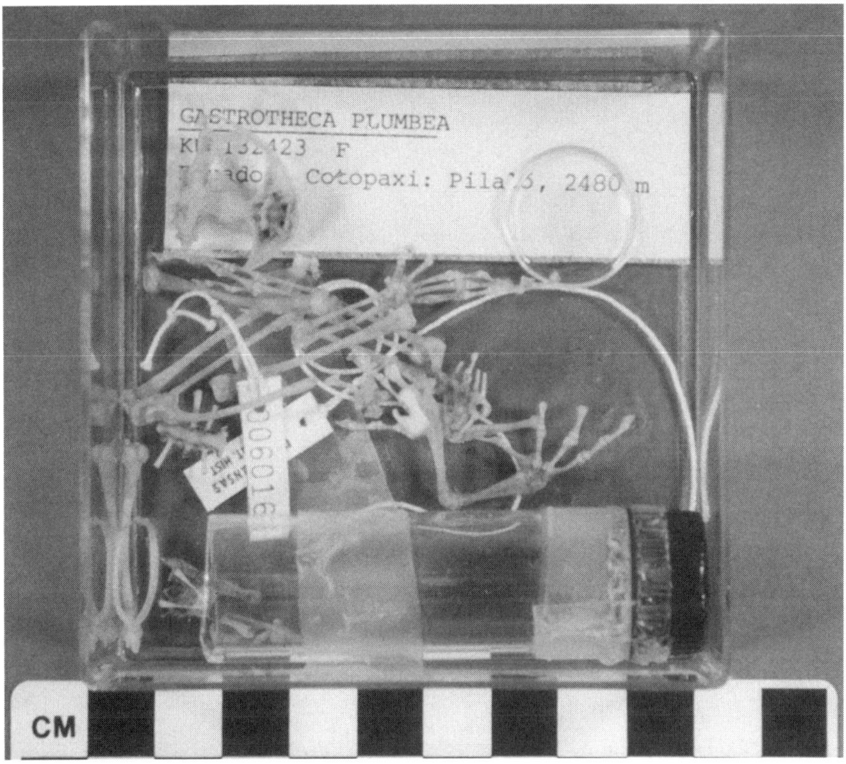

Figure 5.10. Bottom of glass vial broken out as a result of minor variations in storage temperature.

BACTERIAL AND FUNGAL GROWTH IN FLUID-PRESERVED COLLECTIONS—DETECTION AND REMEDIATION

The growth of bacteria or fungi in a container of fluid-preserved specimens is rare, but may occur if the concentration of the preservative falls below a critical level (e.g., ethyl alcohol that drops below 50 percent or so). If not treated immediately, the growth of the bacteria or fungi will damage specimens as both consume the substrate they are growing on. Figure 5.11 shows mold and bacterial growth on long-preserved specimens of iguanid lizards in a stainless steel tank due to a low concentration of preservative alcohol and a bad seal; figure 3.3 shows bacterial growth in a jar with a lowered alcohol level due to repeated topping up with stock alcohol solution.

Specimens from a container inflicted with bacterial or fungal growth should be removed and washed carefully in storage-strength preservative. Bacterial or fungal growth on a specimen may be carefully removed using cotton swabs, taking care to minimize damage to the specimen. The container and closure must be scrubbed with soap and hot water. Once cleaned, the affected specimens should be returned

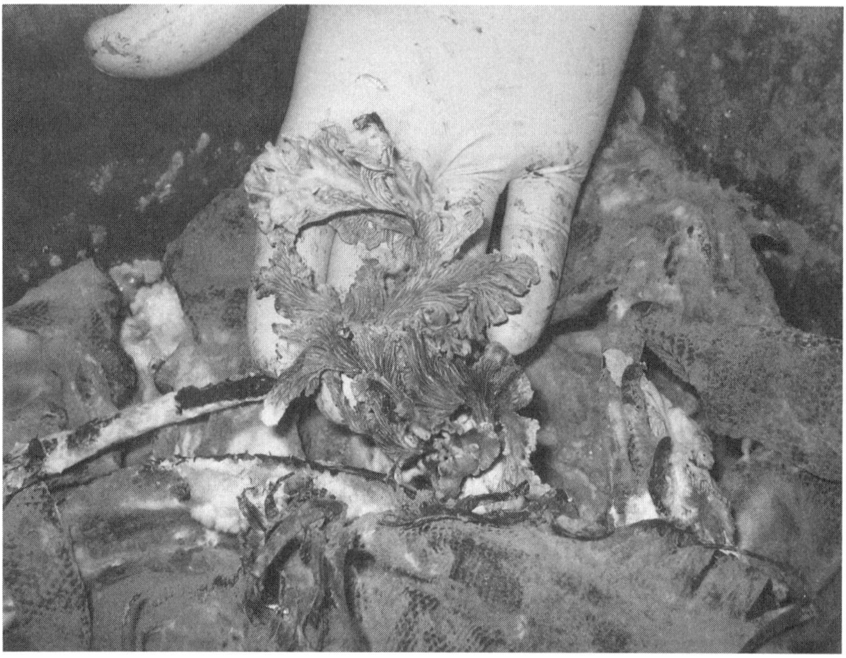

Figure 5.11. Mold and bacterial growth on fluid-preserved lizards in storage tank due to low alcohol concentration.

to storage-strength preservative in a clean container and monitored for several weeks to make sure that the growth does not reappear. Note in the catalog record for the specimen that the growth of bacteria occurred and what cleaning procedures were used. Moore (1999) recommended cleaning fungal growth from specimens with a 5 percent potassium permanganate or copper sulfate solution, then adding a small amount of phenol, menthol, or thymol to the preservative to prevent future growth.

REHYDRATION OF FLUID-PRESERVED SPECIMENS

Rehydration is usually only relatively successful due to the amount of tissue damage that occurs to the specimen during dehydration, as well as the damage that may be caused by the rehydration technique. Successful rehydration depends, in part, on how dehydrated the specimen becomes before rehydration measures are taken, as well as the method of rehydration selected. There are many rehydration techniques described in the literature (see review in table 22). Historically, most published rehydration recipes have been designed for rapid rehydration (a few hours to a few days) using a surfactant. During rehydration, expansion of the cytoplasm and cell membranes often leads to membrane rupture and maceration of tissues. Buffering salts may recrystallize in dehydrated specimens, sometimes causing structural damage to the specimen prior to rehydration (recrystallized buffering salts may be removed by rehydrating the specimens). Caution is urged before using any rehydration technique because some are successful only with selected types of specimens (e.g., small insects or mollusks) and some will cause further damage to specimens or destroy them. In particular, some techniques that work well for hard-bodied small invertebrates do not work well at all for larger invertebrate or vertebrate specimens. Rehydration may be unsuccessful due to the previous preservation history of the specimen, because of the degree of dehydration of the specimen, because of the amount of tissue damage that occurs during the dehydration-rehydration cycle, or due to the chemicals used.

Simon Moore (pers. comm.) has successfully used a 3 percent solution of Decon 90 in warm water (30–40°C) for animals as large as bats, mice, and snakes. Decon 90 is a potassium hydroxide–based alkaline commercial product sold as a laboratory detergent which has a pH of 12–13 when concentrated; dilution produces a pH closer to 8. When rehydrating using Decon 90, specimens should be gently agitated from time to time to reduce the accumulation of air bubbles. The remaining air bubbles can be dispersed with a low vacuum to prevent the specimen from floating (Jeppesen 1998). Moore recommends refixing the specimen after rehydration with careful injections of Steedman's fixative (see table 1).

An alternative to rapid rehydration, particularly for larger specimens with a greater surface-to-volume ratio, is to slowly rehydrate the specimen in a humid atmosphere until the skin has softened enough to absorb water, then stage the specimen from 100 percent water to the appropriate preservative strength (Simmons *forthcoming*; Singer 2014). This technique, although slower than those previously recommended in the literature, seems to be far less damaging to specimens and eliminates the surfactants and other chemicals. Figure 5.12 shows two alcohol-preserved frog specimens that

Figure 5.12. Comparison of two dehydrated frogs; rehydrated individual (left) and dehydrated (right) individual.

were severely dehydrated; the specimen on the left was then rehydrated using this slower technique.

Before beginning any rehydration treatment, the first question to ask is, should the specimen be rehydrated? Even if rehydration is successful, the useful life of the specimen will probably be shortened, the specimen will most likely lose color (most turn some shade of brown), there is a high risk of causing further osmotic damage to the specimen, the specimen may collapse upon itself, and the DNA of a rehydrated specimen will be degraded. As a rule of thumb, I recommend that specimens that are only slightly dehydrated (but still soft and slightly flexible) be carefully rehydrated, but specimens that are very dry are best left as they are (because dehydrated specimens are susceptible to mold and bacterial growth, they should be housed dry in a well-sealed container). If a severely dehydrated specimen is needed for future research, an attempt can be made to rehydrate it at that time, using an appropriate technique. Whatever rehydration technique is chosen should be tested on a similar non-critical specimen first. Once rehydrated, the specimen may be carefully injected with the desired fixative or preservative. The formation of air bubbles is a common problem in dehydrating specimens—these can be removed with a vacuum pump.

MOVING COLLECTIONS

Relocating a fluid-preserved collection to a new storage space is an ideal time to inventory the collection and check on the condition of the specimens, fluid levels, and container seals as each container is handled. It is not unusual to find lost containers or specimens during a collections move. Coetzer et al. (2009) provided a case study for a move of a collection containing more than seventy thousand jars and five thousand large containers. During the move, the collection was inventoried and tracked using an "Inventory, Curation and Reshelving" front-end program designed to work with the collection database; remedial curation was carried out on containers as needed. The inventory, curation, and reshelving program provided selected data fields for easier sorting of information, and displayed the new collection array sequence to allocate the specimens in the new storage area. Remedial curation involved topping up containers and replacing faulty closures or inadequate containers.

EXHIBITION OF FLUID-PRESERVED SPECIMENS

The preferred way to exhibit fluid-preserved specimens is in the most stable, long-term preservative solution (e.g., 70 percent ethyl alcohol or 10 percent buffered formalin). Unfortunately, because alcohol and formaldehyde are both highly regulated in industrial settings, their names tend to catch the attention of the code enforcement authorities, who usually do not allow very large quantities of either fluid on public exhibit. Unfortunately, there are no really satisfactory safer fluids for exhibition.

One alternative is to use a commercially available preservative or holding solution (see table 17), most of which do not have the transparency or clearness of alcohol or formaldehyde. Most proprietary preservatives and holding solutions are glycol-based or glycol- and gluteraldehyde-based (the gluteraldehyde is often listed only as a proprietary ingredient), and quickly become cloudy when specimens are introduced. Although there is no way to predict how quickly a glycol solution will become cloudy, it is usually a matter of just a few days (see discussion in chapter 3). The commercial glycol-containing preservative fluids are marketed as temporary holding solutions for preserved specimens, not as long-term preservatives, because specimens don't hold up well in the long term. The inevitable cloudiness in these solutions may be caused by the glycol, the gluteraldehyde, or both. Although in theory formaldehyde forms water-soluble bonds and thus it should be possible to remove it from a preserved specimen by repeated rinsing, in practice this is extremely difficult to do (see discussion in chapter 2). Glycol compounds (particularly propylene glycol) may turn cloudy when mixed with formaldehyde, and glutaraldehyde may turn milky in saline solutions (such as in the presence of buffered formaldehyde seeping from the preserved specimen). The cloudiness may be caused by the deterioration of the glycol itself.

There are only a few alternative fluids that do not contain glycol (see table 17). One alternative to glycol is phenoxytol, but like glycol it is a cloudy liquid and a poor preservative (see discussion in chapter 3). Other fluids that have been promoted include DMDM-hydantoin, which also seems unsuitable for long-term preservation, and Novec HFE-7100 from 3M Corporation, an engineered fluid that was developed as a cleaner and degreaser (see discussion of these in chapter 3).

A very dilute solution of alcohol or formaldehyde could, in theory, be used for display, assuming that the specimen going on exhibit does not have scientific or archival value. A low concentration of alcohol will result in specimen hydration and put the container contents at risk for bacterial growth, but the risk may be worth it in some exhibit situations. To pursue this option, it would be necessary to discuss the matter with the agency in charge of enforcing the fire code, pointing out that ethanol fumes dissipate so quickly that it rarely poses a hazard in small amounts or low concentrations. Therefore, if the alcohol concentration was reduced to 49 percent (or below) with distilled water, the container and closure thoroughly cleaned, and the container tightly sealed, the specimen might endure at the lower ETOH concentration for a long while (the container would need to be monitored closely for evidence of bacterial or mold growth).

The Armed Forces Institute of Pathology Medical Museum Laboratory (Anon. 1957) proposed a protective coating for friable anatomical specimens on display. The recommendation was to dip the specimen into a mixture of 115 g granular gelatin, 10 g sorbitol, 2–3 gtts (drops) of methyl violet (0.5 percent), and 1,000 cc distilled water. The methyl violet is to keep the water clear, the sorbitol serves as a plasticizer (25 percent polyvinyl alcohol, glycerin, or polyethylene glycol may be substituted). The same paper recommends filling cavities with gelatin or celloidin to displace discolored fluid that may fill them.

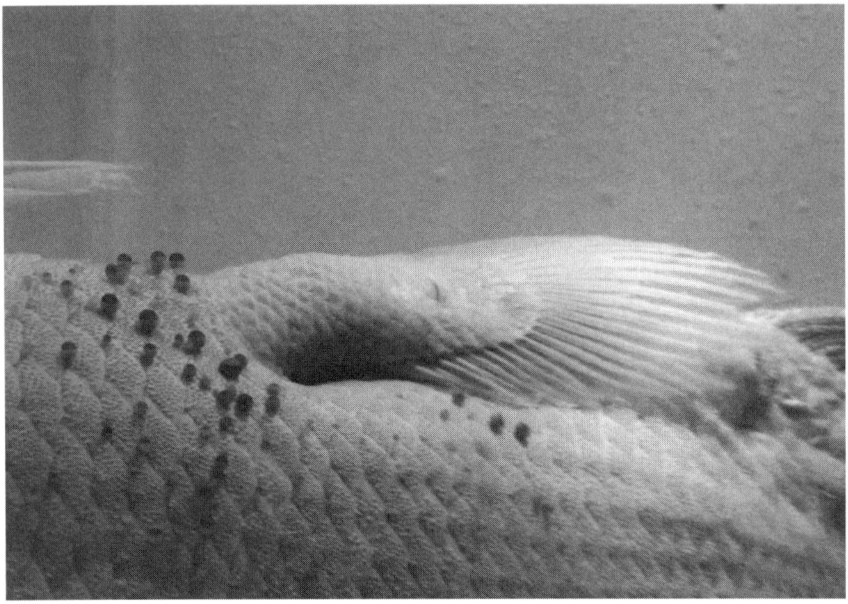

Figure 5.13. Lipid extraction from a coelacanth specimen in Steedman's solution at the Natural History Museum (London).

The Natural History Museum (London) has had a specimen of a coelacanth on exhibit for several years in Steedman's solution (0.5 mL propylene phenoxetol, 4.5 mL propylene glycol, 5 mL formaldehyde, and 90 mL water; see table 1). There is some concern with the quantity of lipids that are being extracted from the specimen by the preservative (see figure 5.13).

DEALING WITH OLD CONTAINERS AND OLD SPECIMENS

When working on old containers of old specimens, an assessment should be made of the historical integrity of the container and the specimen. If the historical value is low, or if the specimen is in imminent danger of being lost, intervention directed toward the specimen can be undertaken. However, if the historical integrity of the container or the specimen is paramount, then great care must be exercised to minimize the evidence of intrusion into the container. Mulder (1997) discusses some of the ethical and aesthetic questions that arise when doing conservation or restoration work on historic specimens.

As discussed earlier, glass containers for fluid-preserved specimens were originally sealed with pig or sheep bladders, often applied over a closure made of glass or other material. It is possible to replace the seals with replicas following the directions of

Moore (1999), who provided a detailed description of a modern bladdering technique: cut open a pig bladder and pin it in a dissecting tray, then fix the bladder with a formaldehyde-based fixative for several days. After fixing, remove the bladder and wash it thoroughly in running water. The bladder can then be stretched over the top of the container and the corners attached temporarily with adhesive tape. Once secured, tie the bladder tightly to the container, just below the rim, using fine twine or thick thread. Coat the twine or thread with varnish to prevent it from coming loose. Once the bladder and binding are dry, the adhesive tape can be removed and the bladder can be trimmed just below the binding. Should the seal be unsuccessful, remove the bladder and rehydrate it in Decon-90. Jar bladders are usually painted black, using a mixture of 20 percent bitumen dissolved in toluene. Bladder life may be extended if it shows signs of becoming brittle by gently applying a lanolin cream to soften it.

Flat glass container closures (whether round or rectangular) should have a small fill-hole (just large enough to admit a hypodermic needle for refilling) drilled near one edge and sealed with a piece of polypropylene rod or covered with a glass microscope slide cover slip.

Lead, parchment, foil, or waxed closures may be removed from old jars by cutting with a scalpel or softening the seal using warm to hot (but not boiling) water (Spencer Jones 1996). Brittle seals on flat glass closures can be removed by carefully cutting the lid free around the circumference with a scalpel. Glass stoppers may be removed after loosening by rocking gently back and forth using the glass tab, by tapping the glass tab gently with a wooden stick, or by submerging the entire jar in warm water to expand the air inside and force the lid off (although this will also heat up the specimen inside). Another method for removing glass stoppers calls for mixing two parts alcohol with one part glycerin and one part sodium chloride and leaving the solution covering the stopper for a few hours, after which the stopper is gently tapped loose (Hiscox 1935). Stubborn screw-top closures may be removed by wrapping the jar lid with a large rubber band or using a leather strap wrench.

Once the container is open, remove the specimen and labels carefully and place them in a holding tray or another container while the jar and closure are cleaned and prepped for refilling. Before resealing the container, clean the sealing surfaces using a scalpel, alcohol and cheesecloth pad, or a soft abrasive dishwashing pad. Specimens once attached to glass or plastic plates or frames that have come loose may be reattached using cotton string, monofilament, or reglued using a 1 percent mixture of necoloidine in a mixture of one part diethyl ether and one part isopropyl alcohol or celloidon.

Some historic seals may be recreated. As discussed previously, a common sealant which came into use in the eighteenth century was made from a combination of Stockholm tar and red lead (lead sesquioxide), which normally lasts about ten years before becoming brittle (Moore 1999). Stockholm tar and red lead was often used with bladders to seal glass containers. The red lead compound is toxic, so must it be handled with care, but it is still available from suppliers of veterinary and agricultural products. To seal a container, the Stockholm tar should be poured onto a glass sheet and the red lead powder carefully mixed in to make a sticky paste (Moore 1999). Apply the paste to the container and closure with a spatula. The closure should be heated in a water bath before being applied to the container and left under a weight for a week while drying. The excess paste can be trimmed with a warm spatula. Gutta-percha (a type of natural latex) seals became common in the early 1900s (Moore 1999; Schorr 1907). When applied properly, gutta-percha forms a grayish white seal that will last thirty to forty years before becoming brittle. Gutta-percha may be applied to the container with a spatula. The closure should be heated in a water bath before being applied to the container and left under a weight for several days while setting. Bitumen, applied while hot, cools to form a black, hard, brittle seal that can only be removed by a combination of scraping with a sharp blade and judicious use of acetone.

Another common nineteenth-century seal was made with gelatin. Frosted gelatin forms a good seal, but if the gelatin has a white mastic appearance or is yellow and flaking the seal will fail. An advantage of gelatin seals is that they are easily reversible by submerging the container and closure in hot water. Moore (1999) described how gelatin seals can be made for glass containers with glass closures using gelatin coignets (very thin sheets of gelatin used in food preparation); powdered or fibrous gelatin is unsuitable for sealing containers due to its melting temperature. About 24 g of thin sheet gelatin should be hydrated in tap water, then melted in a water bath until it has the consistency of syrup. Add 6 mL of glycerol and 3 mL of glacial acetic acid to the gelatin, stir carefully, and then pour the mixture onto a metal tray and allow it to set overnight. Cut the hardened gelatin into squares and store in a well-sealed container until needed. To make a gelatin seal, the contact surfaces of the glass container and closure should be ground so the gelatin will adhere (see earlier discussion). Heat the container closure to about 80°C, and meanwhile heat several squares of the hardened gelatin in a water bath until the gelatin melts. Brush a generous amount of the gelatin on the container rim, being careful not to allow the gelatin to drip down inside the container, then press the closure on the container and leave under a weight overnight.

Flat glass closures on glass jars may be attached with low melting-point microcrystalline paraffin, or with silicon (bearing in mind that silicon is difficult to dissolve and permeable to gas diffusion). The silicon should have a rubbery feel to it, but some silicon compounds solidify over time or may be susceptible to fungal growth (Moore 1999), so use a medium-modulus, neutral-core silicon sealant. Chemical silicon release agents are available commercially that will help clean silicon from glassware. Alternative container sealants include Alsirol (described earlier), and soft, white paraffin. A paraffin seal, applied hot with a spatula, typically lasts for five to ten years before needing replacement.

Acrylic containers, although not recommended (see discussion in Alternatives to Glass Containers), can be sealed with the commercial cements designed for the products or a 1:1 mixture of chloroform or trichloroethylene and glacial acetic acid.

REPAIR OF DAMAGED FLUID-PRESERVED SPECIMENS

Broken specimens can be repaired with glass needles or by using preservative-resistant adhesives such as celloidin (Waterhouse and Graner 2009). Glass needles can be made by heating a glass rod and drawing the ends out until the connecting glass becomes very thin; allow the glass to cool and then snap into the desired lengths, taking care with the very sharp points. Similar needles may also be made of plastic, but will not have the rigidity, durability, or sharp ends that the glass needles do. Specimen repair with glass needles is shown in figure 5.14. Metal wire should not be used to repair fluid-preserved specimens as it will corrode and damage the specimens—copper wire, for example, may react with the preservative to form cupric salts that stain both specimens and preservative.

Celloidin, a mixture of pyroxlin in alcohol, can be used to adhere specimens to a glass or acrylic backing plate (see discussion in chapter 3 and figure 3.5). Moore (2009) recommends leaf gelatin as a reversible adhesive for formaldehyde-preserved specimens (leaf gelatin has longer molecular strands and thus provides better adhesion; do not use powdered gelatin). The gelatin leaves should be hydrated in cold water until soft, then heated in a double boiler until just molten. Add 2.5 mL of glycerol and 1.5 mL of glacial acetic acid to the solution, then pour the liquid on to a glass sheet and allow it to solidify into a soft, sticky substance.

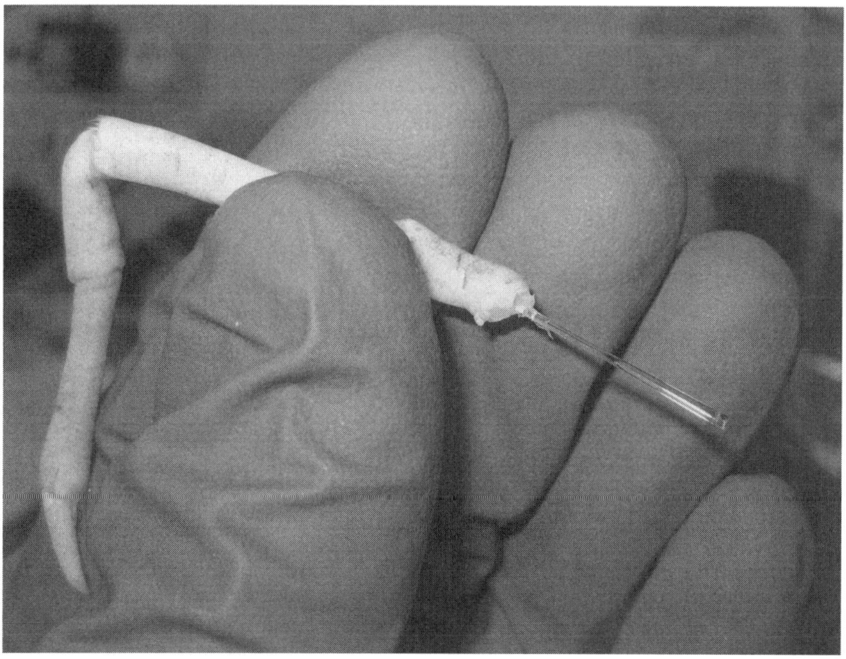

Figure 5.14. Specimen repair using glass needle.

HEALTH AND SAFETY

Standard laboratory practices should protect the people working with fluid preservatives and fixatives from danger, as well as provide procedures for properly handling specimens without causing damage to the specimens (Hawks et al. 2010). Standard procedures include:

1. The use of personal protective equipment, including lab coats; safety glasses, or splash goggles (depending on the chemical in use); and nitrile gloves (which are formaldehyde-resistant). Even when wearing a lab coat, people working with fluid fixatives and preservatives in a laboratory should not wear shorts or short skirts (which leaves the legs exposed), and should not wear open-toed shoes.
2. Clean work habits should be practiced at all times; there should be no consumption of food or beverages or use of tobacco products in the laboratory.
3. Adequate ventilation is critical for working with fluid fixatives and preservatives. Use a bench-top or local exhaust system that draws fumes away from the specimen before the fumes enter the breathing zone of the worker (Burroughs et al. 2006). Ethanol is absorbed through the lungs, mucus membranes, and through the skin (Browning 1965), and may result in "irritation of mucous membranes, headache, inebriation, lack of concentration and somnolence" (Browning 1965, 331). Exhaust systems and fume cabinets must have sufficient make-up air to function efficiently.
4. All chemicals should be properly labeled and stored. Fume cabinets should not be used for storage of chemicals, specimens, or equipment.
5. Standard operating procedures (SOPs) should be written and made available to personnel working with fluid-preserved specimens. SOPs should include instructions for mixing fixatives, preservatives, and other chemical solutions used in the lab; filling containers; accessing the collection and specimens; cleaning glassware; responding to chemical spills; and for any other processes or procedures using chemicals that are conducted in the laboratory.

Laboratory and storage areas for fluid-preserved collections should be equipped with spill kits. A spill kit contains materials necessary for safely cleaning up alcohol or formaldehyde spills and broken containers. A spill kit is usually housed in an HDPE bucket with a good lid. Typical spill kit contents include safety glasses, nitrile gloves, Tyvek overalls, absorbent pads, absorbent pillows, an absorbent boom, and plastic bags for disposal of used absorbent materials and broken or damaged containers. In addition to the spill kit, both laboratory and collection storage areas should be equipped with a wet mop, mop bucket, broom, and dust pan.

For a thorough review of laboratory safety issues and compliance, see Hawks et al. (2010).

FIRE PREVENTION

A fire prevention program for the fluid collection should be developed with a professional fire prevention specialist and the local fire code enforcement authority (Stemen 2010). Ethyl alcohol and most other preservatives are flammable liquids and must be stored and used within approved fire safety guidelines. In the United States, a flammable liquid is defined as one with a flashpoint below 38°C (100°F). The flashpoint of a volatile liquid is the lowest temperature at which it can vaporize and form a mixture with the air that can be ignited. The flashpoint of 96 percent ethyl alcohol is 13°C (55°F); the flashpoint of 70 percent ethyl alcohol is 21°C (70°F) (Stemen 2010).

According to Stemen (2010), fire prevention measures for bulk alcohol (e.g., 96 percent ethyl alcohol in large containers) may include storage in a room that is separated from other areas by fire-rated construction, equipped with explosion-proof electrical connections, and with appropriate ventilation. Fire prevention measures for fluid-preserved collections may include heat and smoke detectors; limited capacity rooms with their own ventilation systems; limits on the amount of alcohol that can be stored per room; fire-rated doors, walls, ceiling, and floors; floor drains and a fire suppression system using wet pipe sprinklers, carbon dioxide gas, foam, or an inert gas mixture. Fire prevention measures may include lowering the temperature in storage to lower the flashpoint of the alcohol (see earlier discussion of preservatives at low temperatures). The design for fluid-collection storage at the Smithsonian Institution includes aggressive ventilation, a hydrocarbon gas detection system, compartmentalization of the facility, automatic sprinklers, run-off control, and maintenance of the temperature at 18.3°C (65°F) to lower the flashpoint of the alcohol (Stemen 2010).

In fluid-preserved collections, a good exhaust system should prevent the accumulation of alcohol vapors. Alcohol vapors disperse rapidly, and the ignition danger drops quickly with increased distance from the source. Bulk alcohol and fluid-preserved collections should not be stored below grade, because alcohol vapors are heavier than air. Possible sources of ignition (sparks, heat, and flame) should be eliminated in the collection storage area (e.g., keep wall plugs at least 1 m above the floor; do not allow the use of electric motors at floor level). All areas where fluid-preserved collections are used should be equipped with wet pipe sprinklers and both heat and smoke detectors. Although lowering the temperature of the collection storage area will lower the flashpoint of alcohol, it is not recommended that the temperature be lower than 18°C (65°F) to protect the quality of the preservative and the specimens, as discussed earlier.

Fires in fluid-preserved collections are rare, but when they do occur, the results can be disastrous. In May 2010, a fire at the Instituto Butantan in Sao Paulo, Brazil, destroyed a historic and valuable collection of eighty thousand preserved snakes, spiders, and scorpions. The building that the collection was housed in was not equipped with a fire suppression system. Although the fire burned very hot (due to the presence of ethanol), according to eyewitnesses and photographs published in the news media, the glass containers of specimens did not explode but broke open due to the heat, allowing the contents to ignite. News photographs of the aftermath of the fire showed the remains of shelving, specimens, and broken containers collapsed in place, a further indication that no explosions occurred.

Fire codes do not directly address the storage of museum specimens in standard preservatives, but typically address the storage of beverage alcohol (which is typically 5–40 percent ETOH) and the storage of large containers of 95 percent ETOH. Neither of these storage situations is comparable with museum specimens in 70 percent ETOH. As a result, local fire officials must adapt the code regulations to museum storage, which means that what is standard practice varies from one place to another. For example, some institutions are required to use explosion-proof safety cabinets, some are not; most institutions are required to have wet pipe sprinkler systems (usually around 0.30 gallons per minute per square foot capacity), but the required output of the system will vary. Most fluid specimen storage facilities are required to be above grade and well ventilated using floor vents because alcohol is heavier than air (ventilation systems are usually 1 cubic foot per minute per square foot). The Natural History Museum of London relocated its collection of more than 450,000 containers of fluid-preserved specimens (approximately half a million liters of alcohol) to a new, purpose-built, energy-efficient facility in 2002 (Brice 2002). The collection storage areas are kept at 13°C, which lowers the flashpoint of the alcohol to the point that sprinklers are not required and standard light fixtures and other electrical equipment can be used. Alcohol (in the form of IMS, or industrial methylated spirits) is piped through the building to the laboratories. The building (which has an area of 10,000 m^2) is divided into sixty fire alarm zones with heat and smoke detectors as well as a gas detection system for alcohol vapors.

Collections managers must work with their local fire marshal or other safety experts to get an interpretation of the regulations that is reasonable and ensures the safety of the collection and those working with it. In general, storage of fluid-preserved collections means a reduced number of ignition sources in fluid storage areas (e.g., few or no wall sockets, no work activities in fluid storage, explosion-proof electrical switches), reduction of the chance of spills (e.g., earthquake bars on shelves), a good ventilation system to prevent the buildup of fumes, and restricted access to fluid collection storage. The design of safe, reasonable facilities for fluid-preserved collections requires the input of a qualified fire protection engineer who has experience with museum fluid-preservation issues and who is willing to work with the museum staff and consultants. Some of the factors to consider when planning storage for fluid-preserved collections include:

- Wet-pipe, water-based sprinkler systems are generally the best option for fire suppression. A very aggressive sprinkler system will probably be required if compactors are used in the space to assure sprinkler penetration. In addition to putting out the fire, water from sprinklers will dilute whatever fixatives and preservatives leak from ruptured containers. Furthermore, water is generally available in a relatively unlimited supply (compared to chemical fire suppression systems, which hold a finite amount of foam or gas in reserve).

- Both smoke and heat detectors should be included in the fire detection system.
- Good closures for specimen containers will limit alcohol vapor evaporation and make the storage area safer.
- Several small rooms are usually required rather than one large room to limit the amount of alcohol in a given area. The maximum room size will be based on the maximum capacity of the fire suppression system, the water or other suppressant supply, and the amount of alcohol to be housed in the space.
- Passive features such as fire walls, fire-rated floors, and fire barriers between storage areas will greatly enhance safety.
- Aggressive ventilation to remove alcohol vapors and prevent a flammable mixture from accumulating will probably be required. Systems with 100 percent air makeup should be avoided, as they are extremely expensive to operate and create a host of problems related to changes in relative humidity and its effects on containers and seals.
- Smoke detectors may be required in return and supply ducts for the ventilation system.
- The HVAC system should be designed to maintain moderate temperatures between 18°C and 21°C (65°F and 75°F), preferably at the low end, with fluctuations of no more than 2°C (5°F), with a relative humidity below 65 percent, but as high as can be reasonably achieved at the temperature set point (it is better to allow relative humidity to drift to maintain a steady temperature).
- There will be restrictions on use of the space for anything other than storage of the collections (collection storage areas should not be used for research, offices, storage, etc.).

FORMALDEHYDE SAFETY

For a good many years following its discovery, formaldehyde was not recognized as a dangerous substance. In 1899, in reference to the irritation of the mucous membranes, Drowne wrote (apparently with no sense of irony) that "one becomes hardened after enduring it for a time and it ceases to annoy much." A 1915 publication called *Formaldehyde the Farmer's Friend* (published by a company that sold formaldehyde) recommended its use "For breaking up colds in the head, place one teaspoon of formaldehyde in a bowl of hot water and inhale the fumes through the nostrils. Repeat from time to time." The pamphlet also recommended using formaldehyde to clean bedding, clothing, floors, cupboards, sinks, and refrigerators. A note published in *Science* magazine in 1931 recommended treating "formalin poisoning" of the fingers with lanolin (Holt 1931). The poisoning was, of course, the fixing of the cells of the epidermis and dermis, so the lanolin would be of minimal help. A note in the widely read *Turtox News* recommended "denaturing" formaldehyde-preserved specimens before working with them by submerging the specimens in a solution of urea and ammonium phosphate (Anon. 1939; Foust et al. 1935).

Formaldehyde affects the respiratory system, is a skin irritant, will fix the tissues in the body of a person exposed to the fumes or fluid, particularly the mucous membranes (Horvath et al. 1988). Formaldehyde will cause allergic contact dermatitis (Browning 1965; Maibach 1983), and has been implicated as a cause of cancer, even at rates as low as 1 ppm (Arts et al. 2006; Perera and Petito 1982). The odor threshold of formaldehyde is so low that if you can smell it, you are already receiving an overdose. For more information on the safety of formaldehyde and other fixative and preservative chemicals, refer to Babin et al. (2010).

SUMMARY OF RECOMMENDATIONS BASED ON WHAT WE KNOW AND DON'T KNOW ABOUT FLUID PRESERVATION

Based on the information compiled in this volume, and considered in the context of the collective experience of many collections care workers, the following recommendations can be made for processing fluid-preserved specimens (see also table 23).

1. A written record of preparation techniques and chemicals used should be included in the collector's field notes.
2. Keep the time interval between death and initial fixation or preservation of specimens as short as possible.
3. Measure solutions carefully when preparing fixatives and preservatives; use deionized or distilled water.
4. Fix specimens at warm (but not hot) temperatures.
5. Use a fluid-to-specimen ratio of at least 7:3.
6. Rinse specimens after fixing.
7. Stage specimens from rinse water to preservative by steps of 20 percent increase in concentration of preservative.
8. Maintain the quality of the preservative fluid. Do not use denatured alcohols.
9. Use containers and closures that are resistant to oxygen permeance and seal well.
10. Fill each container to 90 percent of its volume for ethanol-based preservatives; to 95 percent of its volume for water-based preservatives.
11. Maintain a constant temperature in collection storage areas, preferably a few degrees lower than the place where the containers were sealed.
12. The preferred storage environment is 18–21°C (65–70°F) with relative humidity of 50 percent.
13. Minimize fluctuations in temperature and relative humidity.
14. Protect specimens from light, ultraviolet radiation, and vibrations.
15. Handle specimens carefully; do not allow specimens to dehydrate while they are out of their storage containers.
16. Monitor the collection regularly, particularly for loss of preservative or changes in preservative quality.

6

Fluid-Preserved Collections as Cultural Patrimony

WHY PRESERVE SPECIMENS IN FLUID?

Human beings have an innate impulse to collect objects as a way of orienting themselves in the world, as is demonstrated by the long history of collecting and the history of museums (Simmons 2010). The human desire to preserve objects for the future is directly related to the need to make collections. For centuries, the only organic objects that could be kept in collections were those that could be dehydrated—a process that dramatically alters shapes and colors. The discovery in 1662 that organic materials could be preserved in alcohol for an indefinite length of time changed the nature of collections, enabling a greater variety of objects to be preserved.

There is something about preserving things in fluids that sets them apart from other objects—for example, the reactions that people have when they see a mummy in a museum are not the same as their reactions when they see a human body or body part preserved in fluid. In particular, the image of a human fetus in a bottle elicits strong and tangled emotions of awe, attraction, and revulsion, and is an image has been widely used in literature (as discussed next). A specimen floating in a jar of preservative seems to be suspended between life and death, vivid in detail but shrunken, discolored, perpetually in its aqueous environment. The first public presentation of fluid-preserved specimens before the Royal Society generated a lot of interest, as shown by subsequent accounts in the *Transactions* of members bringing in all sorts of organisms (including a human fetus) that they had preserved in spirits of wine. The anatomical preparations of Frederik Ruysch (see chapter 1) were proclaimed by visitors to the collection as marvelous and extraordinarily life-like. Most contemporary descriptions of the Ruysch specimens seem to describe a standard of preservation that has not been equaled since, although such descriptions should be interpreted in historic context. For example, writing just a hundred years after the death of Ruysch, Jean Nicolas Gannal (1791–1852) opined that:

> It is stated, it is true, that Ruysch, had discovered the means of preserving the dead body with all the appearance of life, without drying, with florid complexion, and supple limbs. But, is this really the fact? and have we not good reasons to doubt such assertions, since no collection of anatomical pieces, prepared by this process, has descended to us, and no explanation has confirmed our knowledge of them? (Gannal 1840, 145)

The very act of putting an object in a bottle endows it with powers it did not previously possess. A specimen in fluid seems forever protected from deterioration, cloaked in the authority conveyed by museum collections—the sealed glass jar and the liquid around the specimen keep it at a respectful distance while simultaneously drawing us closer to it. In the preface to a collection of photographs of museum collections, Frans Schouten describes how a museum object is "supplied with a new significance since it has been uprooted from its origin," and how the photographs "strikingly reflect the alienating effect of the museum" (Thijsen 1990, 1). There is a company in Los Angeles that specializes in costumes and props for movies and theater, and advertises for rent, "A variety of real and replica biological specimens . . . in jars" on their website, claiming that their "prop specimen jar is the largest in the business" and includes "a full term fetal pig" (from 1858), but their most popular specimens are the "all vinyl fetus replicas" (www.bjwinslow.com).

Despite some claims that "A medical museum has little appeal to the general public" (Edwards and Edwards 1959, 2), in fact exhibits of fluid-preserved specimens (particularly medical oddities) have always been a good draw for visitors. Recounting the auction of the Victorian-era Potter Museum, which contained marvelous taxidermy grotesque groups and other specimens collected and preserved by Walter Potter (1835–1918), author Melissa Milgrom described "jars of zoological horrors floating in spirits the color of Kool-Aid" (Milgrom 2010, 174) and observed that, "The more ghastly the specimen, the larger the pre-auction crowd. Nothing drew more people than the Siamese twin fetal piglets in lime green formaldehyde" (Milgrom 2010, 177). The Natural History Museum (London) opened Darwin Centre I in 2002 to house twenty-two million fluid-preserved specimens on 27 km of shelving. Some of the fluid-preserved specimens are on exhibit to the public in an arrangement that gives the museum visitors a glimpse into storage, and guided tours of the collection are very popular with the public. The Berlin Natural History Museum recently opened an exhibit to showcase its specimens of fish, mammals, spiders, amphibians, and reptiles preserved in alcohol (the collection contains more than 276,000 jars in 81,880 L of alcohol on 12.6 km of shelving).

THE FLUID-PRESERVED HUMAN

Human bodies and body parts have long been preserved and often revered, beginning thousands of years before the development of fluid preservation—the process of mummification is labor-intensive and intensely connected to belief in the afterlife, as are the relicts of the saints in Christian tradition, including entire uncorrupted bodies (Cruz 1977, 1984). Medical museums, since their founding, have attracted both the medical community and the general public, but often for very different reasons.

Descriptions of people's reactions to preserved fetuses are varied. One author wrote that "The two-headed fetus swimming in a jar of formaldehyde makes us feel beautiful, psychologists say" (Christy 2008), while the poet Sylvia Plath narrated in her 1963 novel, *The Bell Jar* (Plath 1998) how the protagonist (Esther Greenwood) is shown fetuses preserved in jars by her friend, Buddy Willard, who takes her to a hallway "where they had some big glass bottles full of babies that had died before they were born. The baby in the first bottle had a large white head bent over a tiny curved-up body the size of a frog. The baby in the next bottle was bigger and the baby next to that one was bigger still and the baby in the last bottle was the size of a normal baby and he seemed to be looking at me and smiling a little piggy smile" (Plath 1998, 63). For Ester Greenwood, the fetuses were a metaphor for her mental condition. Later, while looking through a magazine, Ester describes how "The face of Eisenhower beamed up at me, bald and blank as the face of a fetus in a bottle" (Plath 1998, 89). Plath used a similar image in her 1960 poem, "A Life," when she wrote:

> With no attachments, like a foetus in a bottle,
> The obsolete house, the sea, flattened to a picture
> She has one too many dimensions to enter.
> Grief and anger, exorcised,
> Leave her alone now. (Hughes 2008, 150)

Lawrence Durrell invoked another aspect of the image of a fluid-preserved fetus in *Mountolive*, the third book of his *Alexandria Quartet*, when describing an obstinate Egyptian official, "Then he sat with his paws folded over his neat grey waistcoat, glum as a foetus in a bottle, as Mountolive delivered his strongly-worded protest" (Durrell 1958, 251).

The image of a fetus in a bottle is frequently found in other media in popular culture, including the press. In a controversial incident reported by the BBC in 2004, a fourteen-year-old girl, whose baby was stillborn, was given the eleven-week-old fetus in a bottle of formaldehyde to take home when she left the hospital. Her family kept the bottle in the refrigerator before complaining to the County Durham and Darlington Acute Hospitals Trust (BBC, 2004).

In Bangkok one steamy morning, my nephew and I went to see the Museum of Pathology at Sirijit Hospital. The exhibit area is arranged as an old-fashioned teaching institution, which in itself is fascinating (we have lost a lot in museums by making displays that are more eye-catching than they are information-presenting). Most of the pathology specimens were on exhibit in beautiful old glass containers, awash in formaldehyde and other fluid preservatives. Despite the poor lighting and label text mostly in Thai, the pathologies we saw were fascinating—carcinomas, lesions, gout, tissue from brains damaged by stroke. However, the other visitors (almost all

Thais) were most interested in the fetuses. There were a surprising number of fetuses, all with some serious problem (including conjoined twins, spina bifida, anocephalgia, and cleft palette). But these were not fetuses glum in their bottles—despite the terrible deformities (and in some cases, dissections and crude postautopsy stitches), most of the fetuses wore serene expressions as they floated in their jars, hovering quietly just beyond the thin line that separates life and death. Each container that held a fetus was mounted on a platform wide enough to allow the placement of objects around it. The objects that surrounded the fetuses included toys, candy and gum, little images of the Buddha, photos of revered monks, paper airplanes, hair barrettes, coins, plastic bracelets, and handwritten notes. Were these objects left by the parents of children born with similar problems, or perhaps by expectant couples feverishly hoping for a healthy baby? In any case, it was a reminder that what is repulsive and grotesque to one person might be of great interest or be emotionally supportive to another. Each fetus in its glass container compelled you to look closer, to stare, to wonder, simultaneously attracting and repelling as only a grotesque can.

Human body parts preserved in fluids are kept for many reasons. In a short story published in *Granta* magazine, an aging diabetic who lives in a small building on the same property as his divorced wife is trying to cope with the phantom pain from his amputated foot, but "Sometimes he has a panicky thought that they gave it to Jeanne; in a jar, like a tonsil. And that she has it up there in the house, with all his things" (Peelle 2008, 137). In an oddly parallel situation in real life, an incident was reported in Lawrence, Kansas, of an individual who kept his amputated foot in a plastic bucket of formaldehyde and charged his friends to view it (Weslander 2005). When a neighbor complained, police confiscated the foot and held it until they were sure that no crime had been committed. A hospital spokesperson was quoted as saying that it was not unusual for people to take their own body parts home in formaldehyde, noting that in the past, "'They've had women that want their uterus. People take tonsils. They take appendixes. I think it's unusual that someone would want a foot, but it's within their rights because it's theirs." The report went on to quote the owner of the foot as saying that "he cut off two toes, which he was considering giving to friends. He added trinkets to the bucket, including a porcelain horse and a can of Hamm's beer, to make it what he called 'a collage of myself.'" There are many other examples of the fetishism of body parts preserved in fluid, such as an exotic dancer in South Plainfield, New Jersey, who was found to have decorated her apartment with a variety of skulls and a "crudely severed human hand in a mason jar of formaldehyde" that she kept on her dresser; the hand was nicknamed "Freddy" (Associated Press 2006).

In a paper titled "Vision, Headache, and the Halo—Reactions to Stress in the Course of Superego Formation," Greenacre (1947) reports on a patient who clung to the image of body parts in a jar as a repressive coping mechanism:

> Her third year was especially eventful. She had a tonsillectomy under a general anaesthetic; afterwards she was given the excised tonsils in a bottle of preservative to cherish as a momento of the occasion. The patient had been in an adjoining room when her mother miscarried, had gone into her mother's bathroom and discovered something in a pail as well as bloody linen. This made a profound impression and had the effect of an

> overwhelming visual trauma, but was dealt with by fairly quick repression, being fused with the memory of the tonsils in the bottle. (Greenachre 1947, 179–180)

One of the more unusual stories involving a fluid-preserved body part was *Driving Mr. Albert: A Trip across America with Einstein's Brain* (Paterniti 2000), a classic road story but with a Tupperware container of formaldehyde thrown in. Paterniti, a freelance journalist, stumbled on the story of pathologist Thomas Harvey, who performed the autopsy on Albert Einstein (1879–1955). Apparently against the family's wishes, Harvey had removed the brain during the autopsy, preserved and sectioned it, and kept the cubes in a container of formaldehyde for decades, always intending to study it but never getting around to it. Paterniti's book tells the story of how he drove Harvey and the brain in a Buick Skylark from New Jersey to California to hand the preserved brain over to Einstein's granddaughter.

Human body parts are sometimes preserved as a memorial of the dead. During his ill-fated war against the Russian empire, one of Napoleon's soldiers had the heart cut out of the body of his fallen brother and preserved in a container of wine so it could be carried back to the family in France (Talty 2009). After his death from tuberculosis in Paris in 1849 at the age of thirty-nine, and by his prior request, Frederic Chopin's heart was removed and sealed in a vessel described as a crystal urn filled with alcohol (probably brandy or cognac). The heart, in its urn, was later taken to Poland by Chopin's sister, Ludwika. A request to perform a DNA analysis of the heart to determine whether or not Chopin had died of cystic fibrosis instead of tuberculosis was made in 2008, but denied by Polish authorities (Anon. 2008). In Mexico City, in a city park called La Bombilla on the Avenida de los Insurgentes, stands a large tower-like monument that once held a fluid-preserved right arm and hand. The monument commemorates Álvaro Obregón (1880–1928), who—after a distinguished military career—served as president of Mexico from 1920 to 1924. On July 3, 1915, during a battle at Santa Ana de Conde, General Obregón's right arm was shot off. The arm was preserved in fluid (most modern sources say in formaldehyde, but it was probably alcohol) by the attending surgeon. On July 17, 1928, Obregón had just been elected to a second term as president, but had not yet taken office, when he was assassinated at a restaurant near where the monument now stands. Obregón's preserved arm was exhibited in the monument from its dedication in 1935 until around 1999, when the relic was either stolen or removed from view.

One of the most famous preserved bodies is that of Vladimir Lenin (1870–1924). Although technically the body is embalmed rather than preserved in fluid, the son of one of the men who embalmed Lenin wrote that among the preservation options considered was immersion in a "balsamic liquid" made up of glycerin and potassium acetate (Zbarsky and Hutchinson 1997).

A European traveler in 1823 recorded an example of a human body preserved in brandy and displayed to the public:

> Among the delightful villas on the north side of Geneva, there is one at Pregni, once belonging to the empress Josephine. Since her death, it was purchased with the gardens

> and grounds, by our friend, Mr. M., with whom we dined soon after our arrival at Geneva, and had an opportunity of seeing the apartments: they are spacious and commodious, and well suited for the enjoyment of an unostentatious but elegant retirement. A few miles farther east, on the same side of the lake, is situated the château of the late Madame de Stael at Copey. It contains the body of her mother, Madame Neckar, full dressed, and preserved in brandy, by her own particular request. In this singular state it was shown to visitors for several years, but the vault which contained it, was closed a little before we were there. The early attachment of our historian Gibbon to this lady, will preserve her memory much longer than brandy can preserve her body; and the austerity of her temper and singularity of her disposition are sufficiently known by the writings of her daughter. (Bakewell 1823, 68–69)

(Anne Louise Germaine de Staël-Holstein (1766–1817) was a writer and political activist.)

A jar of alcohol containing the head of a man thought to be the Mexican bandit Joaquin Murieta has been sold at auction several times. According to various accounts, after a gun fight, the mutilated hand of one bandit and the head of Murieta were "cut off and preserved in spirits" by rangers who took their trophies to the capital of Sacramento to collect their reward for capturing the bandits (Jackson 1977, 10). In a literary parallel, one of the characters in the 2010 novel *Iron River* by T. Jefferson Parker keeps the head of a Mexican bandit in a jar of alcohol at his home.

One of the more ghoulish fascinations with fluid preservation relates to the fact that such specimens are sometimes consumed. In the World War II novel, *The Good Soldier Švejk*, a visit to a graveyard reveals the final resting place of a person who's death was caused by fluid preservatives when a grave is found with a headstone that reads, "A. Honvéd who plundered the school collections and drank up all the methylated spirit from a jar in which the various reptiles were being preserved now sleeps his last sleep there" (Hašek 1973, 625). This story has a strange parallel in real life—there are accounts of the starving soldiers in the Grand Army of Napoleon, during their disastrous retreat from Moscow in 1812, raiding the medical school at the University of Vilnius (in present-day Lithuania) and consuming the alcohol-preserved specimens (Lobell 2002; Thadeusz 2009). When the serial killer Jeffrey Dahmer was arrested in 1991, it was discovered that he had cooked and eaten parts of some of his victims, and that his apartment contained "a fifty-seven gallon blue plastic drum filled with what appeared to be formaldehyde, in which three headless torsos were marinating" (Baumann 1991, 48) and several bottles of ethyl alcohol and formaldehyde were found in the closet. Other reports detailed how Dahmer dismembered the corpses and saved various parts, primarily the skulls, which he cleaned chemically in his apartment. After his arrest, "A thorough search of Dahmer's Milwaukee apartment turned up body parts of eleven of his victims—some boiled, stripped of flesh, and painted, and others preserved in formaldehyde like lab specimens" (Ewing and McCann 2006, 145).

FOND MEMORIES OF FLUID PRESERVATION

In some cases, fluid-preserved objects are remembered in a more positive light. A case was reported in the press of a specimen of a two-headed piglet, preserved in alcohol, that was stolen in 1999 from an environmental education center in Wisconsin, after being on exhibit for forty years (along with a four-legged pheasant chick and a pair of conjoined raccoons). The center received a number of calls from concerned citizens who were familiar with the exhibit, and at least one offer from a local farmer of a two-bodied, one-headed piglet as a replacement. The specimen was eventually returned, anonymously, to the museum (Balousek 2000).

In his book of essays entitled *Broca's Brain*, scientist and writer Carl Sagan (1979) used the preserved brain of pioneer neurologist Pierre Paul Broca (1824–1880) as an inspiration point to muse on the nature of being human. Broca's brain was preserved along with hundreds of others in the Musée d'Ethnographie du Trocadéro at a time when it was believed that the study of brain anatomy could shed light on how humans think (Broca's brain is now in the Musée de l'Homme in Paris). Although the study of preserved human brains has not revealed the secrets of memory and thought, it has played a critical role in the development of neurology. The Cushing Center, housed in the Yale Medical School Library, contains 50,000 pages of notes and 15,000 photographic negatives documenting more than 2,200 neurological case studies made by the pioneering neurologist Harvey Cushing (1869–1939), along with 650 jars of preserved brains from Cushing's patients (Blair 2010; Epstein 2010; Greenberg 1996). Although the brains currently have minimal research value, they are still useful to show pathologies and surgeries to medical students. The brains and the jars they are in were carefully cleaned and incorporated into the study center, more for their aesthetic value than their teaching value (see figure 6.1).

In a delightful book about his childhood discovery of chemistry, neurologist Oliver Sacks lamented the loss of his pet octopus (killed by the maid who found it soaking in the bath), writing that "I dissected it, sorrowfully, when I got back to London, to learn what I could, and kept its scattered remains in formalin in my bedroom for many years" (Sacks 2001, 234). Later, Sacks and two friends collected cuttlefish which they attempted to preserve in "a little alcohol" so they could take them back to school for their class to dissect (Sacks 2001, 273). Unfortunately, this experiment was not successful:

> A few days later . . . we heard dull thuds emanating from the basement, and going down to investigate, we encountered a grotesque scene: the cuttlefish, insufficiently preserved, had putrefied and fermented, and the gases produced had exploded the jars and blown great lumps of cuttlefish all over the walls and floor; there were even shreds of cuttlefish stuck to the ceiling. The intense smell of putrefaction was awful beyond imagination. We did our best to scrape off the walls and remove the exploded, impacted lumps of cuttlefish and hosed down the basement, gagging, but the stench was not to be removed, and when we opened the windows and doors to air out the basement, it extended outside the house as a sort of miasma for fifty yards in every direction. (Sacks 2001, 273–274)

Figure 6.1. Fluid-preserved brain specimens on exhibit at the Cushing Center, Yale University Medical School Library. Photo by Terry Dagradi, Cushing Center, Yale University.

FLUID PRESERVATION IN VISUAL ART

Images of objects preserved in fluid are relatively rare in visual art, aside from a few depictions of alcohol distillation in the alchemical literature (Roob 2001) and some intriguing jars sealed with bladders in some woodcuts and paintings of alchemical laboratories. There are a few jars of fluid-preserved specimens visible in some of the woodcuts of cabinets of curiosities (e.g., Mauriès 2002). Fluid-preserved specimens are included in a painting by Jean Valette Penot (1710–1777) entitled *Trompe l'Oeil a la Statuette d'Hercule* (a *trompe l'oeil* refers to a technique that uses realistic images to create a three-dimensional optical illusion). The painting (now in the Musée des Beaux-Arts, Remes, France) shows several objects arranged on three shelves in a cabinet, including two jars containing fluid (and presumably, fluid-preserved specimens) on the lower shelf, one sealed with a bladder, the other with a glass stopper. The portrait of Albertus Seba (1665–1736) produced by Jan Maurits Quinkhard (1688-1772) shows Seba standing in front of his collection of fluid-preserved specimens, holding a jar with a snake in it in one hand, while pointing to shells, loose drawings, and an opened copy of his printed catalog with his other hand (Müsch 2001; Simmons and Snider 2012). A 1907 caricature by the anarchist artist Gustave-Henri Jossot (1866–1951), showing a deranged mortician drinking formaldehyde that had been used to preserve a dead fetus, was published in the social protest journal, *L'Assiette au Beurre*.

A few artists have used fluid-preserved objects directly in their work. Joseph Cornell (1903–1972) used old glass jars with glass stoppers, some filled with unknown fluids, in several of his series of works known as *Pharmacy* and *Museum*, created over a ten-year period, beginning in 1943 (Hartigan et al. 2003). The best-known modern artist to use fluid-preserved objects is Damien Hirst, who has used fluid preservation in several art projects, most notably preserving sheep, cows, and sharks in formaldehyde. A work consisting of a large shark in a tank of formaldehyde, *The Physical Impossibility of Death in the Mind of Someone Living*, was commissioned in 1992 for £50,000, and later sold to another collector for £6.5 million. In 2006, it was reported in the press that the shark was badly deteriorating due to poor preservation techniques, including the failure to inject the preservative into the tissues or body cavity of the shark (Ruiz and Harris 2006). As the shark was reported to have "decomposed until its form changed, its skin grew deeply wrinkled and the solution in the tank turned murky," staff members at the Saatchi Gallery inexplicably added bleach to the formaldehyde solution in the tank. When the deteriorating shark was being replaced by his crew, Hirst was quoted as saying that "Three different lengths of needles are being used to inject the shark with formaldehyde. The last shark was never injected, so it decayed from the inside" (Vogel 2006).

Mary Cate Carrol is an American artist who produced a painting titled *American Liberty Upside Down*. The painting shows a man and woman sitting on a couch. A child is depicted on the mother's lap by a red dotted outline. Inside the outline is an actual door that can be opened to see a fetus preserved in formaldehyde. Due to controversy over the presence of the fetus, the artwork was removed from a show at Mary Washington College just before the opening, which ultimately led to a lawsuit that was settled out of court; the college exhibited the piece two years later.

American artist Tracy Hicks has done several installations based on the theme of treasures stored in jars in museum vaults, using a variety objects in different fluids, including alcohol, oils, and beet juice. Participation in a scientific expedition to Central America in 1997 led to Hicks incorporating themes of ecology and disappearing amphibians into this work. In 2005, Hicks exhibited an installation called *Two Cultures: Collection* at the Hall Center for the Humanities at the University of Kansas (see figure 6.2). The installation was composed of 1,300 urethane and silicon casts (treated with phosphorescent pigments and fluorescent dyes) of seventy-nine species of frogs (many endangered, some extinct) from the fluid-preserved scientific collections at the Field Museum (Chicago) and the Biodiversity Institute at the University of Kansas (Anon. 2005; Hill 2005). The project was funded by two grants from the Museum Loan Network (MLN) to Marjorie Swann (professor in the Department of English) and John E. Simmons (Director of the Museum Studies Program and Collections Manager in the Biodiversity Institute) to address how scientific specimens, as objects of cultural patrimony, reflect our interpretations of the diversity of life, and to interrogate the intersection of science and art. By using fluorescent pigments and colored fluids in the jars holding the casts of the frogs, Hicks drew attention to the aesthetics of scientific collections and the mystery of amphibian declines worldwide.

Figure 6.2. Tracy Hicks's installation, *Two Cultures: Collection* at the Hall Center for the Humanities at the University of Kansas. Photo by Tracy Hicks.

FLUID PRESERVATION IN LITERATURE

A very early occurrence of the word alcohol can be found in a fifteenth-century Spanish ballad, *La Misa de Amor* (Menéndez Pidal 1968; Turnbull 1955). At the time, the word alcohol had recently entered Spanish from the Arabic *al kohl*, and referred not to spirit of wine but to a black powder (see Origin of the Name Alcohol in chapter 1):

En la su boca muy linda
lleva un poco de dulzor;
el la su cara tan blanca
un poquito de arrebol,
y un los sus ojuelos garzos
lleva un poco de alcohol

(Menéndez Pidal 1968, 251–252)

References to fluid-preserved objects and specimens abound in literature, often in a context of showing the amorality of scientists and scientific practices. For example, Charles Kingsley parodied the discovery that all primates had a hippocampus in his 1863 children's book, *The Water Babies*, later writing that if a water baby had ever been found, "they would have put it into spirits, or into the *Illustrated News*, or perhaps cut it into two halves, poor dear little thing, and sent one to Professor Owen and one to Professor Huxley, to see what they could each say about it" (Gould 1979, 50). At the time, Richard Owen (1804–1892) and Thomas Henry Huxley (1825–1895) were engaged in a public argument over the validity of the theory of evolution as proposed in Charles Darwin's *On the Origin of Species* (published in 1859).

A more common theme associated with references to fluid preservation is a context of repulsion or horror. In a 1943 novel set in equatorial Africa, *Congo Song* (by Stuart Cloete), the laboratory of the French doctor at a remote agricultural station is described as containing "rows of jars and bottles neatly labeled, filled with gruesome-looking medical specimens bleached white by the alcohol in which they were preserved." In *The History of the Ginger Man*, an autobiographical account by J. P. Donleavy of the writing of his well-known novel, a reference to body parts in formaldehyde is used as an image of repulsion:

> I would, in the cool of the morning, stroll each day, passing the back of the Massachusetts General Hospital, this reminding one of one's mortality. For above the large black morgue doors were the windows of the pathology lab, where one could see steeping in formaldehyde the specimens of brains, liver, lungs, and kidneys in their glass jars. (Donleavy 1994, 205)

In the epilogue to H. G. Wells's 1898 novel, *The War of the Worlds*, the narrator recounts that, "The results of an anatomical examination of the Martians, so far as the prowling dogs had left such an examination possible, I have already given. But everyone is familiar with the magnificent and almost complete specimen in spirits at the Natural History Museum, and the countless drawings that have been made from it; and beyond that the interest of their physiology and structure is purely scientific" (Wells 2005, 178). In Mikhail Bulgakov's (1891–1940) novel, *The White Guard*, the protagonist, Nikolka, interrupts a professor in his laboratory, in the midst of his dissection of a human corpse, to ask permission to search among the bodies in the morgue for his slain commanding officer. The scene illustrates the lack of morality among scientists (Lessing 2003):

> Nikolka took off his cap, noticing the gleaming black blinds drawn low over the windows and a beam of painfully bright light falling on to a desk, behind which was a black beard, a crumpled, exhausted face, and a hooked nose. Then he glanced nervously around the walls at the line of shiny, glass-fronted cabinets containing rows of monstrous things in bottles, brown and yellow, like hideous Chinese faces. (Bulgakov 1971, 269)

A reference to the preservation of Lenin's brain occurs in the 1959 dissident novel, *The Trial Begins*, by Andrei Sinyavsky (1929–1997), set in a Siberian prison camp. The character Globov is guided through the "Puskin Museum of Fine Arts" by another character named "Citizen Rabinovich." Rabinovich takes Globov through a door labeled "A great end gives rise to great energy" where Globov sees "an empty space and, in the middle of it, a glass jar containing a human brain preserved in spirits. It was ridged and furrowed like the crust of the earth. Its two hemispheres throbbed slowly. Around them, a pale green solution flowed through a system of fine tubes and retorts" (Tertz 1982, 69). The scene is reminiscent of a 1930 short story by H. P. Lovecraft, *The Whisperer in Darkness*, which had a slight twist on the preservation in fluid theme—the narrator discovers a number of metal cylinders in a remote farmhouse in the hills of Vermont that contain the surgically extracted brains of several humans, "immersed in an occasionally replenished fluid within an ether-tight cylinder" but kept alive for travel through outer space (Lovecraft 2012, 518).

The Sherwood Anderson (1876–1941) story, *The Egg* (first published in 1921 as *The Triumph of the Egg*), concerns a farmer who became obsessed with his collection of deformed chickens, "preserved in alcohol and each put in its own glass bottle" (Modlin 1992). The farmer and his wife opened a restaurant, and the farmer kept the jars of preserved deformed chickens on display despite his wife's objections because "The grotesques were, he declared, valuable. People, he said, liked to look at strange and wonderful things." The story has a parallel in real life in the form of Henry Thomford, who kept a three-headed chicken in a jar of alcohol to show to travelers at his store (Lorge et al. 2005).

Perhaps the best-known fluid preservation laboratory in fiction was the one that belonged to the character Doc, the marine biologist in John Steinbeck's 1945 novel, *Cannery Row*. The laboratory was described as containing shelves crammed with preserved specimens, an injecting and embalming sink in the basement, and concrete tanks for large specimens outside, all permeated with chemical smells. In the novel, Steinbeck spelled formalin as "formaline," one of several variant spellings based on trade names for commercial formaldehyde solutions (Steinbeck 1973, 22).

In the 1981 novel *Brain* by surgeon and novelist Robin Cook, a neuro-radiologist named Martin Philips discovers an evil plot in a hospital involving connecting disembodied brains to computers. Despite the fact that they are medical professionals, each time a character sees a preserved human brain, he or she becomes nauseated, beginning with Philips when he makes an unauthorized visit to the pathology lab and finds "a series of bookshelves containing glass jars; a whole group of which held brains immersed in preserving fluid" (Cook 1981, 139). Philips takes one of the preserved brains to the X-ray department where "The technician took one look at the brain and turned green" (Cook 1981, 141). Philips places the preserved brain on a paper plate and makes the X-ray himself, but "on all the films it was difficult to make out the internal structure" because "even though the brain had been in formaldehyde, the internal structure must have decomposed enough to blur any radiological definition" (Cook 1981, 141). Philips then carries the brain to his office, where he decides to excise a critical piece of brain tissue and see if it is radioactive enough to fog X-ray film. He asks his secretary, Helen, to get him a knife. "Helen got the one from the peanut butter jar by the coffee urn, marveling at her weird boss. When she returned to this office, she gagged, unprepared for what she saw. Philips was lifting a human brain out of a formaldehyde jar, and putting it on a newspaper, its familiar convolutions glistening in the light from the X-ray viewer. Fighting off a wave of nausea, Helen watched as Philips proceeded to cut a ragged slice from the back of the specimen. After returning the brain to the formaldehyde he headed for the door, carrying the slice of brain on the newspaper" (Cook 1981, 153). Later, after the film is exposed, Philips dumps the "now dried-up slice of brain" (Cook 1981, 158) in a wastebasket.

Fluid preservation is featured in several murder mysteries. In the novel *The Silence of the Lambs* by Thomas Harris (1988) the degree of evil practiced by a psychotic killer is demonstrated by a severed head found in a jar in the killer's car. In *Mr. Smithson's Bones*, a murder mystery set at the Smithsonian Institution, the head of one victim was severed and hidden in a jar of fluid in a collection of preserved anemones. The narrator explains that "I put it in a large jar in alcohol and placed it on the back of a shelf in the invertebrate zoology storage area. . . . It is with the sea anemones. I turned Rebecca's face to the back so all you can see is her hair. It looked rather like an anemone, I thought" (Conroy 1993, 185). The mystery novel *Critique of Criminal Reason* opens with a description of a human head preserved in a jar of alcohol, "like a large conch-shell in the swelling sea" (Gregorio 2006, 10). In *Black Notice*, a medical examiner uses formaldehyde as a defensive weapon when she is attacked by a psychopathic killer. The medical examiner happens to have a jar of formaldehyde in her living room (in which is a preserved piece of skin she removed during an autopsy to attempt to identify a tattoo). When the killer attacks her, the medical examiner opens the jar and throws the liquid in his face. The description of the killer's reaction is an accurate account of what would probably happen: "He shrieked and grabbed his eyes and throat as the chemical burned and made it difficult for him to breathe. He squeezed shut his eyes, shrieking and grabbing at his doused shirt to rip it off, gasping and burning like fire as I ran" (Cornwall 1999, 318). The killer is captured moments later as he rubs snow in his face. In an earlier book by the same author, an expert medical examiner asks to see the brain of a murder victim that was removed by a small-town coroner, and explains that, "It is not uncommon for pathologists to fix brains in a ten percent solution of formaldehyde called formalin. The chemical process preserves and firms tissue. It makes further studies possible, especially in cases involving trauma to this most incredible and least understood of human organs" (Cornwall 1994, 95). However, neither the coroner nor the medical examiner follow good safety procedures—the coroner retrieves the brain from a plastic bucket beneath the sink, lifts out the brain, and lays it on a cutting board, while the narrator describes "fumes from the formaldehyde burning my eyes" (Cornwall 1994, 95).

The capture of a giant squid (*Architeuthis dux*), some 8.62 m in length, in 2004 drew worldwide attention, particularly after it was acquired by The Natural History Museum (London). The British press nicknamed the squid Archie and closely followed its subsequent preservation. The squid had been frozen after capture and remained on ice for its first six months at the Natural History Museum as staff consulted with experts on what to do with it. Ultimately, the squid was thawed, injected with 15 L of buffered 10 percent formalin and installed in a nine-meter-long custom-made tank in a special room in the basement of the museum (Ablett 2012), where it is accessible to the public on special tours. The story of the giant squid led to the publication of a novel about a cult that worships preserved giant squid, *Kraken: An Anatomy* (Miéville 2010). In the novel, Billy Harrow, a curator at the British Museum, takes a group of visitors on a tour of the Darwin Centre through the "specimen maze" so that they can see "The specimens mindlessly concentrated, some posing with their own colourless guts. Flatfish in browning tanks. Jars of huddled mice gone sepia, grotesque mouthfuls like pickled onions. There were sports with excess limbs, fetuses in arcane shapes. They were as carefully shelved as books" (Miéville 2010, 8). In the novel (as in real life), the star attraction of the tour is the 8.62-meter-long giant squid, described as being preserved "in a saline-Formalin mix." When Harrow leads his tour group to the basement, he discovers that the squid has been stolen, nine-meter-long custom tank and all. As the plot thickens, Harrow finds a human stuffed into a glass jar full of fluid preservative. Later in the novel, Harrow is rescued by a "memory angel" from the museum in the form of an animated giant preserving jar with a skull on its lid and whirling skeletal arms that defends him until its glass is shattered by an untimely tossed brick. Harrow realizes that the odd glass-clinking noise he has been hearing faintly for some time is the memory angel keeping an eye on him. At the climax of the story, the stolen giant squid in its tank of formaldehyde is at last found in the back of a moving truck being driven around London by the cultists, who pray to the creature for guidance. A final battle takes place against a villain who wishes to extract the "ink" (a sepia-colored fluid which has, in fact, been used as ink in the past) from the preserved squid. Billy Harrow, the curator, is able to save London from the arch villain by exclaiming that the preserved squid was not a kraken but "a *specimen.* I know. I made it. That's ours" (Miéville 2010, 487), and then claiming that he is no more than "a bottle prophet" (Miéville 2010, 488).

In a Lord Peter Wimsey story, *The Piscatorial Farce of the Stolen Stomach*, a human alimentary canal that is "enclosed in a proper preservative medium in a glass vessel" (Sayers 1972, 195) is stolen from a medical student's cottage (the stomach was bequeathed to the student by his great-uncle). Wimsey figures out that the stomach was stolen because it contained diamonds. Although the jar is broken by the burglar, Wimsey eventually locates the preserved organ in the river, where it "like a drab purse, lay on the shore" (Sayers 1972, 209) being pecked by gulls. Wimsey and his friend slice the stomach open and find most of the diamonds still there.

The protagonist of the novel, *A River Town*, is an Irish immigrant store owner in Australia in 1900 who is haunted by the face of a young woman whose head was kept in a jar of preservative by the police, in the hope that she would eventually be identified (Keneally 1996). This incident has historical precedence—in 1869, a stranger was found murdered in a vacant lot in Spring Valley, New York. A newspaper account stated that "The head and face presented a most ghastly appearance, being mutilated by five ugly wounds. The right eye was entirely gone, and portions of the face had been gnawed off by field mice" (Anon. 1871). Nevertheless, in hopes of identifying the victim, "The mutilated head was severed from the trunk and preserved in alcohol for identification. During the sessions of the inquest thousands of persons inspected the remains, but failed to identify them" (Anon. 1871). Almost a month after the discovery of the body, the victim was identified, and eventually the murderer was caught, brought to trial, and convicted.

FLUID PRESERVATION IN FILM

Fluid-preserved specimens as metaphors for repulsion, horror, the amorality of science, and sometimes just to establish how odd a character is are common themes in films, as in literature. In the 1985 teen romantic comedy, *Better Off Dead* (directed by Savage Steve Holland), Charles DeMar plays a character who spends a lot of class time staring at a pickled pig fetus that he carries around in a jar of formaldehyde.

A more typical use of the fluid preservation metaphor occurs in the 1932 film version of *The Most Dangerous Game* in which a rich hunter, bored with big game, has taken to tracking and shooting human victims. Two shipwrecked strangers who wash up on the hunter's island (Bob Rainsford and Eve Trowbridge) sneak into the secret trophy room of the exiled Russian sportsman Count Zaroff. As the pair tries to hide when they hear the count approaching, Eve bumps into a large glass jar in which a human head is suspended in preservative. The head sways ominously as Count Zaroff enters the room. In *Surviving the Game* (1994, directed by Ernest R. Dickerson), a film based on the same plot device as *The Most Dangerous Game*, the human victim of the hunt, Jack Mason, discovers a number of human heads preserved in jars of alcohol in the hunting club's trophy room. To escape his captors, Mason sets the trophy room ablaze.

The version of *Frankenstein* that is best known today (the 1931 film version from Universal Pictures, directed by James Whale) severely distorts the theme of Mary Shelley's novel (*Frankenstein; or, the Modern Prometheus*, 1818). The monster of Shelley's novel is not inherently evil, but becomes evil in reaction to how he is treated by the humans he encounters. By contrast, in the 1931 film version, the monster is evil due to his biological destiny (because a "criminal brain" was mistakenly implanted in his skull). The idea of the addition of the criminal brain is unique to the film (it is not in the novel). Dr. Frankenstein's hunchbacked assistant, Fritz (played by Dwight Frye) goes to the Goldstadt Medical College to steal a preserved brain. Fritz breaks into the locked lecture room after the professor leaves two brains in jars of fluid on the podium for the "further inspection" of the students. The jar containing the normal brain is in a clear fluid preservative, while the jar containing the criminal brain is in a cloudy fluid preservative. After accidentally dropping the jar with the good brain, Fritz leaves with the jar containing the criminal brain. Mel Brooks's 1974 parody, *Young Frankenstein*, comes much closer to portraying the character of the monster as in the Mary Shelley novel. In *Young Frankenstein*, the hunchbacked assistant (now named Igor), is sent to the "Brain Depository" to fetch a perfect brain. When Igor arrives, the building is closed; a sign on the door reads, "After 5:00 p.m. shove brain through slot in door." Igor breaks in to find row after row of brains in jars and under glass domes, each labeled with the name of a famous donor, including several who died long before fluid preservation was practiced (e.g., Albertus Magus and Cornelius Agrippa). Igor removes the jar containing the brain of Hans Delbruck, "scientist and saint" (the real Hans Delbrück, 1848–1929, was a German scholar who specialized in the history of war) but drops the jar when startled by his own image in a mirror. The brain is destroyed in the fall, so Igor grabs the preserved brain nearest him (labeled "Do not use this brain! Abnormal") and returns to the castle.

Fluid-preserved specimens make an appearance in the 1954 Walt Disney film version of the 1870 Jules Verne novel, *Twenty Thousand Leagues under the Sea*. The sailor Ned Land (Kirk Douglas), wishing to escape captivity on board Captain Nemo's submarine, *Nautilus*, decides to put messages in bottles and throw them overboard, so he orders Conseil (Peter Lorre), a timid scientist, to fetch several jars of preserved specimens. Conseil brings Ned the specimens and announces that, "You've got everything here from the rarest nudibrach to oysters." Ned replies that "Oysters are out of season. Dump them in the sink, all of them. I just want the bottles," to which Conseil replies, "You can't do that, these are priceless." Ned pulls the cork from a bottle and raises the bottle to his nose, and then says, somewhat surprised, "Alcohol," Conseil replies "Pure alcohol," at which point Ned pulls a specimen of a fish from the bottle and drops it in the sink, saying "We'll just drain the pollywogs out and save the grot," and begins pouring the alcohol into a pitcher. The incident with specimens in jars is not in the original text of the novel.

A 1961 film, *Night Tide*, written and directed by Curtis Harrington, used a human hand floating in a jar of preservative to reinforce the threat of evil lurking in the character Captain Samuel Murdock. A sailor named Johnny Drake has fallen in love with Murdock's adopted daughter, Mora. While Murdock is warning Drake that a relationship with Mora will be fatal, he instructs him to open a cabinet to fetch a fresh bottle of gin. When Drake opens the door, he is stunned to see the grisly hand in preservative. Murdock nonchalantly explains "Oh, don't be alarmed, that is just a little Arabian souvenir, the hand of a thief. The Mohammedans punish their thieves by removing the offending portions of the body. Rather gruesome, but logical, don't you think?" to which Drake asks, "How did you ever get that?" Murdock replies that "It was a gift from the Sultan of Marakesh. He knew I collected odd things and he sent it to me. Rather thoughtful of him," to which Drake mumbles, "Yeah, it's very interesting."

In the Icelandic murder mystery *Jar City* by Arnaldur Indridason (2000), a detective locates the preserved brain of a long-dead five-year-old girl in the private collection of a retired pathologist, although it had originally been in a state-owned research collection nicknamed Jar City. In the story, the pathologist explains that the heart of Louis XVII (the son of Louis XVI and Marie Antoinette) was preserved in formaldehyde and later matched by DNA to the royal family, despite the fact that formaldehyde had not been discovered at the time and damages DNA. In 2006 the book was made into a movie, called *Mýrin* (directed by Baltasar Kormákur), which included scenes of fluid-preserved body parts. In fact, Louis XVII's heart was preserved in alcohol after his death in 1795. The specimen was allowed to evaporate eight to ten years later, so that at the time of the DNA testing, the heart had become completely dehydrated (Jehaes et al. 2001).

The 1995 film, *Margaret's Museum* (directed by Mort Ransen) is set in Cape Breton, Nova Scotia, in the 1940s. The story is based on the Sheldon Curie novel, *The Glace Bay Miner's Museum*. The protagonist, Margaret MacNeil (Helena Bonham Carter), opens a museum containing body parts of miners killed on the job in fluid preservative—her husband's tongue, lungs, and fingers; her brother's penis; and her grandfather's lungs. The film depicts Margaret preparing labels for the specimen jars before a visitor, horrified at the exhibit, runs screaming from the museum.

In at least one film, the protagonist is saved by an unusual use of fluid preservative (similar to the medical examiner in the novel *Black Notice*, discussed earlier). *The Relic* (1997, directed by Peter Hyams) is set in a contemporary natural history museum in Chicago. There are several scenes showing racks of fluid-preserved specimens in storage. At one point a curator, Dr. Green, carries a couple of jars by hand from her lab to place them on the shelving. At the climax of the film, a monster (described as a Kothoga) pursues Dr. Greene through a research wing of the museum. To kill the monster, Dr. Greene knocks scores of fluid-preserved specimens to the ground and opens the spigots on several containers of bulk alcohol, then ignites the fluid using an incendiary chemical mixture she has managed to concoct while fleeing the Kothoga. To escape the explosion and fire, Dr. Greene climbs into a maceration tank (presumably filled with water and not yet in use) and closes the lid.

FLUID PRESERVATION IN POPULAR CULTURE

Writers of nonfiction have also employed the image of a preserved specimen to evoke horror and repulsion. A noted big-game hunter of the early twentieth century described how he was badly mauled when attacked by a lion. Back in camp, the hunter related how he dressed his own wounds and "cut off a joint of my right middle finger as it was only just hanging on. I put this amputated finger-joint into alcohol and sent it home to the Museum with the rest of my specimens, and there they noticed the marks of arsenic poisoning under the nail and perceived how busy I had been collecting birds and preparing them for dispatch" (Kittenberger 1989, 27).

During the time of the Jack the Ripper murders in London (the fall of 1888), the president of the Vigilance Committee formed to help catch the murderer received a parcel through the mail that contained a longitudinal section of a kidney, presumably from Jack the Ripper's fourth victim, Catherine Eddowes, whose left kidney had been removed when she was killed and mutilated two weeks before (on the night of September 29–30, 1888). Medical authorities who examined the specimen concluded that the kidney had been preserved in alcohol shortly after the murder (before it was sent to the committee president). However, later writers have cautioned that the kidney may have come from an anatomical specimen unrelated to Jack the Ripper's victims (Wolf 2008). In March 2009 the severed head of King Badu Bonsu II of Ghana, which had been cut off sometime in the 1830s, was discovered preserved in an anatomical collection at the Leiden University Medical Center by Dutch writer Arthur Japin, who was conducting background research for a historic novel. After the discovery, the medical center repatriated the head to Ghana.

Because humans have long consumed alcohol, it seems entirely logical that there would be a convergence of alcohol as a beverage and a preservative—indeed, in many Asian countries, animals (particularly snakes) are often preserved in rice wine which is then consumed as a medicine or tonic, usually to promote longevity and virility (Stutesman 2005). In some formulations, a live snake is submerged in a container of alcohol (usually about 40 percent) along with a selection of herbs and left for several days, after which the snake is removed, gutted, and returned to the bottle. Stutesman (2005, 205) provided a recipe for a Japanese Habu Sake cocktail—coil a specimen of habu (which may refer to one of several species of the vipers *Trimeresurus* or *Ovophis*) in a jar of sake and allow to "ferment" (actually, it would be preserved). By contrast, the recipe provided for the "snakebite" cocktail (Stutesman 2005, 206) does not contain any snake parts at all (shake together 4.5 oz Jack Daniels, 1.5 oz. Cointreau, and 1.5 oz. lime cordial and strain into a tall glass). At the Voodoo Museum in New Orleans, Louisiana, I observed specimens of *Bufo marinus* (cane toads) and *Nerodia* (water snakes) preserved in alcohol that is consumed in religious rituals.

Mescal is an alcoholic beverage that is distilled from the piña (the heart) of various types of agave or maguey plants, principally in the Oaxaca region of Mexico. The piña mash is cooked, sugars added, allowed to ferment, and then distilled twice to yield a mescal that is about 40 percent ethyl alcohol (80 proof). Following distillation, the mescal is aged in wooden barrels to give it color. Most people recognize mescal by the presence of the "worm," which is the larva of either an agave snout weevil or agave moth, both of which are found in the piña after it is harvested. The tradition of adding the worm to mescal is thought to have its origin in providing proof of the alcohol content of the beverage.

The convergence of preservation and consumption has also found its way into popular song. The lyrics to the traditional songs *Lightening Bar Blues* (recorded by a number of artists, including Hoyt Axton on a 1970 album) and *Rambling Boy* (recorded by the Carter Family, the Del McCoury Band, and others) include a request to preserve the narrator's bones in hundred proof alcohol rather than burying the body in a grave.

THE AESTHETICS OF FLUID PRESERVATION

Ever since Dr. Croone showed his "two embryos of puppy-dogs" before the Royal Society in 1662 (see chapter 1), the vast majority of fluid-preserved specimens have been prepared for scientific research, teaching, or museum exhibition, and fluid preservation has been largely a technical and scientific undertaking. Nevertheless, scientific specimens are part of cultural heritage as much as any other objects made or used by human beings. The practice of science is never purely objective—the questions that scientists ask and the ways in which they try to answer them are culturally determined, which means that scientific collections necessarily embody the beliefs and ideals of the cultures which preserve the specimens, creating an intersection of science and culture that has aesthetic value. As can be seen in this brief review of representations of fluid preservation in art, literature, film, and music, objects floating in jars of fluid can evoke our emotions, pique our curiosities, stimulate our imaginations, and represent our deepest fears and horrors.

II

LITERATURE IN THIS FIELD

References

Abbott, M. E. 1918. The British circular memorandum for the collection of war specimens. *Bulletin of the International Association of Medical Museums* 7:15–19.

Ablett, J. 2012. The giant squid, *Architeuthis dux* Steenstrup, 1857 (Mollusca: Cephalopoda): The making of an iconic specimen. *NatSCA News* 23:16–20.

Acton, A., T. Harvey, and M. W. Grow. 2005. An examination of non-formalin-based fixation methods for *Xenopus* embryos. *Developmental Dynamics* 233(4):1464–1469.

Adams, H. R., A. P. Flerchinger, and H. F. Steedman. 1976. Ctenophora fixation and preservation. Pages 270–271 in Steedman, H. F. (editor). *Zooplankton Fixation and Preservation.* Monographs on Oceanographic Methodology 4. Paris: Unesco Press.

Adler, K. 1989. Herpetologists of the past. Pages 5–141 in Adler, K. (editor). *Contributions to the History of Herpetology.* Society for the Study of Amphibians and Reptiles Contributions to Herpetology, No. 5.

Ahlstrom, E. E., and J. R. Thrailkill. 1963. Plankton volume loss with time of preservation. *California Cooperative Oceanic Fisheries Investigations Reports* 9:57–73.

Ahlstrom, E. H. 1976. Maintenance of quality in fish eggs and larvael collected during plankton hauls. Zooplankton Fixation and Preservation. Pages 313–318 in Steedman, H. F. (editor). *Zooplankton Fixation and Preservation.* Monographs on Oceanographic Methodology 4. Paris: Unesco Press.

Aiyappan, A., and S. T. Satyamurti. 1960. *Handbook of Museum Technique.* Madras: Government of Madras.

Alleger, W. W. 1894. Formalin. *Proceedings of the American Microscopical Society* 15 (3): 192–197.

Al-Hassan, L. A. J., J. A. Bujawari, and O. A. El-Silini. 2000. The effect of some preservatives and freezing on certain body dimensions of two species of the family Mullidae collected from Benghazi waters, Libya. *Acta Ichtyologica et Piscatoria* 30(2):127–136.

Amerine, M. A. 1964. Wine. *Scientific American* 211(2):46–56.

Amosov, V. A. 1960. Changes in weight and length of fish after fixation in formalin followed by soaking in water. *Voprosy Ikhtiologii* 16:187–190 [*Fish Research Board of Canada Translation Series* No. 452, 1964].

Anderson, C. L. 1925. Preserved pickled herring. *University of Washington Publications in Fisheries* 1(1):1–64.

Anderson, P. R., F. R. Bacon, and B. W. Byrum. 1975. Effect of surface treatments on the chemical durability and surface composition of soda-lime glass bottles. *Journal of Non-Crystalline Solids* 19:251–262.

Anderson, R. M. 1932. Methods of collecting and preserving vertebrate animals. *National Museums of Canada Bulletin* No. 69, Biological Series No. 18.

Anderson, R. M. 1948. Methods of collecting and preserving vertebrate animals. 2nd rev. ed. *National Museums of Canada Bulletin* No. 69, Biological Series No. 18.

Anderson, R. M. 1965. Methods of collecting and preserving vertebrate animals. 4th rev. ed. *National Museums of Canada Bulletin* No. 69, Biological Series No. 18.

Anderson, S. 1975. Techniques of preservation. *Museums Journal* 74(4):166–168.

Andrei, M. A., and H. H. Genoways. 1999. Changes in pH in museum storage fluids, I—Effects of Resistal paper labels. *Collection Forum* 13(2):63–75.

Andriezen, W. L. 1894. On some of the newer aspects of the pathology of insanity. *Brain* 68:549–692.

Anonymous. n. d. *Suggestions to Collectors of Reptiles and Amphibians*. New York: American Museum of Natural History. (For a discussion of the dates of publication of the six known versions of this pamphlet, see Myers [2000], pages 105–109).

Anonymous. n. d. *Supplement to the Pamphlet "Suggestions to Collectors of Reptiles and Amphibians."* New York: American Museum of Natural History. (For a discussion of the dates of publication of the six known versions of this pamphlet, see Myers [2000], pages 105–109).

Anonymous. n. d. *Suggestions to Collectors of Reptiles and Amphibians*. Chicago: Field Museum of Natural History, 4. (For a discussion of the dates of publication of the six known versions of this pamphlet, see Myers [2000], pages 105–109).

Anonymous. 1793. Sir Ashton Lever's directions for preserving birds, etc. In Anonymous, 1811. *A Selection of Curious Articles from the Gentleman's Magazine*. Vol. 3. 2nd ed. London: Longman, Hurst, Rees, Orme, and Brown.

Anonymous. 1831. *Manual of the Practical Naturalist; or Directions for Collecting, Preparing, and Preserving Subjects of Natural History. Containing instructions and recipes according to the most approved methods for taking and stuffing Quadrupeds, Birds, Fishes, Reptiles. Selecting, preserving and arranging Insects, Minerals, Plants, Shells, &c.* Boston: Lilly and Wait, and Carter, Hendee and Babcock.

Anonymous. 1871. Murder will out. *New York Times*, May 3, 1871.

Anonymous. 1895. Formol as a preserving fluid. *Natural Science* 6(38):229.

Anonymous. 1897. Discussion. *Medical Record* 32:248.

Anonymous. 1905. Theory of killing and preserving. *Bulletin of the College of Charleston Museum* 1(2):15–18.

Anonymous. 1915–1916. *Formaldehyde: The Farmer's Friend. Its' Uses Upon the Farm and in the Home. Treatment of Diseases of Grains, Potatoes and Vegetables*. New York: Perth Amboy Chemical Works.

Anonymous. 1921. *Handbook of Instructions for Collectors, Issued by the British Museum (Natural History)*. 4th ed. London: British Museum (Natural History), viii + 222 pp.

Anonymous, 1935. Preserving herbarium specimens in humid climate. *Turtox News* 13(11):114.

Anonymous. 1938a. A method of displaying snakes. *Turtox News* 15(10):129.

Anonymous. 1938b. Museum specimen labels. *Turtox News* 16(4):75.

Anonymous. 1938c. Maintenance measures for the teaching museum. *Turtox News* 16(4):76–77.

Anonymous. 1939. Are you especially sensitive to formalin fumes? *Turtox News* 17(9):121.

Anonymous. 1943. Notes on the preservation of marine crustaceans. *Turtox News* 21(6):77–78.

Anonymous. 1944. *A Field Collector's Manual in Natural History*. Washington, D. C.: Smithsonian Institution.

Anonymous. 1953. *Instructions for Collectors No. 3. Reptiles, Amphibians and Fishes*. London: British Museum (Natural History).

Anonymous. 1957. *Problem specimens (and cloudy solutions)*. Armed Forces Institute of Pathology Medical Museum Laboratory Technical Outline Series.

Anonymous. 1965. *Instructions for collectors. Number 3. Fishes*. London: British Museum (Natural History).

Anonymous. 1989. Formaldehyde. *Journal of the American Medical Association* 261(8):1183–1187.

Anonymous. 2005. Exhibition. In the croak room. *Nature* 438:27.

Anonymous. 2008. In Poland, Chopin's heart may hold secret of his death. *Agence France-Presse*, June 21, 2008.

Armstrong, J. D., and D. J. Stewart. 1997. The effects of initial length and body curvature on shrinkage of juvenile Atlantic salmon during freezing. *Journal of Fish Biology* 50:903–905.

Arriaza, B. T. 1995. *Beyond Death: The Chincorro Mummies of Ancient Chile*. Washington, D. C.: Smithsonian Institution Press.

Arts, J. H. E., M. A. J. Rennen, and C. de Heer. 2006. Inhaled formaldehyde: Evaluation of sensory irritation in relation to carcinogenicity. *Regulatory Toxicology and Pharmacology* 44:144–160.

Associated Press. 2006. Police find severed hand named "Freddy" in jar at Goth stripper's N. J. home. *Associated Press*, July 25, 2006.

Ayre, C., H. Bishop, H. Fothergill, and J. Freedman. 2008. Collecting life: Field collecting littoral marine animals for museum collections. *NatSCA News* 14:33–40.

Babin, A., D. Hinkamp, K. A. Makos, M. McCann, and M. Pool. 2010. Chemical hazards. Pages 227–349 in Hawks, C. A., M. McCann, K. A. Makos, L. Goldberg, D. Hinkamp, D. C. Ertel, and P. Silence (editors). *Health and Safety for Museum Professionals*. New York: Society for the Preservation of Natural History Collections.

Balick, M. J., L. Rivier, and T. Plowman. 1982. The effects of field preservation on alkaloid content of fresh coca leaves (*Erythroxylum* spp.). *Journal of Ethnopharmacology* 6(3):287–291.

Bailey, J. E. 1952. Life history and ecology of the sculpin, *Cottus bairdi punctulatus*, in southwestern Montana. *Copeia* 1952(4):243–255.

Baird, S. F. 1852. Directions for collecting, preserving, and transporting specimens of natural history. *Smithsonian Contributions to Knowledge* 4:1–23.

Baker, J. R. 1944. The structure and chemical composition of the Golgi element. *Quarterly Journal of Microscopical Science* 85(1):1–17.

Bakewell, R. 1823. *Travels: Comprising Observations Made During a Residence in the Tarentaise and Various Parts of the Grecian and Pennine Alps, and in Switzerland and Auvergne, in the Years 1820, 1821, and 1822*. London: Longmans, Hurst, Rees, Orme and Brown.

Balousek, M. 2000. *Wisconsin Unsolved Mysteries*. Oregon, WI: Badger Books.

Båmstedt, U. 1998. Trophodynamics of *Pleurobrachia pileus* (Ctenophora, Cydippida) and ctenophore summer occurrence off the Norwegian North-West coast. *Sarsia* 83: 169–181.

Bancroft, E. 1769. *An Essay on the Natural History of Guiana, in South America*. London: T. Becket and P. A. DeHondt [Reprinted in 1971 by Arno Press and *The New York Times*, New York, edited by R. M. Goldwyn].

Barker, J., and G. Grigg. 1977. *A Field Guide to Australian Frogs*. Adelaida: Rigby Limited.

Banks, H. J., and D. J. Williams. 1972. Use of surfactant Decon 90 in the preparation and of coccids and other insects for microscopy. *Journal of the Australian Entomological Society* 11:347–348.

Baumann, E. 1991. *Step into My Parlor: The Chilling Story of Serial Killer Jeffrey Dahmer*. Chicago: Bonus Books.

Bayer, J. M., and T. D. Counihan. 2001. Length changes in white sturgeon larvae preserved in ethanol or formaldehyde. *Collection Forum* 15(1–2):57–64.

BBC. 2004. Girl given dead foetus in bottle. A hospital is issuing fresh guidelines to staff after a 14-year-old girl was given her stillborn baby in a specimen bottle and told to take it home. Published 2004/07/22 07:50:57 GMT, http://news. bbc. co. uk/go/pr/fr/-/1/hi/england/3914215. stm.

Bé, A.-W. H., and O. R. Anderson. 1976. Preservation of planktonic Foraminifera and other calcareous plankton. Pages 250–258 in Steedman, H. F. (editor). *Zooplankton Fixation and Preservation*. Monographs on Oceanographic Methodology 4. Paris: Unesco Press.

Bean, T. H. 1881. *Directions for Collecting and Preserving Fish. Proceedings of United States National Museum for 1881*. Washington, D. C.: Government Printing Office, 1882, 4: 235–238.

Bean, T. F. 1895. Directions for collecting and preserving fish. *The Museum* 1(8):236–240.

Beatty, W. 1807. *Authentic Narrative of the Death of Lord Nelson: With the Circumstances Preceding, Attending, and Subsequent to that Event; the Professional Report of His Lordship's Wound; and Several Interesting Anecdotes*. London: T. Candell and W. Davies.

Beers, J. R. 1976a. Sample preparation. Pages 42–48 in Steedman, H. F. (editor). *Zooplankton Fixation and Preservation*. Monographs on Oceanographic Methodology 4. Paris: Unesco Press.

Beers, J. R. 1976b. Preservation and laboratory study of actinopods in plankton samples. Zooplankton Fixation and Preservation. Pages 240–249 in Steedman, H. F. (editor). *Zooplankton Fixation and Preservation*. Monographs on Oceanographic Methodology 4. Paris: Unesco Press.

Beirne, B. P. 1955. *Collecting, Preparing and Preserving Insects*. No. 932. Ottawa: Science Service, Entomology Division, Canada Department of Agriculture.

Ben-Ezra, J., D. A. Johnson, J. Rossi, N. Cook, and A. Wu. 1991. Effect of fixation on the amplification of nucleic acids from paraffin-embedded material by the polymerase chain reaction. *Journal of Histochemistry and Cytochemistry* 39(3):351–354.

Bentley, A. C. 2004. Thermal transfer printers—applications in wet collections. *Society for the Preservation of Natural History Collections Newsletter* 18(2):1–2, 17–18.

Berger, A. J. 1955. Suggestions regarding alcoholic specimens and skeletons of birds. *Auk* 72(3):300–303.

Berger, A. J. 1956. Further notes on alcoholic specimens. *Auk* 73(3):452.

Berland, B. 1984. Basic techniques involved in helminth preservation. *Systematic Parasitology* 6(4):242–245.

Bernard, G. R., and G. C. Wynn. 1964. Weight responses of tissue slices and albumin-gelatine gels during formaldehyde fixation with observations on the effect of pH. *Anatomical Record* 150(4):463–472.

Berry, H. 1944. Antibacterial values of ethylene glycol monophenyl ether (phenoxetol). *Lancet* 247:175–176.

Billy, A. J. 1982. The effects of formalin and isopropyl alcohol on length and weight measurements of *Sarotherodon mossabicus* Trewavas. *Journal of Fish Biology* 21:107–112.

Bininda-Emonds, O. R. P., and A. P. Russell. 1992. Minimization of potential problems associated with the morphometry of spirit-preserved bat wings. *Collection Forum* 8(1):9–14.

Bininda-Emonds, O. R. P., and A. P. Russell. 1993. Effects of preservation on wing morphometry of little brown bats. *Journal of the Zoological Society of London* 230:141–158.

Birch, T. 1968. *The History of the Royal Society of London for Improving of Natural Knowledge from Its First Rise.* Vol. 1. New York: Johnson Reprint Corporation. [Reprint of the 1756–1757 edition published for the Royal Society of London by A. Millar, London].

Bisulca, C., L. Kronthal, L. S. Ford, and M. Arnold. 2006. Corrosion of metal tags in fluid preserved collections. Poster abstract, American Institute of Conservation annual meeting, July 16–19, 2006, Providence, Rhode Island.

Blair, J. 2010. Cushing collection once again open for research. *Yale Medicine* 45(1):4–5.

Blake, E. R. 1949. Preserving birds for study. *Fieldiana: Technique* 7:1–38.

Blaufuss, A. H. 1933. Suggestions for preserving and injecting specimens. *Turtox News* 11(2):18.

Blaxter, J. H. S. 1971. Feeding and condition of Clyde herring larvae. *Rapports et Proces-Verbaux des Réunions/Consell permainent International pour l'Exploration de la Mer* 160:128–136.

Bliss, R. 1872. Appearance of colour in fish kept in alcohol. *Popular Science Review* 11: 335–336.

Bloor, W. R. 1943. *Biochemistry of the Fatty Acids.* American Chemical Society Monograph Series. New York: Reinhold Publishing Company.

Blum, F. 1893a. Der formaldehyd als antisepticum. *Münchener Medicinische Wochenschrift,* 40(32):601–602.

Blum, F. 1893b. Der formaldehyd als härtungsmittel. *Zeitschrift der wissenschaft fur mikroskopie* 10:314–315.

Blum, F. 1894a. Notiz über die anwendung des formaldehyds (formol) als härtungs- und conservirungsmittel. *Anatomischer Anzelger* 9(5):229–231.

Blum, F. 1894b. Ueber formaldehyd. *Münchener Medicinische Wochenschrift,* 41(24):475–477.

Blum, F. 1896. Uber wesen und wert der formolhärtung. *Anatomischer Anzelger* 11:718–727.

Blum, F. 1897. Formol, not Formal. *The American Naturalist* 31(365):464–465.

Blum, J. 1893. Formol als conservierungsflüssigkeit. *Zoologischer Anzeiger* 16:450–452.

Blum, J. 1894. Formol als konservierungsflüssigkeit. *Bericht über die Senckenbergische Naturforschende Gesselschaft in Frankfurt am Main,* June–July, 195–204.

Boas, M. 1958. *Robert Boyle and Seventeenth-Century Chemistry.* Cambridge: Cambridge at the University Press.

Boase, N. A., and R. W. Waller. 1994. The effect of propylene glycol on ethanol concentrations determined by density measurement. *Collection Forum* 19(2):41–49.

Bolnest, E. 1672. *Aurora Chymica: or a Rational Way of Preparing Animals, Vegetables, and Minerals for A Physical Use, by which Preparations they are made most efficacious, safe, and pleasant Medicines for the Preservation and Restoration of the Life of Man.* London: John Starkey.

Borodin, N. S. 1930. Effective method of preserving fishes in natural color. *Museum News* 8(6):11–12.

Borror, D. J., and D. M. DeLong. 1964. *An Introduction to the Study of Insects*. New York: Holt, Rinehart and Winston.

Borror, D. J., D. M. DeLong, and C. A. Triplehorn. 1976. *An Introduction to the Study of Insects*. New York: Holt, Rinehart and Winston.

Borutzky, E. W. 1934. Zur Frage ueber die Technik der quantitativen Untersuchungen der Bodenfauna III. Ueber die Methoden der Bearbeitung des Seebenthos. Ein Vergleich des Frisch-und Roh-(Formol)-gewichts. *Arbeiten der Limnologischen Station zu Kossino der Hydrometeorologischen Administration der USSR* 18:109–132.

Bory de Saint-Vincent, J. B. G. M. 1828. On the modes of preserving the bodies of animals in fluids of different kinds, particularly during voyages and travels. Extracted from the *Dictionaire Classique d'Historie Naturelle* and reprinted in the *Edinburgh New Philosophical Journal*, 160–166.

Bostock, J., and H. T. Riley (translators). 1855. Fifty kinds of German wine. Book XIV, Chapter 8(6) in *The Natural History of Pliny*. Vol. 3. London: Henry G. Bonn.

Boulenger, G. A. 1891. A synopsis of the tadpoles of the European batracians. *Proceedings of the Zoological Society of London* 1891:593–627.

Boyle, R. 1664. *Some Considerations Touching the Usefulness of Experimental Natural Philosophy*. 2nd ed. London: Henry Hall.

Boyle, R. 1665. Observables upon a monstrous head. *Philosophical Transactions of the Royal Society* 1:85–86.

Boyle, R. 1666. A way of preserving birds taken out of the egge, and other small faetus's. *Philosophical Transactions of the Royal Society of London* 1:199–201.

Bracegirdle, B. 1978. *A History of Microtechnique*. Ithaca, NY: Cornell University Press.

Bragg, A. N. 1949. Field preservation of amphibian eggs and larvae. *Turtox News* 27(12):262–263.

Brice, A. 2002. Old museum building is history. *Electrical Times*, April, 20–22.

Bridson, D., and L. Forman (editors). 1998. *The Herbarium Handbook*. 3rd ed. Kew: Royal Botanic Gardens.

British Museum (Natural History). 1953. *Instructions for Collectors No. 3. Reptiles, Amphibians and Fishes*. 6th ed. London: British Museum.

Britt, N. W. 1953. Differences between measurements of living and preserved aquatic nymphs caused by injury and preservatives. *Ecology* 34(4):802–803.

Brock, W. H. 1993. *The Norton History of Chemistry*. New York: W. W. Norton and Company.

Brooks, S. T. 1932. Notes on collecting and preserving some invertebrates. *Museum News* 10(9):6–8.

Browne, M. 1884. *Practical Taxidermy: Manual of instruction to the amateur in collecting, preserving, and setting up natural history specimens of all kinds, to which is added a chapter upon the pictorial arrangement of museums*. London: L. Upcott Gill.

Browning, E. 1965. *Toxicity and Metabolism of Industrial Solvents*. Amsterdam: Elsevier.

Buck, R. A., and J. A. Gilmore (editors). 2010. *Museum Registration Methods*, 5th ed. Washington, D. C.: American Association of Museums.

Bulgakov, M. A. 1971. *The White Guard*. New York: McGraw-Hill.

Burgner, R. L. 1962. Studies of red salmon smolts from the Wood River Lakes, Alaska. Studies of Alaska Red Salmon. *University of Washington Publications in Fisheries* 110:251–297.

Burton, P. T. K. 1969. Two birds specimens probably from Cook's voyages. *Ibis* 111:388–390.

Burroughs, G. E., K. Makos, C. Hawks, and T. J. Ryan. 2006. Exposure of museum staff to formaldehyde during some wet specimen activities. *Collection Forum* 20(1–2):49–54.

Cabodevilla, M. A. (editor). 1998. *El Gran Viaje*. Quito: Ediciones Abya-Yala.
Camacho, A. I., and J. Bedoya. 1994. Evaluation of the effects of different preservative and fixative fluids on aquatic invertebrates from interstitial waters. *Collection Forum* 10(1):20–31.
Cannell, P. F. 1988. Techniques for study of avian syringes. *Wilson Bulletin* 100(2):289–293.
Carter, J. 1994. Use of the DMA-35 digital density meter. *Conservation News* 55:39.
Carter, J. 1996. A comparison of two papers and two inks for use as computer generated labels in fluid preserved collections. *The Biology Curator* 7:5–7.
Carter, J. 1997. An investigation of pH changes in a selection of formaldehyde buffering agents used on a fish, parasitology research collection. *Collection Forum* 13(1):1–10.
Carter, J. 2002. DNA preservation in fluid preserved collections. *Society for the Preservation of Natural History Collections Newsletter* 16(1):14–15.
Carter, J. 2004. Improving molecular preservation in fluid preserved collections. *ICOM Natural History Collections Working Group Newsletter* 13:8–13.
Carter, J. 2009. The effect of pH on ethanol preserved muscle tissue. *Collection Forum* 23(1–2):36–45.
Carter, J. D. 2012. Investigating the effects on tissue preservation of DMDM-Hydantoin using FTR spectroscopy. *Collection Forum* 26(1–2):130–135.
Casas-Andreu, G., G. Valenzuela-López, and A. Ramírez-Bautista. 1991. Cómo hacer una colección de anfibios y reptiles. *Instituto de Biología Universidad Autónoma de México Cuadernos* 10:1–68.
Cato, P. S. 1990. Characteristics of a collection of fluid-preserved mammals and implications for collection management. *Collection Forum* 6(2):53–64.
Cato, P. S. 1991. Summary of a study to evaluate collection manager-type positions. *Collection Forum* 7(2):72–94.
Cato, P. S., and S. L. Williams. 1995. Interaction of research, management, and conservation for serving the long-term interests of natural history collections. *Collection Forum* 11(1):16–27.
Cato, P. S., R. R. Waller, L. Sharp, J. E. Simmons, and S. L. Williams. 1996. *Developing Staff Resources for Managing Collections*. Virginia Museum of Natural History Special Publication Number 4, 71 pp.
Ceska, A., and O. Ceska. 1986. More on the techniques for collecting aquatic and marsh plants. *Annals of the Missouri Botanical Garden* 73(4):825–827.
Chakraborty, A., M. Sakai, and Y. Iwatsuki. 2006. Museum fish specimens and molecular taxonomy: A comparative study on DNA extraction and preservation techniques. *Journal of Applied Ichthyology* 22(2):160–166.
Chamberlain, A. T., and M. P. Pearson. 2001. *Earthly Remains: The History and Science of Preserved Human Bodies*. New York: Oxford University Press.
Chamberlin, J. C. 1925. Heavy mineral oil as a permanent non-volatile preservative for valuable biological material. *Science* 61(1590):634–635.
Champalbert, G., and P. Kerambrun. 1979. Influence du mode de conservation sur la composition chimique elémentaire de Pontella mediterranea. *Marine Biology* 51:357–360.
Chatigny, M. E. 2000. The extraction of DNA from formalin-fixed, ethanol-preserved reptile and amphibian tissues. *Herpetological Review* 31(2):86–87.
Child, R. E. 1994. Labelling of specimens preserved in spirit collections. *Biology Curator's Group Newsletter* 6(4):42.
Christy, B. 2008. *The Lizard King. True Crimes and Passions of the World's Greatest Reptile Smugglers*. New York: Twelve (Hachette Book Group USA).

Ciferri, A. 1971. Swelling and phase transition of insoluble collagen. Pages 101–151 in Elden, H. R. (editor). *Biophysical Properties of the Skin*. New York: Wiley.

Clark, P. F. 1992. Ground glass stoppered jars for fluid collections. Pages 221–223 in Rose, C. L., and A. R. de Torres (editors). *Storage of Natural History Collections: Ideas and Practical Solutions*. New York: Society for the Preservation of Natural History Collections.

Clark, P. F. 1993. Museum storage containers: Back to the future. Pages 347–359 in Palacios, F., C. Martinez, and B. Thomas (editors). *Contributions on the Status, Preservation, and Conservation of Natural History Collections*. Dirección General de Bellas Artes y Archivos, Ministero de Cultura, 2. Madrid: International Symposium and First World Congress on the Preservation and Conservation of Natural History Collections.

Clemens, H. P. 1951. The growth of the burbot Lota lota maculosa (LeSueur) in Lake Erie. *Transactions of the American Fisheries Society* 90(1950):163–173.

Clephane, J. 1753. Experiments by Francis Hume, M. D., on fish and flesh preserved in lime-water. *Philosophical Transactions of the Royal Society* 48:163–164.

Clique, A., J. F. Ponge, Y. D. Sellos, and D. Especel. 2007. Formaldehyde hyperbaric fixation. *Collection Forum* 22(1–2):37–44.

Clutter, R. I., and L. E. Whitesel. 1956. Collection and interpretation of sockeye and salmon scales. *Bulletin of the International Pacific Salmon Fisheries Commission* 9:1–159.

Coetzer, W., O. Gon, and P. H. Skelton. 2009. Moving into the future: Relocating the South African National Fish Collection to a new, dedicated collection facility. *Collection Forum* 23(1–2):1–10.

Cohen, K. S., J. L. Ennis, I. Arons, A. J. Krubsack, R. D. Rowe, and L. G. Lowry. 1979. Caution to contact lens users. *Chemical and Engineering News* 57(4):84.

Cole, F. J. 1921. The history of anatomical injections. Pages 285–343 in Singer, C. J. (editor). *Studies in the History and Method of Science*. Vol. 2. Oxford: Oxford at the Claredon Press.

Cole, F. J. 1944. *A History of Comparative Anatomy from Aristotle to the Eighteenth Century*. London: MacMillan.

Conant, G. 1924. Formalin-acetic-alcohol: a universal killing and fixing agent. *Turtox News* 2(7):25.

Conant, G. 1926. Further remarks on the universal fixative. *Turtox News* 4(7):25–26.

Condit, I. J. 1948. The use of chemicals to prevent molding of herbarium specimens. *Science* 108(2794):58.

Condon, M., and M. D. Whalen. 1983. A plea for collection and preservation of herbivore and pathogen damaged plant materials. *Taxon* 32(1):105–107.

Conroy, R. T. 1993. *Mr. Smithson's Bones: A Mystery at the Smithsonian*. New York: St. Martin's Press.

Cook, F. R. 1965. Collecting and preserving amphibians and reptiles. *National Museum of Canada Bulletin* 69:128–151.

Cornwall, P. 1994. *The Body Farm*. New York: G. F. Putnam's Sons.

Cornwall, P. 1999. *Black Notice*. New York: G. F. Putnam's Sons.

Craig, H. K. 1914. A new method of preparing museum specimens. *Journal of the American Medical Association* 62(116):1241–1242.

Crimmen, O. A. 1989. Phenoxetol: an unsatisfactory preservative for fishes. *Biology Curator's Group Newsletter* 5:26–27.

Crisan, D., E. M. Cadoff, J. C. Mattson, and K. A. Hartle. 1990. Polymerase chain reaction: amplification of DNA from fixed tissue. *Clinical Biochemistry* 23:489–495.

Criscuolo, G. 1992. Extraction and amplification of DNA from wet museum collections. *Ancient DNA Newsletter* 1(1):12–13.

Criscuolo, G. 1994. On the state of preservation of DNA from museum spirit collections. *Biology Curator's Group Newsletter* 6(4):39–41.

Cristobo, F. J., V. Urgorri, M. R. Solorzano, and P. Rios. 1992. Metodos de recogida, estudio y conservación de las colecciones de poriferos. Pages 277–287 in *Contributions on the Status, Preservation, and Conservation of Natural History Collections*. Palacios, F., C. Martinez and B. Thomas. Dirección General de Bellas Artes y Archivos, Ministero de Cultura, 2. Madrid: International Symposium and First World Congress on the Preservation and Conservation of Natural History Collections.

Crombie, A. C. 1952. *Augustine to Galileo. The History of Science A. D. 400–1650*. London: Falcon Educational Books.

Crosland, M. P. 2004. *Historical Studies in the Language of Chemistry*. Mineola, NY: Dover Publications. [Reprint of the 1962 edition].

Cross, F. B. 1962. Collecting and preserving fishes. Pages 41–44 in Hall, E. R. (editor). Collecting and preserving study specimens of vertebrates. *University of Kansas Museum of Natural History Miscellaneous Publications* 30:1–46.

Cruz, J. C. 1977. *The Incorruptibles. A Study of the Incorruption of the Bodies of Various Catholic Saints and Beati*. Rockford, IL: Tan Books and Publishers.

Cruz, J. C. 1984. *Relics*. Huntington, IN: Our Sunday Visitor.

Cunningham, M. K., W. F. Granberry, and K. L. Pope. 2000. Shrinkage of inland silverside larvae preserved in ethanol and formalin. *North American Journal of Fisheries Management* 20(3):816–818.

Cushing, P. E., and J. A. Slowik. 2007. Re-curation of alcohol-preserved specimens: Comparison of gradual versus direct specimen transfer on specimen quality and assessment of specimen value. *Collection Forum* 22(1–2):1–9.

Dabney, J. E. 1974. *Mountain Spirits: A Chronicle of Corn Whiskey from King James' Ulster Plantation to America's Appalachians and the Moonshine Life*. Fairview, NC: Bright Mountain Books.

Dabrowski, K., and R. Bardega. 1982. The changes of fish larvae dimension due to fixation in different preservatives. *Zoologische Jahrbücher. Abteilung für Anatomie und Ontogenie der Tierre* 108:509–516.

Dall, W. H. 1892. Instructions for collecting mollusks, and other useful hints for the conchologist. *Bulletin of the United States National Museum*, No. 39, Part G, pages 1–56.

Davidson, M. H. 1932. Practical suggestions for embryological technique. IV. Frog eggs and tadpoles. *Turtox News* 10(11):203–204.

Davis, C. W. 1978. A note on shrinkage of Atlantic herring larvae preserved in formalin. *National Marine Fisheries Service Northeast Fisheries Center* No. 78–43.

Davis, D. D., and U. R. Gore. 1936. Clearing and staining skeletons of small vertebrates. *Field Museum of Natural History Technique Series* 4:1–15.

Davis, P. 1993. The preservation of fish collections: an historical perspective. Pages 257–287 in Palacios, F., C. Martinez and B. Thomas (editors). *Contributions on the Status, Preservation, and Conservation of Natural History Collections*. Dirección General de Bellas Artes y Archivos, Ministero de Cultura, 2. Madrid: International Symposium and First World Congress on the Preservation and Conservation of Natural History Collections.

Dawson, A. B. 1926. A note on the staining of the skeletons of cleared specimens with alizarin red S. *Stain Technology* 1(4):123–124.

Dean, M. D., and J. W. Ballard. 2001. Factors affecting mitochondrial DNA quality from museum preserved *Drosophila simulans. Entomologia Experimentalis et Applicata* 98:279–283.

Debus, A. G. (editor). 1968. *World Who's Who in Science. A Biographical Dictionary of Notable Scientists from Antiquity to the Present.* Chicago: A. N. Marquis Company.

Deichmann, J. L., J. Boundy, and G. B. Williamson. 2009. Anuran artifacts of preservation: 27 years later. *Phyllomedusa* 8(1):51–58.

Deichmann, J. L., W. E. Duellman, and G. B. Williamson. 2008. Predicting biomass from snount-vent length in New World frogs. *Journal of Herpetology* 42:238–245.

Delafield, F., and T. M. Prudden. 1904. *A Handbook of Pathological Anatomy and Histology.* Seventh edition. New York: William Wood.

de Lafontaine, Y., and W. C. Leggett. 1989. Changes in size and weight of hydromedusae during formalin preservation. *Bulletin of Marine Science* 44(3):1129–1137.

Dempster, W. T. 1960. Rates of penetration of fixing fluids. *American Journal of Anatomy* 107:59–72.

Derry, T. K., and T. I. Williams. 1960. *A Short History of Technology from the Earliest Times to A. D. 1900.* Oxford: Oxford University Press.

Dessauer, H. C., C. J. Cole, and M. S. Hafner. 1996. Collection and storage of tissues. Pages 29–47 in Hillis, D. M., C. Mortiz, and B. K. Mable (editors). *Molecular Systematics.* 2nd ed. Sunderland, MA: Sinauer Associates.

DeWolf, G. P. 1968. Notes on making an herbarium. *Arnoldia* 28:69–111.

De Villiers, C. G. S. 1929. Some features of the early development of Breviceps. *Annals of the Transvaal Museum* 13(3):142–151.

Dingerkus, G. 1982. Preliminary observations on acidification of alcohol in museum specimen jars. *Curation Newsletter* 5:1–3.

Dingerkus, G., and L. D. Uhler. 1977. Differential staining of bone and cartilage in cleared and stained fish using alcian blue to stain cartilage and enzymes for clearing flesh. *Journal of Stain Technology* 52(4):229–232.

DiStefano, R. J., M. J. Roell, B. A. Wagner, and J. J. Decoske. 1994. Relative performances of four preservatives on fish and crayfish. *Transactions of the American Fisheries Society* 123:817–823.

Dobell, C. 1932. *Antony van Leeuwenhoek and His "Little Animals."* New York: Harcourt, Brace and Company.

Dobson, J. 1970. Historical introduction. Pages xi–xvi in Tompsett, D. H. *Anatomical Techniques.* 2nd ed. Edinburg: E. & S. Livingstone.

Donald, G. L., and C. G. Patterson. 1977. Effect of preservation on wet weight biomass of chironomid larvae. *Hydrobiologia* 53:75–80.

Donleavy, J. P. 1994. *The History of The Ginger Man.* Boston: Houghton Mifflin/Seymour Lawrence.

Doolittle, A. A. 1925. Waterproofing labels for wet specimens. *Turtox News* 3(5):19.

Down, R. 1989. "Old" preservative methods. Pages 27–32 in Horie, C. V. (editor). *Conservation of Natural History Specimens: Spirit Collections.* Manchester: The Manchester Museum and Department of Environmental Biology, The University of Manchester.

Drahl, C. 2008. Seeking an eternal solution. Fluorinated fluid is the protagonist of an ongoing experiment in preserving biological specimens. *Chemical and Engineering News* 86(44):25–28.

Drowne, F. P. 1899. Formalin as a preservative. *The Museum* 5(12):179–181.

Drury, R. A. B., and E. A. Wallington. 1980. *Carleton's Histological Technique*. 5th ed. Oxford: Oxford University Press.

Dubeau, L., L. A. Chandler, J. R. Gralow, P. W. Nichols, and P. A. Jones. 1986. Southern blot analysis of DNA extracted from formalin-fixed pathology specimens. *Cancer Research* 46:2964–2969.

Dubeau, L., K. Weinberg, P. A. Jones, and P. W. Nichols. 1988. Studies on immunoglobulin gene rearrangement in formalin-fixed, paraffin-embedded pathology specimens. *American Journal of Pathology* 130(3):588–594.

Durrell, L. 1958. *Mountolive*. London: Faber and Faber.

Ealy, J. B. 1991. An alternative to formaldehyde. *The Science Teacher* 58(4):26–27.

Edmonds, G. F., S. L. Jensen, and L. Berner. 1976. *The Mayflies of North and Central America*. Minneapolis: University of Minnesota Press.

Edwards, M. S., T. F. Turner, and Z. D. Sharp. 2002. Short- and long-term effects of fixation and preservation on stable isotope values ($\delta^{13}C$, $\delta^{15}N$, $\delta^{34}S$) of fluid-preserved museum specimens. *Copeia* 2002(4):1106–1112.

Edwards, J. J., and M. J. Edwards. 1959. *Medical Museum Technology*. London: Oxford University Press.

Eekhout, X. 2010. Sampling amphibians and reptiles. Pages 530–557 in Eymann, J., J. Degreef, C. Häuser, J. C. Monje, Y. Samyn, and D. VandenSpiegel (editors). *Manual on Field Recording Techniques and Protocols for All Taxa Biodiversity Inventories*. Belgium: Belgian National Focal Point to the Global Taxonomy Initiative.

Egginton, S., and S. Cordiner. 1995. Effect of fixation protocols on muscle preservation and *in situ* diffusion distances. *Journal of Fish Biology* 47:59–69.

Ehmann, H. 1989. Neutral paraformaldehyde solution for field use. *Herpetological Review* 20(2):44–45.

Ellis, D. V. 1987. Biomass loss in wet-preserved reference collections. *Collection Forum* 3(1–2):6–8.

El Mahdy, C. 1989. *Mummies, Myth and Magic in Ancient Egypt*. New York: Thames and Hudson.

Eltoum, I., J. Fredenburgh, R. B. Myers, and W. E. Grizzle. 2001a. Introduction to the theory and practice of fixation of tissues. *Journal of Histotechnology* 24(3):173–190.

Eltoum, I., J. Fredenburgh, R. B. Myers, and W. E. Grizzle. 2001b. Advanced concepts in fixation: 1. Effects of fixation on immunohistochemistry, reversibility of fixation and recovery of proteins, nucleic acids, and other molecules from fixed and processed tissues. 2. Developmental methods of fixation. *Journal of Histotechnology* 24(3):201–210.

Emerson, W. K., and A. Ross. 1965. Invertebrate collections: trash or treasure? *Curator* 8(4):333–347.

Engel, S. 1974. Effects of formalin and freezing on length, weight, and condition of cisco and yellow perch. *Transactions of the American Fisheries Society* 103:136–138.

Epstein, R. H. 2010. Inside neurosurgery's rise. *New York Times*, August 24, 2010, D1.

Estep, A. D. 2004. Use of polycarbonate food storage pails in wet collections. *Society for the Preservation of Natural History Collections Newsletter* 18(1):1–2.

Ewing, C. P., and J. T. McCann. 2006. *Minds on Trial: Great Cases in Law and Psychology*. New York: Oxford University Press.

Evans, D. 1958. Storage of preserved insect specimens. *The Canadian Entomologist* 90:461–463.

Evans, H. E. 1948. Clearing and staining small vertebrates in toto, for demonstrating ossification. *Turtox News* 26(2):42–47.

Fancett, M. S. 1986. Species composition and abundance of scyphomedusae in Port Phillips Bay, Victoria. *Australian Journal of Marine and Freshwater Research* 37:379–384.

Fang, S-G., Q-H. Wan, and N. Fujihara. 2002. Formalin removal from archival tissue by critical point drying. *BioTechniques* 33(3):604–611.

Farris. D. A. 1963. Shrinkage of sardine (Sardinops caerulea) larvae upon preservation in buffered formalin. *Copeia* 1963(1):185–186.

Favret, C., K. S. Cummings, R. J. McGinley, E. J. Heske, K. P. Johnson, C. A. Philips, L. R. Phillippe, M. E. Retzer, C. A. Taylor, and M. J. Wetzel. 2007. Profiling natural history collections: A method for quantitative and comparative health assessment. *Collection Forum* 22(1–2):53–65.

Fenaux, R. 1976. The appendiculatrians. Pages 309–312 in Steedman, H. F. (editor). *Zooplankton Fixation and Preservation*. Monographs on Oceanographic Methodology 4. Paris: Unesco Press.

Fink, W. L., K. E. Hartel, W. G. Saul, E. M. Koon, and E. O. Wiley. 1979. *A report on current supplies and practices used in curation of ichthyological collections*. American Society of Ichthyologists and Herpetologists ad hoc Subcommittee Report, 63 pages.

Fish, P. A. 1893. Brain preservation, with a résumé of some old and new methods. Pages 385–400 in *The Wilder Quarter-Century Book*. Ithaca, NY: Comstock Publishing.

Fish, P. A. 1895. The use of formalin in neurology. *Proceedings of the American Microscopical Society* 17:319–330.

Fisher, R. D., L. K. Gordon, and C. Hawks. 1995. Division of Mammals fluid assessment project report, United States National Museum, unpublished, 18 pages.

Fisk, E. L., and R. M. Addoms. 1935. *A Laboratory Manual of General Botany*. Rev. ed. New York: MacMillan.

Ford, L. S., and J. E. Simmons 1997. The diffusion of knowledge: Agassiz (1807–1873), Ruthven (1882–1971), and the growth of herpetological collections. Pages 577–593 in Pietsch, T. W., and W. D. Anderson (editors). *Collection Building in Ichthyology and Herpetology*. American Society of Ichthyologists and Herpetologists Special Publication Number 3.

Forbes, R. J. 1948. *Short History of the Art of Distillation from the Beginnings up to the Death of Cellier Blumenthal*. Leiden: E. J. Brill.

Forman, L., and D. Bridson. 1989. *The Herbarium Handbook*. Rev. ed. Kew: Royal Botanic Gardens.

Fosberg, F. R. 1947. Formaldehyde in plant collecting. *Science* 106(2750):250–251.

Foust, H. L., T. S. Leith, H. M. Tabbut, and L. Bowstead. 1935. Denaturizing formaldehyde. *Science* 82(2134):498.

Fowler, G. M., and S. J. Smith. 1983. Length changes in silver hake (*Merluccius bilinearis*) larvae: Effects of formalin, ethanol, and freezing. *Canadian Journal of Fisheries and Aquatic Science* 40(7):866–870.

Fox, C. H., and C. Benton. 1987. Formaldehyde: the fixative. *Journal of Histotechnology* 10(3):199–201.

Fox, C. H., F. B. Johnson, J. Whiting, and P. P. Roller. 1985. Formaldehyde fixation. *Journal of Histochemistry and Cytochemistry* 33(8):845–853.

Fraser, J. H. 1961. Use of polyethylene glycol in biology. *Nature* 189(4760):241–242.

Freeman, B. L., E. K. Moyer, and A. M. Lassek. 1955. The pH of fixing fluids during fixation of tissues. *Anatomical Record* 121(3):593–600.

French, R. 1994. *Ancient Natural History. Histories of Nature*. New York: Routledge.

Fry, C. H. 1985. The effect of alcohol immersion on the plumage colours of bee-eaters. *Bulletin of the British Ornithologists' Club* 105(2):78–79.

Fudge, H. 1968. Biochemical analysis of preserved zooplankton. *Nature* 219(27 July):380–381.

Furnestin, M.-L. 1976. Fixation and preservation of Chaetogratha. Pages 272–278 in Steedman, H. F. (editor). *Zooplankton Fixation and Preservation*. Monographs on Oceanographic Methodology 4. Paris: Unesco Press.

Galigher, A. E., and E. N. Kozloff. 1971. *Essentials of Practical Microtechnique*. Philadelphia: Lea and Febiger.

Gamble, J. 1983. Effects of formaldehyde on the respiratory system. Pages 175–197 in Gibson, J. E. (editor). *Formaldehyde Toxicity*. Washington, D.C.: Hemisphere Publishing Company.

Gannal, J. N., and R. Harlan. 1840. *History of Embalming and of Preparations in Anatomy, Pathology and Natural History*. Philadelphia: Judah Dobson.

Geiger, D. L., B. A. Marshall, W. F. Ponder, T. Sasaki, and A. Warén. 2007. Techniques for collecting, handling, preparing, storing and examining small molluscan specimens. *Molluscan Research* 27(1):1–50.

Geldmacher, A., and K. Wieland. 1999. Implications of mechanical deformation and formaldehyde preservation for the identification of stage-specific characteristics of Baltic Cod Eggs. *Journal of Applied Ichthyology* 15:75–79.

Gere, C. 2003. A brief history of brain archiving. *Journal of the History of Neurosciences* 12(4):396–410.

Gerrick, D. J. 1968. A comparative study of antioxidants in color preservation of fish. *Ohio Journal of Science* 68(4):239–240.

Gibb, T. J., and C. Y. Oseto. 2006. *Arthropod Collection and Identification: Laboratory and Field Techniques*. New York: Academic Press.

Gibbs, P. E. 1976. Fixation and preservation of planktonic polychaeta. Pages 279–280 in Steedman, H. F. (editor). *Zooplankton Fixation and Preservation*. Monographs on Oceanographic Methodology 4. Paris: Unesco Press.

Gibbs, R. H., E. Jarosewich, and H. L. Windom. 1974. Heavy metal concentrations in museum fish specimens: effects of preservatives and time. *Science* 184(4135):475–477.

Giguère, L. A., J.-F. St-Pierre, B. Bermier, A. F. Vézina, and J.-G. Rondeau. 1989. Can we estimate the true weight of zooplankton samples after chemical preservation? *Canadian Journal of Fisheries and Aquatic Sciences* 46:522–527.

Gillette, J. W. 2004. Jar sealing tape for use with fluid-preserved collections: a summary of recommendations and sources. *Society for the Preservation of Natural History Collections Newsletter* 18(2):15.

Gillespie, J. W., C. J. M. Best, V. E. Bichel, K. A. Cole, et al. 2002. Evaluation of non-formalin tissue fixation for molecular profiling studies. *American Journal of Pathology* 160(2):449–457.

Gloyd, H. K. 1938. Methods of preserving and labeling amphibians and reptiles for scientific study. *Turtox News* 16(3):49–53; 66–67.

Goates, C. Y., and K. Knutson. 1994. Enhanced permeation of polar compounds through human epidermis. I. Permeability and membrane structural changes in the presence of shorter chain alcohols. *Biochimica et Biophysica Acta* 1195:169–179.

Goelz, S., S. R. Hamilton, and B. Vogelstein. 1985. Purification of DNA from formaldehyde fixed and paraffin embedded human tissue. *Biochemical and Biophysical Research Communications* 130(1):118–126.

Gould, S. J. 1979. *Ever Since Darwin. Reflections on Natural History*. Harmondworth, England: Penguin Books.
Grandperrin, R., and C. Caboche. 1968. Aperçu sur l'action des procédés de conservation sur la biomasse d'organismes micronectoniques et macroplanctoniques. *Journal du Consell International pour l'Exploration de la Mer* 32:209–215.
Graves, G. 1817. *The Naturalist's Pocket-Book, or Tourist's Companion, being a brief introduction to the different branches of Natural History, with approved methods for collecting and preserving the various Productions of Nature*. London: W. and S. Graves.
Graves, G. R., and M. J. Braun. 1992. Museums: storehouses of DNA? *Science* 255(5050):1335–1336.
Greenacre, P. 1947. Vision, headache, and the halo. Reactions to stress in the course of superego formation. *Psychoanalytic Quarterly* 16(2):177–194.
Greensburg, B. 1996. Med student unlocks surgical pioneer's trove. *Philadelphia Inquirer*, March 3, 1996, A12.
Greer, C. E., S. L. Peterson, N. B. Kiviat, and M. M. Manos. 1991. PCR amplification from paraffin-embedded tissues. Effects of fixative and fixation time. *American Journal of Clinical Pathology* 95:117–124.
Gregersen, K. M. 2008. *Liquid preservation in the 19th century*. Master's thesis, School of Conservation, Royal Academy of Fine Arts, Copenhagen.
Gregorio, M. 2006. *Critique of Criminal Reason*. London: Faber and Faber.
Grene, D. [translator]. 1987. *Herodotus. The History*. Chicago: University of Chicago Press.
Grew, N. 1681. *Musaeum Regalis Societatis, or a Catalogue and Description of the Natural and Artificial Rarities belonging to the Royal Society and preserved at Gresham Colledge*. London: W. Rawlins.
Griffin, L. F. 1922. Practicable substitutes for grain alcohol. *Science* 55(1419):262–263.
Griffiths, F. B., A. Fleminger, B. Kimor, and M. Vannucci. 1976. Shipboard and curating techniques. Pages 17–31 in Steedman, H. F. (editor). *Zooplankton Fixation and Preservation*. Monographs on Oceanographic Methodology 4. Paris: Unesco Press.
Guillette, L. J., M. S. Rand, V. DeMarco, and K. Etheridge. 1988. Fixative and alcohol-induced weight change in eggs from the lizard, *Sceloporus undulatus*. *Journal of Herpetology* 22(1):115–118.
Gunther, R. T. 1927. *The Diary and Will of Elias Ashmole, Edited and Extended from the Original Manuscripts*. Oxford: Oxford University Press.
Haly, A. 1892. A medium for preserving the colours of fish and other animals. *Nature* 45:212.
Hammel, B. E. 1987. The origami of botany: A guide to collecting and mounting specimens of Cyclanthaceae. *Annals of the Missouri Botanical Garden* 74(4): 897–902.
Hangay, C., and M. Dingley. 1985a. *Biological Museum Methods. Volume 1. Vertebrates*. Sydney, Australia: Academic Press.
Hangay, C., and M. Dingley. 1985b. *Biological Museum Methods: Plants, Invertebrates, and Techniques*. Vol. 2. Sydney, Australia: Academic Press.
Hargrave, C. W., K. G. Sutherland, E. Marsh-Matthews, and W. J. Matthews. 2005. Multiple interacting factors affect pH in museum storage solutions. *Collection Forum* 19(1–2):23–31.
Harrington, H. D. 1947. Preserving plants in formaldehyde fumes. *Turtox News* 25(12):238.
Harris, J. B. 1968. Practical aspects of lighting as related to conservation. Pages 133–138 in Thompson, G. (editor). *Contributions to the London Conference on Museum Climatology*. London: International Institute for Conservation of Historic and Artistic Works.
Harris, R. H. 1960. Alizarine transparencies. *Museums Journal* 60(4):99–101.

Harris, R. H. 1979. The conservation of one of the earliest known examples of a fluid-preserved injection dissection. *Museums Journal* 79(2):71–72.

Hartigan, L. R., W. Hopps, R. Vine, and R. Lehrman. 2003. *Joseph Cornell. Shadowplay . . . Eterniday*. New York: Thames and Hudson.

Haselkorn, R., and P. Doty. 1961. The reaction of formaldehyde with polynucleotides. *Journal of Biological Chemistry* 236(10):2738–2745.

Hašek, J. 1973. *The Good Soldier Švejk and His Fortunes in the World War*. London: Penguin and Heinemann.

Hatch, L. F. 1961. *Isopropyl Alcohol*. New York: McGraw-Hill.

Hawks, C. A. 2003. Re-use of ethanol in processing biological specimens. *Conserve O Gram* 11/5:1–4.

Hawks, C., M. McCann, K. A. Makos, L. Goldberg, D. Hinkamp, D. C. Ertel Jr., and P. Silence (editors). 2010. *Health and Safety for Museum Professionals*. New York: Society for the Preservation of Natural History Collections and the Health and Safety Committee of the American Institute for Conservation of Historic and Artistic Works.

Hawks, C. A., and S. L. Williams. 1986. Care of specimen labels in vertebrate research collections. Pages 105–108 in Waddington, J. and D. M. Rudkin (editors). *Proceedings of the 1985 Workshop on Care and Maintenance of Natural History Collections*. Toronto: Life Sciences Miscellaneous Publications of the Royal Ontario Museum.

Hay, D. E. 1981. Effects of capture and fixation on gut contents and body size of Pacific herring larvae. The early life history of fish II. *Rapports et Process-Verbaux des Reúnions Conseil Permanent Exploration de la Mer* 178:395–400.

Hay, D. E. 1982. Fixation shrinkage of herring larvae: Effects of salinity, formalin concentration, and other factors. *Canadian Journal of Fisheries and Aquatic Sciences* 39:1138–1143.

Hay, D. E. 1984. Weight loss and change of condition factor during fixation of Pacific herring, *Clupea harengus pallasi*, eggs and larvae. *Journal of Fish Biology* 25:421–433.

Haynes, R. R. 1984. Techniques for collecting aquatic and marsh plants. *Annals of the Missouri Botanical Garden* 71(1):229–231.

Heming, T. A., and R. P. Preston. 1981. Differential effect of formalin preservation on yolk and tissue of young chinook salmon (*Onchorynchus tschwaytscha*). *Canadian Journal of Zoology* 59:1608–1611.

Henderson, N. E. 1963. Influence of light and temperature on the reproductive cycle of the Eastern brook trout, *Salvelinus fontinalis* (Mitchell). *Journal of the Fisheries Research Board of Canada* 20(4):859–897.

Hendrickx, F., J.-P. Maelfait, A. D. Mayer, F. M. G. Tack, and M. G. Verloo. 2003. Storage mediums affect metal concentration in woodlice (Isopoda). *Environmental Pollution* 21:87–93.

Higgins, A. Z., and J. O. M. Karlsson. 2013. Effects of intercellular junction protein expression on intracellular ice formation in mouse insulinoma cells. *Biophysical Journal* 105:2006–2015.

Hildebrand, M. 1968. *Anatomical Preparations*. Berkeley: University of California Press, Berkeley.

Hile, R. 1936. Age and growth of the cisco, *Leuichthys artedi* (Le Sueur), in the lakes of Northeastern Highlands, Wisconsin. *Bulletin of the United States Bureau of Fisheries* 48(19):211–317.

Hile, R. 1941. Age and growth of the rock bass, *Amblophlites rupestris* (Rafinesque), in Nebish Lake, Wisconsin. *Transactions of the Wisconsin Academy of Science, Arts, and Letters* 33:189–337.

Hill, S. 2005. Fluorescent frogs and technicolor toads. An artist explores the links between science and art in (almost) living color. *Kansas Alumni* 2005(6):64–65.

Hillcourt, W. 1970. *The New Field Book of Nature Activities and Hobbies.* Revised version of Field Book of Nature Activities and Conservation. New York: G. P. Putnam's Sons.

Hiscox, G. D. (editor). 1935. *Henley's Twentieth Century Book of Formulas, Processes and Trade Secrets.* Revised and enlarged, by T. O'Conner Sloane. London: Scientific and Commercial Publishing Company.

Hiscox, G. D., and T. O. Sloan (editors). 1939. *Fortunes in Formulas for Home, Farm, and Workshop.* New York: Books, Inc.

Hoar, W. S. 1939. The weight-length relationship of the Atlantic salmon. *Journal of the Fisheries Research Board of Canada* 4:441–460.

Hodge, W. H. 1947. The use of alcohol in plant collecting. *Rhodora* 49:207–210.

Hofmann, A. W. 1867. Contributions to the history of methylic aldehyde. *Proceedings of the Royal Society of London* 16:156–159.

Hoffmann, A., J. Decher, F. Rovero, J. Schaer, C. Voigt, and G. Wibbelt. 2010. Field methods and techniques for monitoring mammals. Pages 482–529 in Eymann, J., J. Degreef, C. Häuser, J. C. Monje, Y. Samyn, and D. VandenSpiegel (editors). *Manual on Field Recording Techniques and Protocols for All Taxa Biodiversity Inventories.* Belgium: Belgian National Focal Point to the Global Taxonomy Initiative.

Hollister, G. 1934. Clearing and dyeing fish for bone study. *Zoologica* 12(10):89–101.

Holm, Å. 1957. Specimina linnaeana I Uppsala bevarade zoogiska samlingar från Linnes tid. *Uppsala Universitets Årsskrift* 6:1–68.

Holmyard, E. J. 1957. *Alchemy.* Harmondsworth, Middlesex, England: Penguin Books.

Holst, B. 2003. The Marie Selby Botanical Gardens spirit collections: A valuable botanical resource. *Plant Science Bulletin* 49(4):141.

Holt, V. 1931. Cure for formaldehyde poisoning. *Science* 73(1903):675.

Hoogmoed, M. S. 1973. Het verzamelen en conserveren van ambibieën voor museumgebruik. *Museologia* No. 1–X:18–22.

Hopkins, T. L. 1968. Carbon and nitrogen content of fresh and preserved *Nematoscelis difficilis*, a euphausiid crustacean. *Journal du Consell permanent International pour l'Exploration de la Mer* 31(3):300–304.

Hopwood, D. 1967. Some aspects of fixation with glutaraldehyde. A biochemical and histochemical comparison of the effects of formaldehyde and glutaraldehyde fixation on various enzymes and glycogen, with a note on penetration of glutaraldehyde into liver. *Journal of Anatomy* 101(1):83–92.

Hopwood, D. 1973. Theoretical and practical aspects of glutaraldehyde fixation. Pages 47–83 in Stoward, P. J. (editor). *Fixation in Histochemistry.* London: Chapman and Hall.

Horie, V. 1983. Sealing of museum jars with silicone mastic. *Conservation News* 20:13–14.

Horie, V. 1994. Environmental control for spirit collections. *Biology Curator's Group Newsletter* 6(6):43–44.

Hornaday, W. T. 1905. *Taxidermy and Zoological Collecting. A Complete Handbook for the Amateur Taxidermist, Collector, Osteologist, Museum-Builder, Sportsman, and Traveler.* New York: Charles Scribner's Sons.

Hornell, J. 1895. The use of formalin as a preservative medium for marine animals. *Natural Science*, 7(46):416–420.

Hornell, J. 1900. The respective values of formalin and alcohol as preservatives for zoological specimens. *Laboratorium et Museum et Clinicum* 5:85–89.

Horvath, E. P., H. Anderson, W. E. Pierce, L. Hanrahan, J. D. Wendlick. 1988. Effects of formaldehyde on the mucous membranes and lungs. A study of an industrial problem. *Journal of the American Medical Association* 259(5):701–707.

Houde, P., and M. J. Braun. 1988. Museum collections as a source of DNA from studies of avian phylogeny. *The Auk* 105:773–776.

Howlett, P., and J. Horák. 2006. Integrated collections management at the National Museum Wales. *Collection Forum* 21(1–2):58–63.

Howmiller, R. P. 1972. Effect of preservatives on weights of some common macrobenthic invertebrates. *Transactions of the American Fisheries Society* 101:743–746.

Hughes, G. W., and J. A. Cosgrove. 1990. pH change in a formalin borax solution with inferences about uses of neutralized formalin in vertebrate collections. *Collection Forum* 6(1):21–26.

Hughes, T. (editor). 2008. *The Collected Poems of Sylvia Plath*. New York: Harper Perennial.

Hughey, J. R., P. C. Silva, and M. H. Hommersand. 2001. Solving taxonomic and nomenclatural problems in Pacific gigartinaceae (Rhodophyta) using DNA from type material. *Journal of Phycology* 37:1091–1109.

Huheey, J. E. 1971. Concerning the use of paraformaldehyde as a field preservative. *Copeia* 1963(1):192–193.

Humphries, D. 2003. The Kunstkamera, St. Petersburg, Russia. *Fortean Times* 172:76–78.

Huxley, R. (editor). 2007. *The Great Naturalists*. London: Thames & Hudson.

Huxley, R., and J. Bryant. 2007. How to prepare seaweed specimens. Society for the Preservation of Natural History Collections How to Series No. 2, supplement to *SPNHC Newsletter* 21(1), 2 pages.

Impraim, C. C., R. K. Saiki, H. A. Erlich, and R. L. Teplitz. 1987. Analysis of DNA extracted from formalin-fixed, paraffin-embedded tissues by enzymatic amplifications and hybridization with sequence-specific oligonucleotides. *Biochemical and Biophysical Research Communications* 142(3):710–716.

Jackson, F. S. 1919. The preservation of fresh-water bryozoa. *Transactions of the American Microscopical Society* 38(3):217–220.

Jackson, J. H. 1977. *Bad Company: The Story of California's Legendary and Actual Stage-Robbers, Bandits, Highwaymen and Outlaws from the Fifties to the Eighties*. Lincoln: University of Nebraska Press.

Jackson, V. 1978. Studies on histone organization in the nucleosome using formaldehyde as a reversible cross-linking agent. *Cell* 15(3):945–954.

Jacques, H. E. 1939. *How to Know the Insects*. Pictured Key Nature Series, Mt. Pleasant.

Jawad, L. A. 2003. The effect of formalin, alcohol and freezing on some body proportions of *Alepes djeddaba* (Pisces: Carangidae) collected from the Red Sea coast of Yemen. *Revista de Biologia Marina y Oceanografía* 38(2):77–80.

Jehaes, E., H. Pfeiffer, K. Toprak, R. Decorte, B. Brinkmann, and J. J. Cassiman. 2001. Mitochondrial DNA analysis of the putative heart of Louis XVII, son of Louis XVI and Marie-Antoinette. *European Journal of Human Genetics* 9(3):185–190.

Jennings, S. 1991. The effects of capture, net retention, and preservation upon lengths of larval and juvenile bass, *Dicentrachus labrax* L. *Journal of Fish Biology* 38:349–357.

Jensh, R. P., and R. L. Brent. 1966. Rapid schedules for KOH clearing and alizarin red S staining of fetal rat bone. *Stain Technology* 41(3):179–183.

Jeppesen, P. H. 1998. Use of vacuum in rehydration of biological tissue, with a review of liquids used. *Crustaceana* 55(3):268–273.

Johnson, A. G. 1948. Effect of formaldehyde on Picea and Tsuga herbarium specimens. *Science* 107(2777):294.

Johnson, A. R. 1971. Extraction and purification of lipids. Pages 131–136 in Johnson, A. R. and J. B. Davenport (editors). *Biochemistry and Methodology of Lipids*. New York: Wiley-Interscience.

Jones, B. C., and G. H. Geen. 1977. Morphometric changes in an elasmobranch (*Squalus acantias*) after preservation. *Canadian Journal of Zoology* 55:1060–1062.

Jones, D. 1973. Reactions of aldehydes with unsaturated fatty acids during histological fixation. Pages 1–45 in Stoward, P. J. (editor). *Fixation in Histochemistry*. London: Chapman and Hall.

Jones, D. 1976. Chemistry of fixation and preservation with aldehydes. Pages 155–171 in Steedman, H. F. (editor). *Zooplankton Fixation and Preservation*. Monographs on Oceanographic Methodology 4. Paris: Unesco Press.

Jones, E. M., and R. D. Owen. 1987. Fluid preservation of specimens. Pages 51–63 in Genoways, H. H., C. Jones, and O. L. Rossolimo (editors). *Mammal Collection Management*. Lubbock: Texas Tech University Press.

Jones, M. L. 2001. To fix, to harden, to preserve—Fixation: a brief history. *Journal of Histotechnology* 24(3):155–162.

Joram, P. R. 2005. A new sealing compound for museum jars. *The Anatomical Record* 148(4):631–632.

Jordan, D. S. 1905. *Fishes*. New York: Henry Holt and Company.

Jordan, D. S. 1907. *A Guide to the Study of Fishes*. Vol. 1. New York: Henry Holt and Company.

Jørgensen, P. M., J. E. Lawesson, and L. B. Holm-Nielsen. 1984. A guide to collecting passion flowers. *Annals of the Missouri Botanical Garden* 71(4):1172–1174.

Judah, E. L. 1918. On the technique of the preservation of war material. *Bulletin of the International Association of Medical Museums* 7:21–29.

Judah, E. L. 1922. Personal modifications in the techniques of the Kaiserling Methods of colour preservation. *Bulletin of the International Association of Medical Museums* 8:62–64.

Kaiserling, C. 1896. Über die conservirung von sammlungspräparaten mit erhaltung der natürlichen farben. *Berliner Klinische Wochenschrift* 33(35):775–777.

Kaiserling, C. 1897. Weitere mittheilungen über die herstellung möglichst naturgefreuer sammlungspräparate. *Virchows Archiv für Pathologische Anatomie und Physiologie und für Klinische Medizin*, 147(3):389–417.

Kaiserling, C. 1900. Über die konservirung und aufstellung pathologisch-anatomischer präparate für schau- und lehrsammlungen. *Verhandlungen der Deutschen Pathologischen Gellschaft* 2:203–207.

Kaplan, B. B. 1993. *"Divulging of Useful Truths in Physick." The Medical Agenda of Robert Boyle*. Baltimore: Johns Hopkins University Press.

Karns, D. R. 1986. *Field Herpetology. Methods for the Study of Amphibians and Reptiles in Minnesota*. James Ford Bell Museum of Natural History Occasional Paper 18:1–88.

Keel, W. G., W. Moser, J. Giaccai, A. Ormos, J. Tanner, and L. A. Weigt. 2011. Alcohol recycling at the Smithsonian Institution, National Museum of Natural History (NMNH). *Collection Forum* 25(1–2):10–21.

Kellicott, D. S. 1896. Formalin in the zoölogical and histological laboratory. *Transactions of the American Microscopical Society*, 17(1):331–335.

Kendrick. W. R. 1969. Preservation of fleshy fungi for taxonomy. *Mycologia* 61(2):392–395.

Keneally, T. 1996. *A River Town*. Mishawaka, IN: Plume Books.

Kenyon, F. C. 1895. Formol as a preserving fluid. *The American Naturalist* 29(337):82–91.

Kenyon, F. C. 1897. Formol or Formalin. *The American Naturalist* 31(365):466.

Kessler, R. 1989. Farberhaltende fixations- und konservierrungstechnik fur feuchtpräparate. *Der Präparator* 35(4):161–170.

Kiernan, J. A. 1990. *Histological and Histochemical Methods. Theory and Practice*. 2nd ed. Oxford: Pergamon Press.

Kilburn, K. H., B. C. Seidman, and R. Warshaw. 1985. Neurobehavioral and respiratory symptoms of formaldehyde and xylene exposure in histology technicians. *Archives of Environmental Health* 40(4):229–233.

Kimor, B. 1976. The fixation and preservation of marine Protozoa: some problems and general considerations. Pages 231–235 in Steedman, H. F. (editor). *Zooplankton Fixation and Preservation*. Monographs on Oceanographic Methodology 4. Paris: Unesco Press.

King, J. R., and S. D. Porter. 2004. Recommendations on the use of alcohols for preservation of ant specimens (Hymenoptera, Formicidae). *Insect Sociaux* 51(2):197–202.

Kishimani, C. H. 1989. Archival storage of disintegrating labels from fluid-preserved specimens. *Collection Forum* 5(1):1–4.

Kishinami, C. H. 1992. Storage system for deteriorating fluid specimen labels. Pages 259–260 in Rose, C. L., and A. R. de Torres (editors). *Storage of Natural History Collections: Ideas and Practical Solutions*. York, PA: Society for the Preservation of Natural History Collections.

Kittenberger, K. 1989. *Big Game Hunting and Collecting in East Africa, 1903–1926*. New York: St. Martin's Press.

Klanten, S. O., L. van Herwerden, and J. H. Choat. 2003. Acquiring reef fish DNA sequences from formalin-fixed museum specimens. *Bulletin of Marine Science* 73(3):771–776.

Klauber, L. M. 1935. Notes on herpetological field collecting. 3rd rev. *San Diego Society of Natural History Leaflet* no. 1, 10 pages.

Klauber, L. M. 1943. Tail-length differences in snakes, with notes on sexual dimorphism and the coefficient of divergence. *Bulletin of the Zoological Society of San Diego* 18:1–60.

Klauber, L. M. 1972. *Rattlesnakes. Their Habits, Life Histories, and Influence on Mankind*. Vol. 2. 2nd ed. Berkeley: University of California.

Knudsen, J. W. 1966. *Biological Techniques. Collecting, Preserving, and Illustrating Plants and Animals*. New York: Harper and Row.

Knudsen, J. W. 1972. *Collecting and Preserving Plants and Animals*. New York: Harper and Row.

Köhler, G. 2001. *Anfibios y Reptiles de Nicaragua*. Offenbach, Germany: Herpeton, Verlag Elke Köhler.

Kotrba, M., and K. Golbig. 2009. A new approach to stabilize the pH in fluid-preserved natural history collections. *Collection Forum* 23(1–2):18–22.

Kotrba, M., and K. Golbig. 2011. Effect of container glass quality on pH in natural history wet collections. *Collection Forum* 25(1):129–135.

Kristoffersen, J. B., and A. G. V. Salvanes. 1998. Effects of formaldehyde and ethanol preservation on body and otoliths of *Maurolicus muelleri* and *Benthosema galaciale*. *Sarsia* 83(2):95–102.

Krogmann, L., and J. Holstein. 2010. Preserving and specimen handling: Insects and other invertebrates. Pages 463–481 in Eymann, J., J. Degreef, C. Häuser, J. C. Monje, Y. Samyn, and D. VandenSpiegel (editors). *Manual on Field Recording Techniques and Protocols for All Taxa Biodiversity Inventories*. Belgium: Belgian National Focal Point to the Global Taxonomy Initiative, 653 pages.

Kruse, G. H., and E. L. Dalley. 1990. Length changes in capelin, *Mallotus villosus* (Müller), larvae due to preservation in formalin and anhydrous alcohol. *Journal of Fish Biology* 36:619–621.

Kuch, U., M. Pfenniger, and A. Bahl. 1999. Laundry detergent effectively preserves amphibian and reptile blood and tissue for DNA isolation. *Herpetological Review* 30(2):80–82.

Kuckhan, T. S. 1771. Four letters from Mr. T. S. Kuckhan, to the president and members of the Royal Society, on the preservation of dead birds. *Philosophical Transactions of the Royal Society* 60:302–320.

Kunz, G., and G. Wilcke. 1991. Untersuchungen zur konservierung biologischen materials unter dem aspect der formaldehyde-reduzierung. *Der Präparator* 38(2):63–74.

Laframboise, S., R. M. Rankin, and M. M. L. Steigerwald. 1993. Managing change: alcohol transfer at the Canadian Museum of Nature. Pages 28–33 in Snyder, A. M. (editor). *The 1992 American Society of Ichthyologists and Herpetologists Workshop on Collections Care and Management Issues.*

Lai, Y.-C. 1963. *Effects of several preservatives on proportional measurements of the fat-headed minnows,* Pimephales promelas. Master's Thesis, University of Kansas, 40 pages.

Lall, V. K., M. R. Suseela, and S. Devi. 1996. Fixation artifacts and inadequate preparation procedures leading to misinterpretation of structures in plant specimens. *Phytomorphology* 46(1):19–23.

Lamph, S. A., and W. E. Bennit. 2009. Evidence review: non-formalin fixatives. *Centre for Evidence-based Purchasing*, Policy and Innovation Directorate, NHS Purchasing and Supply Agency.

Lappalainen, A., and P. Kangas. 1975. Littoral benthos of the northern Baltic Sea II. Interrelationships of wet, dry and ash-free dry weights of macrofauna in the Tvärminne area. *Internationale Revue der Gesamten Hydrobiologie* 60(3):297–312.

Lavenberg, R. J., G. E. McGowen, and R. E. Woodsum. 1984. Preservation and curation. Pages 57–59 in Moser, H. G., W. J. Richards, D. M. Cohen, M. P. Fahay, A. W. Kendall, and S. L. Richardson (editors). *Ontogeny and Systematics of Fishes.* American Society of Ichthyologists and Herpetologists Special Publication No. 1.

Lazell, J. D., Jr. 1972. The anoles (Sauria, Iguanidae) of the Lesser Antilles. *Bulletin of the Museum of Comparative Zoology* 143(1):1–115.

Leach, L. C. 1995. The preparation of good (useful) herbarium specimens of succulent plants especially of the spiny euphorbias and of the stapeliads and aloes. *Asklepios* 64:24–26.

Leal-Klevezas, D. S., I. O. Martínez-Vázquez, B. Cuevas-Hernández, and J. P. Martínez-Soriano. 2000. Antifreeze solution improves DNA recovery by preserving the integrity of pathogen-infected blood and other tissues. *Clinical and Diagnostic Laboratory Immunology* 7(6):945–946.

Lebedeva, L. I., and Y. I. Kozlova. 1969. Size-weight characterization of live and formalin-fixed freshwater Cladocera. *Hydrobiolgia* 5:59–65.

Lee, J. C. 1982. Accuracy in anuran morphometrics: artifacts of preservation. *Systematic Zoology* 31(3):266–281.

Lefler, H. T. (editor). 1967. *A New Voyage to Carolina by John Lawson.* Chapel Hill: University of North Carolina Press.

Leicester, H. M. 1956. *The Historical Background of Chemistry.* Toronto: Wiley.

Lendrum, A. C. 1941. On the fixation of post-mortem material. *Journal of Pathology and Bacteriology* 52(1):132–137.

Leslie, J. K. 1983. Observations on length changes in preserved young freshwater fish. *Canadian Technical Report of Fisheries and Aquatic Sciences*, no. 1236.

Leslie, J. K., and J. E. Moore. 1986. Changes in length of fixed and preserved young freshwater fish. *Canadian Journal of Fisheries and Aquatic Sciences* 43:1079–1081.

Lessing, D. 2003. Foreword. Pages vii–xii in Bulgakov, M. *The Fatal Eggs*. London: Modern Voices, Hesperus Press.

Leuven, R. S. E., T. C. M. Brock, and H. A. M. van Druten. 1985. Effects of preservation on dry-and ash-free dry weight biomass of some common aquatic macro-invertebrates. *Hydrobiolgia* 127:151–159.

Levey, M. 1960. The earliest stages in the evolution of the still. *Isis* 51(1):31–34.

Levi, H. W. 1966. The care of alcoholic collections of small invertebrates. *Systematic Zoology* 15(3):183–188.

Levine, R. J., R. D. Dal Corso, P. B. Blunden, and M. C. Battigelli. 1984. The effects of occupational exposure on the respiratory health of West Virginia morticians. *Journal of Occupational Medicine* 26(2):91–98.

Levison, L. A. 1904. A case of fatal formaldehyde poisoning. *Journal of the American Medical Association* 42:1492.

Lichine, A. 1989. Distilled spirits. Pages 188–190 in *The Encyclopedia Americana*. Vol. 9. Danbury: Grolier Incorporated.

Ligname, A. H. 1964. Fixation method for amphibian eggs. *Journal of the Royal Microscopical Society* 83(4):481–482.

Lillie, R. D. 1954. *Histopathologic Technique and Practical Histochemistry*. New York: McGraw-Hill.

Lillie, R. D. 1965. *Histopathologic Technique and Practical Histochemistry*. 3rd ed. New York: McGraw-Hill.

Lincoln, R. J., and J. G. Sheals. 1979. *Invertebrate Animals: Collection and Preservation*. Cambridge: British Museum (Natural History) and Cambridge University Press.

Lobell, J. A. 2002. Digging Napoleon's dead. *Archaeology* 55(5).

Lo Bianco, S. 1899. The methods employed at the Naples Zoological Station for the preservation of marine animals. *Bulletin of the United States National Museum*, No. 39, Part M, pages 1–42.

Lockwood, S. J. 1973. Weight and length changes of O-group (*Pleuronectes platessa*) after preservation in 4% neutral formalin. *Journal du Consell International pour l'Exploration de la Mer* 35(1):100–101.

Lockwood, S. J., and C. de B. Daly. 1975. Further observations on the effects of preservation in 4% neutral formalin on the length and weight of O-group flatfish. *Journal du Consell International pour l'Exploration de la Mer* 36(2):170–175.

Lorge, L., D. Weston, and R. Phelps. 2005. *Castro Valley*. Mt. Pleasant, NC: Arcadia Publishing Company.

Lovecraft, H. P. 2012. *Great Tales of Terror*. New York: Fall River Press.

Lucas, F. A. 1896. Biological Society of Washington—258th Meeting, Saturday, March 21. *Science* 3(68):603–604.

Lundvall, W. 1905. Weiteres über demonstration embryonaler skellette. *Anatomischer Anzeiger* 27:521.

Mabee, P. M., E. Aldridge, E. Warren, and K. Helenurm. 1998. Effect of clearing and staining on fish length. *Copeia* 1998(2):346–353.

Macleod, K. 1975. Museum lighting. *Canadian Conservation Institute Technical Bulletin* 2:1–13.

Madin, L. P., C. M. Cetta, and V. L. McAlister. 1981. Elemental and biochemical composition of salps (Tunicata: Thailiacea). *Marine Biology* 63:217–226.

Maibach, H. 1983. Formaldehyde: effects on animal and human skin. Pages 166–174 in Gibson, J. E. (editor). *Formaldehyde Toxicity*. Washington, D.C.: Hemisphere Publishing.

Maisano, J. A. 2008. A protocol for clearing and double-staining squamate specimens. *Herpetological Review* 39(1):52–54.

Mallory, F. B., and J. H. Wright. 1918. *Pathological Technique. A Practical Manual for Workers in Pathological Histology and Bacteriology including Directions for the Performance of Autopsies and for Clinical Diagnosis by Laboratory Methods*. 7th ed., revised and enlarged. Philadelphia: W. B. Saunders.

Manchester, S., M. E. Collinson, and K. Goth. 1994. Fruits of the Juglandaceae from the Eocene of Messel, Germany, and implications for early Tertiary phytogeographic exchange between Europe and Western North America. *International Journal of Plant Sciences* 155(3):388–394.

Mandal, A. K. 1988. Preparation and storage of fluid-preserved specimens. Pages 303–313 in B. K. Tikader (editor). *Management of Mammal Collections in Tropical Environments*. Calcutta, India: Zoological Survey of India, Technical and General Press.

Mandrioli, M., F. Borsatti, and L. Mola. 2006. Factors affecting DNA preservation from museum-collected lepidopteran specimens. *Entomologica Experimentis et Applicata* 120:239–244.

Marhue, L. 1983. Techniques to restore dried-up invertebrate specimens. *Syllogeus* 44:175–177.

Markle, D. F. 1984. Phosphate buffered formalin for long term preservation of formalin fixed ichthyoplankton. *Copeia* 1984(2):525–528.

Martin, J. E. H. 1977. *The insects and arachnids of Canada. Part 1. Collecting, preparing, and preserving insects, mites and spiders*. Canada Department of Agriculture Research Branch (Publication 1643).

Martin, M., and M. Terrence. 1962. A solution for the preservation of color in fishes. *Turtox News* 49(4):122–123.

Martin, R. F. 1978. Clutch weight/total body weight ratios of lizards (Reptilia, Lacertilia, Iguanidae): Preservation induced variation. *Journal of Herpetology* 12:248–251.

Matson, F. R. 1949. Hygroscopicity of soda-lime-silica container glasses. *The Journal of the American Ceramic Society* 32(4):121–128.

Mauriès, P. 2002. *Cabinets of Curiosities*. London: Thames and Hudson Ltd.

Maynard, C. J. 1870. *The Naturalist's Guide in Collecting and Preserving Objects of Natural History with A Complete Catalogue of the Birds of Eastern Massachusetts*. Boston: Fields, Osgood.

McDiarmid, R. W. 1994. Preparing amphibians as scientific specimens. Pages 289–297 in Heyer, W. R., M. A. Donnelly, R. W. McDiarmid, L-A. C. Hayek, and M. S. Foster (editors). *Measuring and Monitoring Biological Diversity. Standard Methods for Amphibians*. Washington, D.C.: Smithsonian Institution Press.

McDiarmid, R. W., and R. Altig. 1999. Research: materials and techniques. Pages 7–23 in McDiarmid, R. W., and R. Altig (editors). *Tadpoles: The Biology of Anuran Larvae*. Chicago: University of Chicago Press.

McGinley, R. J. 1989. Entomological collections management—are we really managing? *Insect Collection News* 2:19–24.

McGinley, R. J. 1993. Where's the management in collections management? Planning for improved care, greater use, and growth of collections. Pp. 309–338 in Rose, C. L., S. L. Williams, and J. Gisbert (editors). *Congreso Mundial Sobre Preservación y Conservación de*

Colecciones de Historia Natural. Volume 3. Temas de Actualidad, Iniciativas y Direcciones Futuras sobre Preservación y Conservación de Colecciones de Historia Natural. Madrid: Dirección General de Bellas Artes y Archivos.

McGovern, P. E. 2009. *Uncorking the Past: The Quest for Wine, Beer, and Other Alcoholic Beverages.* Berkeley: University of California Press.

McMahon, T. E., and J. C. Tash. 1979. Effects of formalin (buffered and unbuffered) and hydrochloric acid on fish otoliths. *Copeia* 1979(1):155–156.

Menéndez Pidal, R. 1968. *Flor Nueva de Romances Viejos Que Recogió de la Tradición Antigua y Moderna.* Madrid: Espasa-Calpe, S. A.

Merritt, E. E. 2008. *National Standards and Best Practices for U.S. Museums.* Washington, D.C.: American Association of Museums.

Michaelis, F. B. 1981. Preservation of freshwater macrophytes using glycerol. *Aquatic Botany* 11:389.

Michel, R. H., P. E. McGovern, and V. R. Badler. 1992. Chemical evidence for ancient beer. *Nature* 360:24.

Middleton, N. 2007. *Make Your Own Herbarium Specimens.* University of Melbourne Herbarium, School of Botany.

Miéville, C. 2010. *Kraken. An Anatomy.* New York: Ballantine Books.

Milgrom, M. 2010. *Still Life. Adventures in Taxidermy.* Boston: Houghton Mifflin Harcourt.

Miller, G. S. 1899. Directions for preparing study specimens of small mammals. *Bulletin of the U.S. National Museum*, Part N, 39:1–10.

Mills, A. 1975. Measuring changes that occur during frozen storage of fish: A review. *Journal of Food Technology* 10:483–496.

Mills, E. L., K. Pittman, and B. Munroe. 1982. Effect of preservation on the weight of marine benthic invertebrates. *Canadian Journal of Fisheries and Aquatic Science* 39:221–224.

Modigh, M., and S. Castaldo. 2005. Effects of fixatives on ciliates as related to cell size. *Journal of Plankton Research* 27(8):845–849.

Modlin, C. E. (editor). 1992. *Certain Things Last. The Selected Short Stories of Sherwood Anderson.* New York: Four Walls Eight Windows, Village Station.

Moku, M., K. Mori, and Y. Watanabe. 2004. Shrinkage in the body length of myctophid fish (*Diaphus* slender-type spp.) larvae with various preservatives. *Copeia* 2004(3):647–651.

Möller, H. 1980. Population dynamics of *Aurelia aurita* medusae in Kiel Bight, Germany (FRG). *Marine Biology* 60:123–128.

Monick, J. A. 1968. *Alcohols: Their Chemistry, Properties and Manufacture.* New York: Reinhold Book Corporation.

Moore, H. E. 1950. A substitute for formaldehyde and alcohol in plant collecting. *Rhodosora* 52:123–124.

Moore, S. J. 1983. A clean and safe gravimetric method to differentiate spirit, formalin and other fluid preserving media. *Conservation News* 20:113.

Moore, S. J. 1994. What fluid is in this bottle? *Biology Curator's Group Newsletter* 6(4): 44–45.

Moore, S. J. 1997. Phenoxetol, friend or foe? (A personal history). *Natural Sciences Conservation Group Newsletter* 6:24–25.

Moore, S. J. 1999. Fluid preservation. Pages 92–132 in D. Carter and A. Walker (editors), *Care and Conservation of Natural History Collections.* London: Butterworth-Heinemann.

Moore, S. J. 2001. Transferring biological specimens from formalin to alcohol. *Natural Sciences Conservation Group Newsletter* number 17:43–45.

Moore, S. J. 2002a. Leaching and degradation of lipids in zoological fluid-preserved specimens. *Biology Curator* 22:44–46.

Moore, S. J. 2002b. A brief history of fluid preservation, with some basic facts about it, including labels and inks. *Natural Sciences Conservation Group Newsletter* number 18:21–24.

Moore, S. 2005a. Conservation of a foetal elephant's hind limb, Cole Museum of Zoology, Reading University. Natural Sciences Collections Association, *NatSCA News* 5:21–24.

Moore, S. 2005b. Problems with lipid and fluid-preserved specimens. Natural Science Collections Association *NatSCA News* 7:14–15.

Moore, S. 2006. Conserving transparencies. *NatSCA News* 8:12–14.

Moore, S. 2009a. The future of fluid preservation—fixation in the balance: the case for using formalin. *NatSCA News* 16:18–22.

Moore, S. 2009b. Adhesives for fluid-preserved specimens. *NatSCA News* 16:32–35.

Moore, S. 2010. Preservation of botanical specimens in fluid. *NatSCA News* 19:85–88.

Mori, S., and G. T. Prance. 1987. A guide to collecting Lecythidaceae. *Annals of the Missouri Botanical Garden* 74(2): 321–30.

Morkert, S. B., and R. A. Bergstedt. 1990. Shrinkage of sea lamprey larvae preserved in formalin. *North American Journal of Fisheries Management* 10:484–486.

Morris, R. J. 1972. The preservation of some oceanic animals for lipid analysis. *Journal of the Fisheries Research Board of Canada* 29:1303–1307.

Morton, J. E. 1950. Collecting and preserving zoological specimens. *Tuatara* 3(3):104–114.

Motoda, S., R. Marumo, and T. Tokioka. 1976. Fixation and preservation experiments on marine zooplankton in Japan: outline of experiments and results. Pages 325–332 in Steedman, H. F. (editor). *Zooplankton Fixation and Preservation*. Monographs on Oceanographic Methodology 4. Paris: Unesco Press.

Mukherjee, D. P., and A. S. Hoffman. 1971. Physical and mechanical properties of elaastin. Pages 219–241 in Elden, H. R. (editor). *Biophysical Properties of the Skin*. New York: Wiley.

Mulder, W. J. 1997. Choices in conservation and restoration of historic specimens, an ethical approach. *Bulletin of the European Association of Museums of the History of Medical Sciences* 23:30–38.

Müsch, I. Albertus Seba's collection of natural specimens and its pictorial inventory. Pages 6–24 in Seba, Albertus. *Cabinet of Natural Curiosities. The Complete Plates in Colour 1734–1765*. Köln: Taschen.

Mutlu, E. 1995. Effect of formaldehyde on the gelatinous zooplankton (*Pleurobrachia pileus, Aurelia aurita*) during preservation. *Turkish Journal of Zoology* 20:423–426.

Mutlu, E., and F. Bingel. 1999. Distribution and abundance of ctenophores, and their zooplankton food in the Black Sea. I. *Pleurobrachia pileus*. *Marine Biology* 135:589–601.

Myers, C. W. 2000. A history of herpetology at the American Museum of Natural History. *Bulletin of the American Museum of Natural History* 252:1–232.

Myers, G. S. 1956. Brief directions for preserving and shipping specimens of fishes, amphibians and reptiles. 2nd ed. *Natural History Museum of Stanford University Circular* 5:1–3.

Nagorsen, D. W., and R. L. Peterson. 1980. *Mammal Collector's Manual. A Guide for Collecting, Documenting, and Preparing Mammal Specimens for Scientific Research*. Life Sciences Miscellaneous Publications, Royal Ontario Museum.

Nakanishi, M., A. C. Wilson, R. A. Nolan, G. C. Gorman, and G. S. Bailey. 1969. Phenoxyethanol: Protein preservative for taxonomists. *Science* 163:681–683.

Necker, W. L. 1940. Short directions for preserving reptiles and amphibians. *Chicago Academy of Sciences Leaflet* 15:1–2.

Neill, W. T. 1950. *How to preserve reptiles and amphibians for scientific study*. Ross Allen Reptile Institute, Silver Springs, Florida, Special Publication 2:1–15.

Nelson, C. E. 1971. Effects of different preservation on dermal postularity of the microhylid frogs *Gastrophryne* and *Hypopachus*. *Herpetological Review* 3(5):93–94.

Nelson, D. W., and J. Sparks. 1999. Paraformaldehyde/Alconox problems. *Curation Newsletter* 12:1–2.

Neumann, D. 2010. Preservation of freshwater fishes in the field. Pages 587–631 in Eymann, J., J. Degreef, C. Häuser, J. C. Monje, Y. Samyn, and D. VandenSpiegel (editors). *Manual on Field Recording Techniques and Protocols for All Taxa Biodiversity Inventories*. Belgium: Belgian National Focal Point to the Global Taxonomy Initiative.

Newcombe, F. C. 1894. Proceedings of the Botanical Club, A. A. A. S., Brooklyn Meeting, August 15–22, 1894. *Bulletin of the Torrey Botanical Club*, vol. 21(9):406–411.

Nicola, L. 1771. An easy method of preserving subjects in spirits. *Transactions of the American Philosophical Society* 1:244–246.

Nieuwland, J. A., and A. D. Slavin. 1928. Preservation of monotropa and similar plants without discoloration. *Proceedings of the Indiana Academy of Science* 38:103–104.

NIOSH. 1976. *Criteria for a recommended standard . . . occupational exposure to formaldehyde*. Department of Health, Education and Welfare Publication No. 77–126.

NIOSH. 1991. Carcinogenicity of acetaldehyde and malonaldehyde, and mutagenicity of related low-molecular-weight aldehydes. *National Institute for Occupational Safety and Health (NIOSH) Current Intelligence Bulletin* 55.

Nishikawa, J., and M. Terazaki. 1996. Tissue shrinkage of two gelatinous zooplankton, *Thalia democratica* and *Dolioletta gegenbauri* (Tunicata: Thaliacea) in preservative. *Bulletin of the Plankton Society of Japan* 43(1):1–7.

Nitz, O. W. 1989. Alcohol. Pages 513–516 in *The Encyclopedia Americana*, Vol. 1. Danbury, CT: Grolier Incorporated.

Noble, G. K. 1917. The systematic status of some batrachians from South America. *Bulletin of the American Museum of Natural History* 37:793–814.

Notton, D. G. 2010. Maintaining concentration: A new practical method for profiling and topping up alcohol-preserved collections. *Collection Forum* 24(1–2):1–27.

Nowak, M., and W. Popek. 2012. Morphological alterations in a small-sized cyprinid fish, Leucaspius delineates, caused by preservation with different ethanol solutions. *Pices Hungarici* 6:101–105.

Nuovo, G. J., and R. M. Richart. 1989. Buffered formalin is the superior fixative for the detection of HPV DNA by *In Situ* hybridization analysis. *American Journal of Pathology* 134(4):807–842.

Nuovo, G. J., and S. J. Silverstein. 1988. Comparison of formalin, buffered formalin, and Bouin's fixation on the detection of human papillomavirus deoxyribonucleic acid from genital lesions. *Laboratory Investigation* 59(5):720–724.

Oldroyd, H. 1958. *Collecting, Preserving and Studying Insects*. New York: MacMillan.

Omori, M. 1978. Some factors affecting on dry weight, organic weight and concentrations of carbon and nitrogen in freshly prepared and in preserved zooplankton. *Internationale Revue der Gesamten Hydrobiologie* 63:261–269.

Omori, M., and A. Fleminger. 1976. Laboratory methods for processing crustacean zooplankton. Pages 281–286 in Steedman, H. F. (editor). *Zooplankton Fixation and Preservation*. Monographs on Oceanographic Methodology 4. Paris: Unesco Press.

Oorthuyzen, W., and C. Sadée. 1982. *Voorkomen en groei van* Pleurobrachia pileus *en* Aurelia aurita *in de westelijke Waddenzee en hun mogelijke rol als predator van platvislarven*. Texel: Netherlands Institute of Sea Research.

Ouwens, P. A. 1907. On a new method of preserving snakes and other reptiles. *Bulletin du Département de l'Agriculture aux Indes-Néerlandaises* 8:1–3.

Owen, G., and H. F. Steedman. 1956. Preservation of animal tissues, with a note on staining solutions. *Quarterly Journal of Microscopical Science* 97(3):319–321.

Owens, J. S., and E. C. Emanuel. 1942. Effect of storage conditions on weathering of commercial glass containers. *Journal of the American Ceramic Society* 25(13):359–371.

Owens, S. 2007. Maria Sybilla Merian. Pages 138–175 in Attenborough, D. *Amazing Rare Things: The Art of Natural History in the Age of Discovery*. London: Royal Collection Enterprises.

Pabst, R. 1987. Exposure to formaldehyde in anatomy: An occupational health hazard? *Anatomical Record* 219:109–112.

Palmer, L. 1996. Assessing preservation fluids in a wet collection: How important is container style? *Natural History Conservation* 10:10–12.

Parker, G. H., and R. Floyd. 1896. The preservation of mammalian brains by means of formol and alcohol. *Anatomischer Anzeiger* 11(5):156–158.

Parker, R. P. 1963. Effects of formalin on length and weight of fishes. *Journal of the Fisheries Research Board of Canada* 20:1441–1455.

Parsons, U. 1831. *Directions for Making Anatomical Preparations, Formed on the Basis of Pole, Marjolin and Breschet, and Including the New Method of Mr. Swan*. Philadelphia: Carey and Lea.

Partington, J. R. 1989. *A Short History of Chemistry*. London: Macmillan. [Dover reprint of the 1957 edition].

Patten, B. M., and R. Philpott. 1921. The shrinkage of embryos in the process preparatory to sectioning. *Anatomical Record* 20:393–413.

Paterniti, M. 2000. *Driving Mr. Albert: A Trip across American with Einstein's Brain*. New York: Dial Books.

Paulet, T. G., and H. Kaiser. 2004. Formalin-induced shrinkage in preserved goldfish (*Carassius auratus* L.) larvae. *Journal of Applied Ichthyology* 20:551–552.

Payne, L. M., L. G. Wilson, and H. Hartley. 1960. William Croone, F. R. S. (1633–1684). *Notes and Records of the Royal Society of London* 15:211–219.

Pearcy, W. G. 1962. Ecology of an estuarine population of winter flounder, *Pseudopleuronectes americanus* (Walbaum). III. Distribution, abundance, growth, and production of juveniles: survival of larvae and juveniles. *Bulletin of the Bingham Oceanographic Collection* 18(1):39–64.

Pearse, A. G. E. 1980. *Histochemistry: Theoretical and Applied: Preparative and Optical Technology*. Vol. 1. Edinburg: Churchill Livingstone.

Peele, L. 2008. Phantom pain. *Granta* 102:131–151.

Penrith, M. J. 1971. Notes on new products of interest to curators of wet biological collections. *Southern African Museums Association Bulletin* 9:499–507.

Penzig, G. 1894. La formalina come liquido conservatore dei preparati vegetali. *Malpighia* 8:331–336.

Perera, F., and C. Petito. 1982. Formaldehyde: a question of cancer policy? *Science* 216(4552):1285–1291.

Perry, R., and B. A. Thomas. 1992. Herbarium practice. Pages 448–454 in Thompson, J. M. A. (editor). *Manual of Curatorship. A Guide to Museum Practice*. Oxford: Butterworth-Heinemann.

Persing, D. H., S. R. Telford, P. N. Rys, D. E. Dodge, T. J. White, S. E. Malawista, and A. Spielman. 1990. Detection of *Borrelia burgdorferi* DNA in museum specimens of *Ixodes dammini* ticks. *Science* 249(4975):1420–1423.

Petersen, K. W. 1976. Fixation and preservation of planktonic coelenterates. Pages 268–269 in Steedman, H. F. (editor). *Zooplankton Fixation and Preservation*. Monographs on Oceanographic Methodology 4. Paris: Unesco Press.

Pettingill, O. S. 1970. *Ornithology in Laboratory and Field*. 4th ed. Minneapolis, MN: Burgess Publishing.

Pettitt, C. 1976. Label materials for wet-preserved biological specimens. *Museums Journal* 75(4):175–176.

Phillips, C. J. 1984. Field fixation and storage of fluid preserved specimens for museum mammal collections. Pages 283–301 in B. K. Tikader (editor). *Management of Mammal Collections in Tropical Environments*. Calcutta, India: Zoological Survey of India, Technical and General Press.

Pickering, J. 1997. A survey of ethanol concentrations in the collections at the Oxford University Museum of Natural History. *The Biology Curator* 10:1–5.

Piechocki, R., and J. Händel. 2007. *Makroskopische Präparations-technik. Wirbellose. Leitfaden für das Sammeln, Präparieren und Konservieren*. Stuttgart, Germany: E. Schweizerbart'sche Verlagsbuchhandlung.

Pilkington, R. 1959. *Robert Boyle: Father of Chemistry*. London: John Murray.

Pinzl, A. *forthcoming*-a. Annotated Bibliography: References on Herbarium Curation, Methodology and Management.

Pinzl, A. *forthcoming*-b. Index to References on Herbarium Curation, Methodology and Management.

Pisani, G. R. 1973. *A Guide to Preservation Techniques for Amphibians and Reptiles*. Society for the Study of Amphibians and Reptiles Herpetological Circular 1:1–22.

Pitkin, B. 1995. Labelling specimens in the life science departments at the Natural History Museum, London, using computers. *The Biology Curator* 1995(4):24–26.

Plath, S. 1998. *The Bell Jar*. New York: HarperCollins.

Pole, T. 1790. *The Anatomical Instructor; or, An Illustration of the Modern and Most Approved Methods of Preparing and Preserving the Different Parts of the Human Body, and of Quadrupeds, by Injection, Corrosion, Maceration, Distention, Articulation, Modelling, Etc*. London: Couchman and Fry.

Pole, T. 1813. *The Anatomical Instructor; or, An Illustration of the Modern and Most Approved Methods of Preparing and Preserving the Different Parts of the Human Body, and of Quadrupeds, by Injection, Corrosion, Maceration, Distention, Articulation, Modelling, Etc. A New Edition with Additional Notes*. London: J. Callow, Medical Bookseller.

Pope, C. H. 1940. *China's Animal Frontier*. New York: Viking Press.

Pope, P. H. 1928. Isopropyl alcohol as a preservative. *Science* 68(1768):487–488.

Pope, P. H. 1931. Formalin poisoning. *Science* 73(1897):495–496.

Post, G. 1967. *Methods of Sampling and Preserving Field Specimens for Laboratory Examination or Laboratory Analysis*. Boulder, CO: Pruett Press.

Potthoff, T. 1984. Clearing and staining techniques. Pages 35–37 in Moser, H. G, W. J. Richards, D. M. Cohen, M. P. Fahay, A. W. Kendall, and S. L. Richardson (editors). *Ontogeny and Systematics of Fishes*. American Society of Ichthyologists and Herpetologists, No. 1.

Presnell, J. K., and M. P. Schreibman. 1997. *Humason's Animal Tissue Techniques*. Baltimore: Johns Hopkins University Press.

Price, J. C., and G. R. Fitzgerald. 1996. Categories of specimens: a collections management tool. *Collection Forum* 12(1):8–13.

Prum, R. O., R. L. Morrison, and G. R. Ten Eyck. 1994. Structural color production by constructive reflection from ordered collagen arrays in a bird (*Philepitta castanea*: Eurylaimidae). *Journal of Morphology* 222:61–72.

Providence Franklin Society. 1829. *General Directions, for Collecting and Preserving Articles in the Various Departments of Natural History: Respectfully Submitted by the Franklin Society of Providence, R. I. to the Attention of Travellers, Seafaring Men, and All Lovers of Nature and of Nature's Handy Works, Who Are Willing to Lend Their Aid in Promoting the Cause of Science, and Advancing the Progress of the Arts*. J. C. Parmenter, Printer, 8 pages.

Pruess, P. W., R. L. Dailey, and E. S. Lehman. 1985. Exposure to formaldehyde. Pages 247–259 in Turoski, V. (editor). *Formaldehyde. Analytical Chemistry and Toxicity*. Washington, D.C.: American Chemical Society.

Puchtler, H., and S. N. Meloan. 1985. On the chemistry of formaldehyde fixation and its effects on immunohistochemical reactions. *Histochemistry* 82:201–204.

Quester, R., and R. Schröder. 1997. The shrinkage of the human brain stem during formalin fixation and embedding in paraffin. *Journal of Neuroscience Methods* 75:81–89.

Quay, W. B. 1974. Bird and mammal specimens in fluid—objectives and method. *Curator* 17(2):91–104.

Radtke, R. L. 1989. Larval fish age, growth, and body shrinkage: Information available from otoliths. *Canadian Journal of Fisheries and Aquatic Science* 46:1884–1894.

Radtke, R. L., and W. G. Waiwood. 1980. Otolith formation and body shrinkage due to fixation in larval cod (*Gadus morhua*). *Canadian Technical Reports Fisheries and Aquatic Science* 929:1–10.

Ransome, H. M. 1937. *The Sacred Bee in Ancient Times and Folklore*. London: George Allen and Unwin.

Réaumur, R. A. F. 1748. Diverse means for preserving from corruption dead birds, intended to be sent to remote countries, so that they may arrive there in good condition. Some of the same means may be employed for preserving quadrupeds, reptiles, fishes and insects. *Philosophical Transactions of the Royal Society of London* 45:304–320.

Redenbaugh, W. A. 1895. Preservation of some marine animals. *The American Naturalist* 29(340):399–401.

Reed, R. N. 2001. Effects of museum preservation techniques on length and mass of snakes. *Amphibia-Reptilia* 22(4):488–491.

Reese, N. E., W. M. Boyce, I. A. Gardner, and D. M. Nelson. 1996. Fixation affects morphometric characters of *Psoroptes cuniculi* mites (Acari: Psoroptidae). *Journal of Medical Entomology* 33(5):835–838.

Reid, D. G. 2000. Preservation and curation of marine molluscan specimens. *The Biology Curator* 18:11–17.

Reid, G. 1994. The preparation and preservation of collections. Pages 28–69 in Stansfield, G., J. Mathias, and G. Reid (editors). *Manual of Natural History Curatorship*. London: HMSO.

Redenbaugh, W. A. 1895. Preservation of some marine animals. *The American Naturalist* 29(340):399–401.

Riley, C. V. 1892. Directions for colleting and preserving insects. *Bulletin of the United States National Museum*, No. 39, Part F, pages 1–147.

Robins, C. R., and R. H. Backus. 1952. Tri-sodium phosphate as a storage medium for otoliths. *Copeia* 1952(3):207.

Rogers, B. B., L. C. Alpert, E. A. S. Hine, and G. J. Buffone. 1990. Analysis of DNA in fresh and fixed tissue by the polymerase chain reaction. *American Journal of Pathology* 136(3):541–548.

Roob, A. 2001. *Alchemy and Mysticism. The Hermetic Museum*. Köln: Taschen.

Roper, C. F. E., and M. J. Sweeney. Techniques for fixation, preservation,and curation of cephalopods. *Memoirs of the National Museum Victoria* 44:29–48.

Rosenthal, H., D. Kuhlmann, and O. Fukuhara. 1978. Shrinkage of newly hatched larvae of the Red Sea bream (Chrysophrys major) preserved in formalin. *Archiv für Fischereiwissenschaft* 29(1/2):59–63.

Rosenthal, H., and H. von Westernhagen. 1976. Effect of formalin fixation on length, weight and condition factor of three siganid species. *Archive für Fischereiwissenschaft* 27(1):37–43.

Ross, G. C. 1961. Labels for biological material. *Museums Journal* 61(3):177–179.

Ross, K. F. A. 1953. Cell shrinkage caused by fixatives and paraffin-wax embedding in ordinary cytological preparations. *Quarterly Journal of Microscopical Science* 94:125–139.

Ruiz, C., and G. Haris. 2006. Damien Hirst in talks to replace rotting shark. *The Art Newspaper*, accessed June 28, 2006.

Russ, J. L., C. J. Kolman, S. L. Jewett, and N. Tuross. 1997. *The whites of their eyes: Criteria for selection of fluid stored fish for molecular studies.* Poster abstract, annual meeting of the Society for the Preservation of Natural History Collections, University of Wisconsin-Madison, July 8–13, 1997.

Ruthven, A. G. 1912. Directions for collecting and preserving specimens of reptiles and amphibians for museum purposes. *University of Michigan Museum of Natural History 14th Report, Michigan Academy of Sciences* 165–177.

Sacks, O. 2001. *Uncle Tungsten. Memories of a Chemical Boyhood.* New York: Vintage Books.

Sagan, C. 1979. *Broca's Brain. Reflections on the Romance of Science.* New York: Random House.

Sagnes, P. 1997. Potential artefacts in morphometric analysis of fish: effects of formalin preservation on O+ grayling. *Journal of Fish Biology* 50:910–914.

Sameoto, D. D. 1971. The distribution of herring (*Clupea harengus* L.) larvae along the southern coast of Nova Scotia with some observations on the ecology of herring larvae and the biomass of macrozooplankton on the Scotian shelf. *Fisheries Research Board of Canada Technical Report No. 252.*

Sayers, D. L. 1972. *Lord Peter. A Collection of All the Lord Peter Wimsey Stories.* New York: Perennial Library, Harper and Row.

Schaefer, H., A. Zesch, and G. Stuttgen. 1982. *Skin Permeability.* Berlin: Springer-Verlag.

Schmid, R. 1981. Polyethylene bags for permanently sealing jars of pickled specimens. *Taxon* 30(1):369–370.

Schmidt, K. P. 1932. Museum tags of chemical proof paper. *Science* 75(1931):23–24.

Schnack, D., and H. Rosenthal. 1978. Shrinkage of Pacific herring larvae due to formalin fixation and preservation. *Berichte der Deutschen Wissenschaftlichen Kommission für Meeresforschung* 26(3/4):222–226.

Schram, L. A., G. R. Ploskey, and E. H. Schmitz. 1981. Dry weight loss in Ceriodaphnia lacustris following formalin preservation. *Transactions of the American Microscopical Society* 100(3):326–329.

Schultes, R. E. 1947. The use of formaldehyde in plant collecting. *Rhodora* 49:54–60.

Schultz, A. H. 1924. Preparation and preservation of anatomical and embryological material in the field. *Journal of Mammalogy* 5(1):16–24.

Schultze, O. 1897. Über herstellung und conserverung durchsichtiger embryonen zum stadium der skellettbildung. *Anatomischer Anzeiger* 13:3–5,

Scott, N. J., and A. L. Aquino-Shuster. 1989. The effects of freezing on formalin preservation of specimens of frogs and snakes. *Collection Forum* 5(2):41–46.

Scrocchi, G., and S. Kretzschmar. 1996. *Guía de métodos de captura y preparación de anfibios y reptiles para estudios científicos y manejo de colecciones herpetológicas.* Fundación Miguel Lillo Miscelanea 102:1–44.

Scully, F. J. 1937. Preservation of plant material in natural colors. *Rhodora* 39:16–19.

Sendall, K., and G. W. Hughes. 1997. Correcting alcohol concentrations. *Society for the Preservation of Natural History Collections Newsletter* 11(1):6–7.

Serns, S. L. 1982. Influence of formalin on size of walleye eggs. *Progressive Fish-Culturist* 44(3):149.

Seutin, G., B. N. White, and P. T. Boag. 1991. Preservation of avian blood and tissue samples for DNA analysis. *Canadian Journal of Zoology* 69:82–90.

Seymour, R. B., and G. B. Kauffman. 1992. Formaldehyde: A simple compound with many uses. *Journal of Chemical Education* 69(6):457–458.

Sheim, M. 1951. Color restoration and preservation of gross museum specimens. *Bulletin of the International Association of Medical Museums* 32:117–123.

Shephard, S. 2000. *Pickled, Potted, and Canned. How the Art and Science of Food Preserving Changed the World.* New York: Simon and Schuster.

Shetter, D. S. 1936. Shrinkage of trout at death and on preservation. *Copeia* 1936(1):60–61.

Shipman, F. M., and A. T. Thomas. 1991. Beverage production. *The New Encyclopaedia Britannica.* Chicago: Encyclopaedia Britannica, Inc., 14:734–754.

Sigler, W. F. 1949. Life history of the white bass, *Lepibema chrysops* (Rafinesque), of Spirit Lake, Iowa. *Iowa Agricultural Experimental Station Research Bulletin* 366:203–244.

Silverster, C. F. 1922. Retrospect: Preserving fluids for gross pathological specimens with review of the literature. *Bulletin of the International Association of Medical Museums* 8:53–61.

Simmons, J. E. 1993. Natural history collections management in North America. *Journal of Biological Curation* 1(3/4):1–17.

Simmons, J. E. 1999a. Storage concerns for fluid-preserved collections. *Conserve O Gram* 11(3):1–4.

Simmons, J. E. 1999b. Storage containers and labels for fluid-preserved collections. *Conserve O Gram* 11(4):1–4.

Simmons, J. E. 2002. *Herpetological Collecting and Collections Management.* Rev. ed. Society for the Study of Amphibians and Reptiles Herpetological Circular 3. New York: Taylor and Francis.

Simmons, J. E. 2010. History of Museums. Pages 2096–2106 in Bates, M. J., and M. N. Maack (editors). *Encyclopedia of Library and Information Sciences*, 3rd ed. New York: Taylor and Francis.

Simmons, J. E. 2010. Collections management. Pages 515–549 in Hawks, C. A., M. McCann, K. A. Makos, L. Goldberg, D. Hinkamp, D. C. Ertel, and P. Silence (editors). *Health and Safety for Museum Professionals.* New York: Society for the Preservation of Natural History Collections.

Simmons, J. E. 2013. Application of preventive conservation to solve the coming crisis in collections management. *Collection Forum* 27(1–2):89–101.

Simmons, J. E. 2014. Collection care and management. In McCarthy, C. (editor). *Museum Practice: Critical Debates in the Museum Sector.* Vol. 4. New York: International Handbook of Museum Studies, Wiley-Blackwell.

Simmons, J. E. *forthcoming*. Conservation and re-housing of human body slice exhibit.

Simmons, J. E., M. Arthur, D. S. Mcleod, and T. Quinn. *forthcoming*. Evaluation of a rehydration method for fluid preserved specimens.

Simmons, J. E., A. C. Campbell, and E. Greenbaum. 2007. Presence of elemental mercury in fluid preserved specimens. *Collection Forum* 22(1–2):32–36.

Simmons, J. E., and J. Snider. 2012. Observation and distillation—preservation, depiction, and the perception of nature. *Bibliographica Herpetologica* 9(1–2):115–134.

Singer, C. J. (editor). 1921. *Studies in the History and Method of Science.* Oxford: Oxford at the Clarendon Press.

Singer, R. A. 2014. Are dehydrated specimens a lost cause? A case-study to reclaim dehydrated fluid preserved specimens. *Collection Forum* 28.

Slevin, J. R. 1927. The making of a scientific collection of reptiles and amphibians. *Proceedings of the California Academy of Sciences* 16(9):231–259.

Smith, A. G. 1951. *Making a collection of amphibians and reptiles*. Contributions of the Department of Biological Sciences, Loyola University number 6, Leaflet No. 1:1–4.

Smith, D. G. 1995. Preservation of color in larval fishes. *Curation Newsletter* 11:5–6.

Smith, H. M. 1955. Effect of preservatives upon pattern in a Mexican garter snake. *Herpetologica* 11(3):165–168.

Smith, H. M. 1975. A case for field notes from life. *Bulletin of the Maryland Herpetological Society* 11(1):8–13.

Smith, J. L. B. 1947. A neutral solution of formaldehyde for biological purposes. *Transactions of the Royal Society of South Africa* 31(3):279–282.

Snyder, A. M. 1999. Mechanical and manual production of labels for collections stored in fluids: A few examples of papers, ink and production protocols. *Curation Newsletter* 12:4–10.

Soderstrom, T. R., and S. M. Young. 1983. A guide to collecting bamboos. *Annals of the Missouri Botanical Garden* 70(1):128–136.

Spence, T. F. 1967. *Teaching and Display Techniques in Anatomy and Zoology*. Oxford: Pergamon Press.

Spencer Jones, M. E. 1996. Procedure for curating old glass museum jars. *The Biology Curator* 7:3–5.

Springer, V. G., and G. D. Johnson. 2000. Use and advantages of ethanol solution of alizarin red S dye for staining bone in fishes. *Copeia* 2000(1):300–301.

Stackhouse, J. 1800. Observations on preserving specimens of plants. *Transactions of the Linnean Society*, 5:2123.

Stahl, J. 1893. Formalin. *Pharmaceutische Zeitung* 38(22):173-–176.

Stanford, J. A. 1973. A centrifuge method for determining live weights of aquatic insect larvae, with a note on weight loss in preservative. *Ecology* 54:449-–451.

Stearns, R. P. 1953. James Petiver: promoter of natural science, c. 1663-–1718. *Proceedings of the American Antiquarian Society* 62:243-–365.

Steedman, H. F. 1976a. *Zooplankton Fixation and Preservation*. Monographs on Oceanographic Methodology 4, The UNESCO Press, Paris, 350 pages.

Steedman, H. F. 1976b. General and applied data on formaldehyde fixation and preservation of marine zooplankton. Pages 103–154 in Steedman, H. F. (editor). *Zooplankton Fixation and Preservation*. Monographs on Oceanographic Methodology 4. Paris: Unesco Press.

Steedman, H. F. 1976c. Osmotic pressurers in fixation and preservation. Pages 186–195 in Steedman, H. F. (editor). *Zooplankton Fixation and Preservation*. Monographs on Oceanographic Methodology 4. Paris: Unesco Press.

Steedman, H. F. 1976d. Cell products: calcium salts. Pages 209–221 in Steedman, H. F. (editor). *Zooplankton Fixation and Preservation*. Monographs on Oceanographic Methodology 4. Paris: Unesco Press.

Steigerwald, M., and S. Laframboise. 1996. Tape application: a jar sealing method for reducing ethanol evaporation in fluid-preserved collections. *Collection Forum* 12(2):45–54.

Steinbeck, J. 1973. *Cannery Row*. New York: Viking Press.

Stejneger, L. 1891. Directions for collecting reptiles and batrachians. *Bulletin of the United States National Museum* 39:1–13.

Stejneger, L. 1911. Directions for preserving small herpetological specimens in formalin. *Supplementary to Part E of the Bulletin of the United States National Museum*, number 39:1.

Stejneger, L. 1917. Cuban amphibians and reptiles collected from the United States National Museum from 1899 to 1902. *Proceedings of the United States National Museum* 53(2205):259–291.

Stemen, B. L. 2010. Fire safety program management. Pages 77–103 in Hawks, C. A., M. McCann, K. A. Makos, L. Goldberg, D. Hinkamp, D. C. Ertel, and P. Silence (editors). *Health and Safety for Museum Professionals*. New York: Society for the Preservation of Natural History Collections.

Stephenson, A. B., and J. L. Riley. 1995. Fixation and preservation of museum marine collections using formaldehyde/glutaraldehyde mixes. *Collection Forum* 11(2):58–68.

Stevenson, J. C. 1962. Distribution and survival of herring larvae (*Clupea pallasii* Valenciennes) in British Columbia waters. *Journal of the Fisheries Research Board of Canada* 19(5):735–810.

Stickland, N. C. 1975. A detailed analysis of the effects of various fixatives on animal tissue with particular reference to muscle tissue. *Stain Technology* 50:255–264.

Stobo, W. T. 1972. Effects of formalin on the length and weight of yellow perch. *Transactions of the American Fisheries Society* 101(2):362–364.

Stockdale, G. F., and F. V. Tooey. 1950. Effect of humid conditions on glass surfaces studied by photographic and transmission techniques. *Journal of the American Ceramic Society* 33(1):11–16.

Stoddard, R. W. 1989. Fixatives and preservatives: their effects on tissue. Pages 1–25 in Horie, C. V. (editor). *Conservation of Natural History Specimens: Spirit Collections*. Manchester: Manchester Museum and Department of Environmental Biology, University of Manchester.

Stoecker, D. K., D. J. Gifford, and M. Putt. 1994. Preservation of marine planktonic ciliates: losses and cell shrinkage during fixation. *Marine Ecology Progress Series* 110:293–299.

Stone, B. C. 1983. A guide to collecting Pandanaceae (Pandanus, Freycinetia, and Sararanga). *Annals of the Missouri Botanical Garden* 70(1):137–145.

Stone, G. E. 1899. Formalin as a preservative for botanical specimens. *Journal of Applied Microscopy* 11(10):537–540.

Strayer, J. L. 1965. Preserving snakes in glass tubing. *Turtox News* 43(7):166–167.

Stroud, B. B. 1897a. Formal, (Formaldehyde 40 per cent. solution). *The American Naturalist* 31(361):92–96.

Stroud, B. B. 1897b. The name of Formal. *The American Naturalist* 31(365):465–466.

Stuart, B. L., K. A. Dugan, M. W. Allard, and M. Kearney. 2006. Extraction of nuclear DNA from bone of skeletonised and fluid-preserved museum specimens. *Systematics and Biodiversity* 4(2):133–136.

Stuart, J. N. 1995. Observations on formalin-induced darkening of herpetological specimens. *Collection Forum* 11(2):39–45.

Stutesman, D. 2005. *Snake*. London: Reaktion Books.

Suzumoto, A. Y. 1992. New materials for sealing old crocks. *Collection Forum* 8(2):68–72.

Swingle, C. F. 1930. Oxyquinoline sulphate as a preservative for plant tissues. *The Botanical Gazette* 90:333–334.

Talty, S. 2009. *The Illustrious Dead. The Terrifying Story of How Typhus Killed Napoleon's Greatest Army*. New York: Crown Publishers.

Takizawa, K. Y. Fujita, Y. Ogushi, and S. Matsuno. 1994. Relative change in body length and weight in several fish larvae due to formalin fixation and preservation. *Fisheries Science* 60(4):355–359.

Tarkhan, A. A. 1931. The effects of fixatives and other reagents on cell-size and tissue-bulk. *Journal of the Royal Microscopical Society* 51, Ser. II: 387–399.

Tate, R. 1883. Land and freshwater shells, etc. Pages 102–116 in Taylor, J. E. (editor). *Notes on Collecting and Preserving Natural History Objects*. London: W. H. Allen.

Taub, A. M. 1962. The use of paraformaldehyde as a field preservative. *Copeia* 1962(1):209–210.

Taylor, F. J. R. 1976. Flagellates. Pages 259–264 in Steedman, H. F. (editor). *Zooplankton Fixation and Preservation*. Monographs on Oceanographic Methodology 4. Paris: Unesco Press.

Taylor, W. R. 1967a. An enzyme method of clearing and staining small vertebrates. *Proceedings of the United States National Museum* 122(3596):1–17.

Taylor, W. R. 1967b. Outline of a method of clearing and staining tissues with pancreatic enzymes and staining bones of small vertebrates. *Turtox News* 45(12):308–309.

Taylor, W. R. 1977. Observations on specimen fixation. *Proceedings of the Biological Society of Washington* 90(4):753–763.

Taylor, W. R. 1981. Preservative practices: water in tissues, specimen volume, and alcohol concentration. *Curation Newsletter* 2:1–3.

Taylor, W. R., and G. C. Van Dyke. 1985. Revised procedures for staining and clearing small fishes and other vertebrates for bone and cartilage studies. *Cybium* 9(2):107–119.

Tredwell, W. 2006. Botanical spirit collections. *NatSCA News* 9:2–3.

Terentjev, P. V., and S. A. Chernov. 1949. *Key to Amphibians and Reptiles*. (1965 translation from Russian by L. Kochva.) Jerusalem: Israel Program for Scientific Translations.

Tertz, A. 1982. *The Trial Begins* and *On Socialist Realism*. Berkeley: University of California Press.

Thadeusz, F. 2009. Geschichte. Marsch der maladen. *Der Spiegel* 31:119.

Thede, C. E. 1996. Survey of a fluid-preserved collection. *Natural History Conservation* 10:12–17.

Theilacker, G. H. 1980. Changes in body measurements of larval northern anchovy, *Engraulis mordax*, and other fishes due to handling and preservation. *Fishery Bulletin* 78(3):685–692.

Thibault-Botha, D., and T. Bowen. 2004. Impact of Formalin Preservation on Pleurobrachia bachei (Ctenophora). *Journal of Experimental Marine Biology and Ecology* 303:11–17.

Thijsen, M. 1990. *Grotesque. Natural Historical & Formaldehyde Photography*. Amsterdam: Fragment Uitgeverij.

Thomas, O. 1892. On the probable identity of certain specimens, formerly in the Lidth de Jeude collection, and now in the British Museum, with those figured by Albert Seba in his "Thesaurus" of 1734. *Proceedings of the Zoological Society* 22:309–317.

Tomlinson, P. B. 1965. Collecting botanical material in fluid preservatives with special reference to the tropics. Pages 117–119 in Fosberg, F. R., and M-H. Sachet. *Manual for Tropical Herbaria*. Vol. 39, Appendix IE. Utrecht, the Netherlands: International Bureau for Plant Taxonomy and Nomenclature, Regnum Vegetabile.

Thompson, J. R., M. H. Thompson, and S. Drummond. 1966. A method for restoring dried crustacean specimens to taxonomically usable condition. *Crustaceana* 10(1):189.

Tokioka, T. 1976. Fixation, preservation and taxonomic aspects of Thaliacea. Pages 305–308 in Steedman, H. F. (editor). *Zooplankton Fixation and Preservation*. Monographs on Oceanographic Methodology 4. Paris: Unesco Press.

Treadwell, E. 2006. Botanical spirit collections. *NatSCA News* 9:2–3.

Treasurer, J. W. 1992. Length and weight changes in O+ perch, *Perca fluviatilis* L., following fixation in formaldehyde. *Journal of Fish Biology* 41:1033–1036.

Trillat, A. 1892. Sur les propriétés antiseptiques de la formaldéhyde. *Comptes Reundus Hebdomadaires des Séances de l'Académie des Science* 114(May):1278–1281.

Tucker, J. W., and A. J. Chester. 1984. Effects of salinity, formalin concentration and buffer on quality of preservation of southern flounder (Paralichthys lethostigma) larvae. *Copeia* 1984(4):981–988.

Turnbull, E. L. 1955. *Ten Centuries of Spanish Poetry: An Anthology in English Verse with Original Texts, from the XIth Century to the Generation of 1898*. Baltimore: Johns Hopkins Press.

Turner, R. D. 1976. Fixation and preservation of molluscan zooplankton. Pages 290–300 in Steedman, H. F. (editor). *Zooplankton Fixation and Preservation*. Monographs on Oceanographic Methodology 4. Paris: Unesco Press.

Tyler, M. J. 1962. On the preservation of anuran tadpoles. *The Australian Journal of Science* 25:222.

Valdecasas, A. G., and A. Baltanas. 1989. A note on the use of Angelier's fluid for freshwater invertebrates. *Archiv für Hydrobiologie* 115(2):313–316.

van Cleave, H. J., and J. A. Ross. 1947. A method for reclaiming zoological specimens. *Science* 105:318.

van Dam, A. J. 1997. Conservation of fluid preserved specimens, properties of sealants and their effect on preservation quality. *Bulletin of the European Association of Museums of the History of Medical Sciences* 23:22–28.

van Dam, A. J. 2000. The interactions of preservative fluid, specimen container, and sealant in a fluid collection. *Collection Forum* 14(1–2):78–92.

van Dam, A. J. 2003. DMDM-hydantoin: The promising result of a search for an alternative in fluid preservation of biological specimens. *Collection Forum* 18(1–2):104–115.

van Dam, A. J., J. P. M. van der Ploeg, G. J. M. Koper, and D. Bedeaux. 2000. The warping and cracking of Plexiglas™ specimen containers. *Collection Forum* 14(1–2):47–56.

van Dijk, D. E. 1966. Systematics and field keys to the families, genera and described species of Southern African anuran tadpoles. *Annals of the Natal Museum* 18(2):231–286.

Van der Veer, H. W., and W. Oorthuysen. 1985. Abundance and food demand of the scyphomedusa *Aurelia aurita* in the western Dutch Wadden Sea. *Netherlands Journal of Sea Research* 19:38–44.

Van Guelpen, L. 1999. Larval fish preservation: ethanol acidity from Byron Weston Resistall label paper. *Curation Newsletter* 12:12–14.

Van Guelpen, L., J. F. McElman, and K. J. Sulak. 1987. Additional label papers for specimens in aqueous preservatives. *Curation Newsletter* 9:3–4.

Van Oosten, J. 1929. Life history of the lake herrings (*Leucichuthys artedi* Le Sueur) of Lake Huron as revealed by its scales, with a critique of the scale method. *Bulletin of the United States Bureau of Fisheries* 44:265–428.

van Steenis, C. G. G. 1935. On the preservation of botanical specimens in liquid. *Chronica Botanica* 1:341–342.

Vink, C. J., S. M. Thomas, P. Paquin, C. Y. Hayashi, and M. Hedin. 2005. The effects of preservatives and temperatures on arachnid DNA. *Invertebrate Systematics* 19:99–104.

Vinther, J., D. E. G. Briggs, J. Clarke, G. Mayr, and R. O. Prum. 2010. Structural coloration in a fossil feather. *Biology Letters* 6:128–131.

Viswanathan, M. V., H. B. Singh, and P. R. Bhagwat. 1991. A novel wet preservation method for plants. *Journal of Scientific and Industrial Research* 50(12):989–990.

Vitt, L. J., J. M. Howland, and A. R. Dunham. 1985. The effect of formalin fixation on weight of lizard eggs. *Journal of Herpetology* 19(2):298–299.

Vogel, C. 2006. Swimming with famous dead sharks. *New York Times*, October 1, 2006.

Vogt, K. D. 1991. Reconstituting dehydrated museum specimens. *Curator* 34(2):125–131.

Vogt, K. D. 1998. The reconstitution of dehydrated museum specimens II. *The Biology Curator* 13:6.
Vogt, K. D. 2001. The reconstitution of dehydrated museum specimens III. *The Biology Curator* 21:16–17.
Voight, J. R. 2001. Morphological deformation in preserved specimens of the deep-sea octopus Graneledone. *Journal of Molluscan Studies* 67:95–102.
Volk, W. A., and M. F. Wheeler. 1984. *Basic Microbiology*. 5th ed. New York: Harper and Row.
Von Endt, D. 1994. Spirit collections: a preliminary analysis of some organic materials found in the storage fluid of mammals. *Collection Forum* 10(1):10–19.
Von Endt, D. W. 2000. Staying in shape: The stability of structural proteins in natural history museum storage fluids. *Polymer Preprints* 41(2):1794–1795.
Von Endt, D. W., C. A. Ross, and P. E. Hare. 1999. Initial results from cleaning small vertebrate skeletons using the enzyme trypsin. *Collection Forum* 13(2):51–62.
Von Endt, D. W., E. R. Yourd, and P. E. Hare. 2000. Spirit collections: accelerated aging studies concerning the stability of keratin in ethanol and formalin. *Collection Forum* 14(1–2):66–77.
von Humboldt, A. 1889. *Personal Narrative of Travels to the Equinoctial Regions of America, during the Years 1799–1804*. Vol. 1. London: George Bell.
von Schiller, D., and A. G. Solimini. 2005. Differential effects of preservation on the estimation of biomass of two common mayfly species. *Archiv für Hydrobiologie* 164(3):325–334.
Waddington, J. 1989. Natural history collections management at the Royal Ontario Museum. *Collection Forum* 5(1):27–31.
Wagstaffe, R., and J. H. Fidler. 1955. *The Preservation of Natural History Specimens: Invertebrates*. Vol. 1. London: H. F. and G. Witherby.
Wagstaffe, R., and J. H. Fidler. 1968. *The Preservation of Natural History Specimens. Part Two—Zoology—Vertebrates; Part Three—Botany; Part Four—Geology*. London: H. F. and G. Witherby.
Walker, H. J., G. H. Burgess, A. Y. Suzumoto, C. I. Klepadio, L. VanGuelpen, and J. M. Humphries. 1995. Isopropanol revisited. *Curation Newsletter* 11:3–4.
Walker, H. J., C. I. Klepadlo, and J. E. Simmons. 1999. An alternative to glass? Yes! (For some cases). *Curation Newsletter* 12:3–4.
Walker, J. 1811. Sir Ashton Lever's directions for preserving birds, etc., 1773. Pages 502–503 in *A Selection of Curious Articles from the Gentleman's Magazine*. 2nd ed. Vol. 3. London: Longman, Hurst, Rees, Orme, and Brown.
Walker, J. F. 1964. *Formaldehyde*. American Chemical Society Monograph Series. New York: Reinhold Publishing.
Waller, R. 1992. Temperature- and humidity-sensitive mineralogical and petrological specimens. Pages 25–50 in Howie, F. M. (editor). *The Care and Conservation of Geological Material: Minerals, Rocks, Meteorites and Lunar Finds*. Oxford: Butterworth-Heinmann.
Waller, R., and T. J. K. Strang. 1996. Physical chemical properties of preservative solutions—I. Ethanol-water solutions. *Collection Forum* 12(2):70–85.
Waller, R., and D. E. McAllister. 1986. A spot test for distinguishing formalin from alcohol solutions. Pages. 93–99 in Wadington, J. and D. M. Rudkin (eds.). *Proceedings of the 1985 Workshop on the Care and Maintenance of Natural History Collections*. Life Sciences Miscellaneous Publications, Royal Ontario Museum.
Waller, R., and J. E. Simmons. 2003. An exploratory assessment of the state of a fluid-preserved herpetological collection. *Collection Forum* 18(1–2):1–37.

Waller, R. A., and W. N. Eschmeyer. 1965. A method for preserving color in biological specimens. *Bioscience* 15(5):361.

Ward, R. 1890. *The Sportsman's Handbook to Practical Collecting, Preserving, and Artistic Setting-up of Trophies and Specimens to which is added a Synoptical Guide to the Hunting Grounds of the World*. London: Simpkin, Marshall, Hamilton, Kent.

Warén, A., E. Dock-Åkerman, and H. Taylor. 2010. Metal-lid jars in museum collections? *Collection Forum* 24(1–2):52–61.

Warford, A., J. H. Pringle, J. Hay, S. D. Henderson, and I. Lauder. 1988. Southern blot analysis of DNA extracted from formol-saline fixed and paraffin wax embedded tissue. *Journal of Pathology* 154:313–320.

Warmington, A. R., J. M. Wilkinson, and C. B. Riley. 2000. Evaluation of ethnol-based fixatives as a substitute for formalin in diagnostic clinical laboratorires. *Journal of Histotechnology* 23(4):299–308.

Wassersug, R. J. 1976. A procedure for differential staining of cartilage and bone in whole formalin-fixed vertebrates. *Stain Technology* 51(2):131–134.

Waterhouse, D., and B. Graner. 2009. Fluid preservation course 7th–10th April 2008 Shirehall Study Centre, Norwich. *NatSCA News* 16:11–17.

Weber, C., T. Jaccoud, and A. de Chambrier. 1984. A temporary field fixing and preserving solution for ornithological collecting. *Curator* 7(4):281–286.

Wells, H. G. 2005. *The War of the Worlds*. London: Penguin Books.

Wentworth, J. E. 1939. The hydrosulphite method of preserving museum specimens. *Bulletin of the International Association of Medical Museums* 19:79–82.

Wentworth, J. E. 1942. The preservation of museum specimens in war-time. *Journal of Pathology and Bacteriology* 54(1):137–138.

Westlander, E. 2005. Amputee happy to be reunited with foot. Police had seized limb after complaints from neighbors about public spectacle. *Lawrence Journal-World*, July 26, 2005.

White, D. A., and E. J. Peters. 1969. A method of preserving color in aquatic vertebrates and invertebrates. *Turtox News* 47(9):296–297.

Wiederholm, T., and L. Ericksson. 1977. Effects of alcohol-preservation on the weight of some benthic invertebrates. *Zoon* 5:29–31.

Williams, E. E. 1962. The anoles of the northern Leewards, Anguilla to Monserrat: New data and a new species. *Bulletin of the Museum of Comparative Zoology* 127(9):453–465.

Williams, R., and D. B. Robins. 1982. Effects of preservation on wet weight, dry weight, nitrogen and carbon contents of *Calanus helgolandicus* (Crustacea: Copepoda). *Marine Biology* 71:271–281.

Williams, S. L. 2005. Policy theory and application for museums. *Collection Forum* 19(1–2):32–44.

Williams, S. L., and Cato, P. S. 1995. Interaction of research, management, and conservation for serving the long-term interests of natural history collections. *Collection Forum* 11(1):16–2.

Williams, S. L., R. R. Monk, and J. Arroyo-Cabrales. 1996. Applying McGinley's model for collections assessment to collections of Recent vertebrates. *Collection Forum* 12(1):21–35.

Williams, S. T. 2007. Safe and legal shipment of tissue samples: Does it affect DNA quality? *Journal of Molluscan Studies* 73(4):416–418.

Williamson, D. I., and G. Russell. 1965. Ethylene glycol as a preservative for marine organisms. *Nature* 206(4991):1370–1371.

Winokur, R. M., and S. Hillyard. 1992. Pelvic cutaneous musculture in toads of the genus *Bufo*. *Copeia* 1992(3):760–769.

Windsor, A. S. 1934. Alcohol notes. *Turtox News* 12(11):215–217.

Windsor, D. A. 1971. Ammonium sulfate as a preservative which does not remove color from frogs. *Copeia* 1971(2):356–357.

Wineski, L. E., and A. W. English. 1989. Phenoxyethanol as a nontoxic preservative in the dissection laboratory. *Acta Anatomica* 136:155–158.

Withering, W. 1801. *Systematic Arrangement of British Plants, with an easy Introduction to the Study of Botany*. 4th ed., vol. 1. London: T. Cadell.

Wolf, G. 2008. A kidney from hell? A nephrological view of the Whitechapel murders in 1888. *Nephrology Dialysis Transplantation* 23:3343–3349.

Yamamoto, T. 1985. The Royal Ontario Museum system of collections management. *International Journal of Museum Management and Curatorship* 4:273–298.

Yeh, C. F., and R. G. Hodson. 1975. Effects of formalin on length and weight of bluegill and white crappie from Lake Nasworthy, Texas. *Southwestern Naturalist* 20:315–321.

Yip, S. Y. 1982. A note on the effect of preserving ctenophores in formaldehyde-seawater. *Irish Naturalist's Journal* 20(10):416–419.

Zala, K., N. D. Pentcheff, and R. Wetzer. 2005. Laser-printed labels in wet collections: Will they hold up? *Collection Forum* 19(1–2):49–56.

Zbarsky, I., and S. Hutchinson. 1997. *Lenin's Embalmers*. London: Harvill Press.

Zinke, O. 1991. Naßkonservierung in ethyleneglycol—eine möglichkeit zur bewahrung des silberglanzes bei fischen. *Der Präparator* 37(1):37–41.

Zweifel, R. G. 1966. Guidelines for the care of a herpetological collection. *Curator* 9(1):24–35.

Bibliography

Adams, M., and J. Bailey. 1994. The reorganisation of the natural history museum's avian spirit collection. *Biology Curator Group Newsletter* 6(4):41–42.

Ahlander, E., S. O. Kullander, and B. Fernholm. 1997. Ichthyological collection building at the Swedish Museum of Natural History, Stockholm. Pages 13–25 in Pietsch, T. W., and W. D. Anderson Jr. (editors). *Collection Building in Ichthyology and Herpetology*. American Society of Ichthyologists and Herpetologists Special Publication Number 3.

Anonymous. n.d. Preserving zoological specimens. *Turtox Service Leaflet*.

Anonymous. n.d. *Oxford University Museum formulas*. Oxford University Museum [unpublished manuscript].

Anonymous. 1928. Crayfish troubles. *Turtox News* 6(10):63.

Anonymous. 1931. How long will an embalmed specimen remain in good condition for dissection? *Turtox News* 9(11):89–90.

Anonymous. 1943. Alcohol solutions. *Turtox News* 21(8):105.

Anonymous. 1982. *Investigation of exposure to formaldehyde from preserved biological specimens*. United States Consumer Product Safety Commission.

Anonymous. 1996. Procedure for curation old glass museum jars. *The Biology Curator* 7:3–5.

Anonymous. 1998. Working safely with glutaraldehyde. *Techlines* 5(1):1–6.

Arndt, W. 1937. Aloholfragen in naturkunde museum. *Museumskunde, Neue Folge*. 9:61–106.

Audette, L. G. 1965. A method of constructing plastic display jars. *The Anatomical Record* 153(4):383–388.

Balachandran, T. 1976. Fixation and preservation experiments on marine zooplankton at Cochin: outline of experiments and results. Pages 333–339 in Steedman, H. F. (editor). *Zooplankton Fixation and Preservation*. Monographs on Oceanographic Methodology 4. Paris: Unesco Press.

Barriga, R., and M. Olalla. 1983. Conservación y preparación de peces. *Manual Para Museos. Publicaciones del Museo Ecuatoriano de Ciencias Naturales, serie Miscelaneas* 4(2):61–67.

Bates, M. F. 1990. The hidden value of a reptile wet collection with special reference to the National Museum, Bloemfontein. Pages 139–143 in Herholdt, E. M. (editor). *Natural History Collections: Their Management and Value.* Pretoria: Transvaal Museum.

Batson, O. V. 1921. The differential staining of bone. *Anatomical Record* 22:159–164.

Bayless, J., and C. Shepherd. 1993. Removing wet specimens from long-term storage in formalin. *Conserve O Gram* 11(1):1–4.

Beccaloni, J. 2002. A comparison of trisodium phosphate and Decon 90 as rehydrating agents for Arachnida and Myripoda dry specimens. *The Biology Curator* 22:15–23.

Beebe, W. 1929. The architectural beauty of fish. New adaptations of an old method. *Bulletin of the New York Zoological Society* 32:67–69.

Bell, R. 1977. The museum preparation of herpetological specimens. *Cotswold Herpetological Symposium Report* 3:2–8.

Bello, R. A., J. H. Luft, and G. M. Pigott. 1981. Improved histological procedure for microscopic demonstration of related changes in fish muscle tissue structure during holding and freezing. *Journal of Food Science* 46:733–740.

Black, L. F. 1991. *From Liquid to Vapor and Back. Four Centuries of the First Chemical Separation Process, a Checklist of an Exhibition at the Hugh M. Morris Library.* Newark: University of Delaware Library.

Blanton, P. L., and N. L. Biggs. 1968. Density of fresh and embalmed human compact and cancellous bone. *American Journal of Physical Anthropology* 29:39–44.

Blaufuss, A. H. 1943. Notes on the preservation of marine crustaceans. *Turtox News* 21(6):77–78.

Burdi, A. R. 1965. Toluidine blue-alizarin red S staining of cartilage and bone in whole-mount skeletons in vitro. *Stain Technology* 40:45–48.

Burgess, G. H. 1983. Acetone in isopropanol. *Curation Newsletter* 6:1.

Burgess, G. H. 1984. Additional comments on the above. *Curation Newsletter* 7:4.

Cannell, P. F., M. R. Bakst, and C. S. Asa. 1988. Suggestions regarding alcoholic bird collections. *The Condor* 90:500–503.

Carter, J. 2002. Fixation: an overview. *Natural Science Conservation Group Newsletter* 18: 27–32.

Castle, P. H. J. 1976. Criteria of satisfactory fixation and preservation of eel larvae (Leptocephali). Zooplankton Fixation and Preservation. Pages 319–320 in Steedman, H. F. (editor). *Zooplankton Fixation and Preservation.* Monographs on Oceanographic Methodology 4. Paris: Unesco Press.

Cater, C. W. 1963. The evaluation of aldehyde of other difunctional compounds as cross-linking agents for collagen. *Laboratory Reports of the British Leather Manufacturer's Research Association* 62:69–84.

Cato, P. S. 1986. Guidelines for managing bird collections. *Museology* 7:1–78.

Chen, T. P., and C. R. Fellers. 1926. Fish preservation by hypochlorites. *University of Washington Publications in Fisheries* 1:1–10.

Clark, P. F., O. A. Crimmen, F. C. Naggs, A. D. Wahl, and M. C. Mansfield. 1994. Transportation of fluid-preserved natural history specimens stored in glass containers: new solutions to an old problem. *Collection Forum* 10(1):1–9.

Cleverton, M. A. 1943. A new fixative for animal tissues. *Science* 97(2511):168.

Congdon, J. D., L. J. Vitt, J. E. Platz, and A. C. Hulse. 1975. A portable rapid injector for preserving fish, amphibians and reptile specimens in the field. *Copeia* 1975(1):183.

Cosgrove, J. A. 1992. Tub and divider system for fluid collections. Pages 205–206 in Rose, C. L., and A. R. de Torres (editors). *Storage of Natural History Collections: Ideas and Practical Solutions.* New York: Society for the Preservation of Natural History Collections.

Crary, D. D. 1962. Modified benzyl alcohol clearing of alizarin-stained specimens without loss of flexibility. *Stain Technology* 37:124–125.

Cumley, R. W., J. F. Crow, and A. B. Griffin. 1939. Clearing specimens for demonstration of bone. *Stain Technology* 14:7–11.

Dang, P. T. 1983. Glycerine as a preservative for immature insects. Pages 69–71 in Faber, D. J. (editor). Proceedings of the 1981 Workshop on Care and Maintenance of Natural History Collections. *Syllogeus* 44:1–196.

Darovec, J. E. 1979. Concrete blocks and "Xerox" machines as tools for curating fishes. *Association of Systematic Collections Newsletter* 7(3):30.

Davies, K. C., and J. Hull. 1976. *The Zoological Collections of the Oxford University Museum: A Historical Review and General Account, with Comprehensive Donor Index to the Year 1975.* Oxford: Oxford University Museum.

Davies, M. 1993. Collection and preservation of the anura. Pages 113–14 in Glasby, C. J., G. J. B. Ross, and P. L. Beesley (editors). *Fauna of Australia. Amphibia & Reptilia.* Canberra: Australian Government Publishing Service.

de la Torre, L. 1951. A method for cleaning skulls of specimens preserved in alcohol. *Journal of Mammalogy* 32(2):230–231.

de Moor, F. C. 1990. Containers for wet collections—problems and solutions. Pages 27–36 in Herholdt, E. M. (editor). *Natural History Collections: Their Management and Value.* Pretoria: Transvaal Museum.

Diefenbacher, E. H., and V. L. Wiggins. 2008. Rapid new protocol for clearing and staining of specimens for osteological examination. *Herpetological Review* 39(1):54–55.

Donaldson, H. H. 1894. Preliminary observations on some changes caused in the nervous tissues by reagents commonly employed to harden them. *Journal of Morphology* 9:123–166.

Duellman, W. E. 1962. Collecting and preserving amphibians and reptiles. Pages 37–40 in Hall, E. R. (editor). *Collecting and Preserving Study Specimens of Vert*ebrates. University of Kansas Museum of Natural History Miscellaneous Publications 30:1–46.

Dundee, H. A. 1962. A low-cost storage tank for preserved animals. *Turtox News* 40(12):298–299.

Edwards, S. R., W. J. Paule, and C. M. Pomerat. 1964. The effect of various fixatives and dehydrating fluids on tissue culture cells. *Anatomical Record* 148:278.

Elkan, E. 1970. Embalming frogs. *British Journal of Herpetology* 4(6):155–156.

Ely, F. A. 1910. Formaldehyde poisoning. *Journal of the American Medical Association* 54:1140–1141.

Ezzell, J., B. Richter, and E. Francis. 1996. Selective extraction of polychlorinated biphenyls from fish tissue using accelerated solvent extraction. *American Environmental Laboratory* 12/96:12–13.

Faber, D. J. 1983. A storage system to reduce the exposure of animal collections to light. Pages 71–73 in Faber, D. J. (editor). Proceedings of the 1981 Workshop on Care and Maintenance of Natural History Collections. *Syllogeus* 44:1–196.

Faber, D. J. 1986. Unbuffered formaldehyde for long-term storage of larval fishes. Page 92 in Waddington, J., and D. M. Rudkin (editors). *Proceedings of the 1985 Workshop on Care and Maintenance of Natural History Collections.* Toronto: Life Sciences Miscellaneous Publications of the Royal Ontario Museum.

Fikioris, M. A. 1992. Vapor permeable covers for storage units. Pages 131–132 in Rose, C. L., and A. R. de Torres (editors). *Storage of Natural History Collections: Ideas and Practical Solutions.* New York: Society for the Preservation of Natural History Collections.

Forstner, M. R., J. R. Dixon, and D. E. Starkey. 1997. An improved technique for formalin fixation of large vertebrates. *Herpetological Review* 28(2):77.

Fosberg, F. R., and M.-H. Sachet. 1965. *Manual for Tropical Herbaria*. Utrecht, the Netherlands: Regnum Vegetabile. International Bureau for Plant Taxonomy and Nomenclature, 39:1–132.

Fraser, J. H. 1961. Use of polyethylene glycol in biology. *Nature* 189(4760):241–242.

Fremling, C. R., and D. L. Hemming. 1965. A new method of taxidermy using polyethylene glycol as an impregnation medium. *The American Biology Teacher* 27(9):697–701.

Fuller, T., A. Blount, and C. Bossert. 1992. Support system for collections stored in jars. Pages 231–232 in Rose, C. L., and A. R. de Torres (editors). *Storage of Natural History Collections: Ideas and Practical Solutions*. New York: Society for the Preservation of Natural History Collections.

Gale, F. C., and J. J. R. Geary. 1956. A method of embalming human cadavers for anatomical dissection and subsequent histological study. *Anatomical Record* 124(1):95–99.

Gamble, D. L., and M. M. Wells. 1924. The preservation of mammals for dissection. *Turtox News* 2(9):33–34.

Gibson, J. E. 1983. *Formaldehyde Toxicity*. Chemical Industry Institute of Toxicology Series, Washington, D.C.: Hemisphere Publishing.

Gisbert, J., F. Palacios, and R. Garcia-Perea. 1990. Labeling vertebrate collections with Tyvek synthetic paper. *Collection Forum* 6(1):35–37.

Gisbert, J., R. Garcia-Perea, and F. Palacios. 1992. Durable specimen labels. Pages 265–266 in Rose, C. L., and A. R. de Torres (editors). *Storage of Natural History Collections: Ideas and Practical Solutions*. New York: Society for the Preservation of Natural History Collections.

Gisbert, J., and R. Garcia-Perea. 1992. Double-jar system for types in fluid collections. Pages 225–226 in Rose, C. L., and A. R. de Torres (editors). *Storage of Natural History Collections: Ideas and Practical Solutions*. New York: Society for the Preservation of Natural History Collections.

Goebel, A. M., and J. E. Simmons. 1993. Preserving DNA in liquid-preserved museum specimens. Pages 34–52 in Snyder, A. M. (editor). *1992 ASIH Workshop on Collection Care and Management Issues in Herpetology and Ichthyology*. Albuquerque, NM: American Society of Ichthyologists and Herpetologists.

Gold, K. 1976. Methods for preserving Tintinnida. Pages 236–239 31 in Steedman, H. F. (editor). *Zooplankton Fixation and Preservation*. Monographs on Oceanographic Methodology 4. Paris: Unesco Press.

Goldspink, G. 1961. Fixation of muscle. *Nature* 192(4809):1305–1306.

Goldspink, G., S. Gelder, L. Clapison, and P. Overfield. 1973. Pre- and post-rigor fixation of muscle. *Journal of Anatomy* 114:1–6.

Gordon, F. J. 1934. Preservation of small amphibians in gelatin. *Science* 80:475.

Gough, J., H. Berry, and B. M. Still. 1944. Phenoxetol in the treatment of pyocyanea infections. *Lancet* 247:176–178.

Gough, J., and J. E. Wentworth. 1950 The use of thin sections of entire organs in morbid anatomical studies. *Journal of the Royal Microscopical Society* 69(4):231–235.

Graves, G. 1818. *The Naturalist's Pocket-Book: or Tourist's Companion, Being a Brief Introduction to the Different Branches of Natural History, with Approved Methods for Collecting and Preserving the Various Products of Nature*. London: Longman.

Gray, P. 1928. The preparation of alizarine transparencies. *Museums Journal*, (28):341–342.

Green, M. C. 1952. A rapid method of clearing and staining specimens for the demonstration of bone. *Ohio Journal of Science* 52(1):31–33.

Greer, C. E., J. K. Lund, and M. M. Manos. 1991. PCR amplification from paraffin-embedded tissues. Recommendations on fixatives for long-term storage and prospective studies. *Genomic Research* 1:46–50.

Habenstein, R. W., and W. M. Lamers. 1962. *The History of American Funeral Directing.* Milwaukee, WI: Bulfin Printers.

Hahn, D. E. 1959. New methods and suggestions on preserving herptiles. *Journal of the Ohio Herpetological Society* 2(1):3–4.

Hall, E. R. 1962. Collecting and preparing study specimens of vertebrates. *University of Kansas Museum of Natural History Miscellaneous Publication* 30:1–46.

Halliday, R. G., and B. Roscoe. 1976. The effects of icing and freezing on the length and weight of groundfish species. *International Commission for the Northwest Atlantic Fisheries Research Document* 69/2.

Hanken, J., and R. Wassersug. 1981. The visible skeleton. A new double-stain technique reveals the nature of the "hard" tissues. *Functional Photography* 16:22–26.

Harris, R. H. 1959. Small vertebrate skeletons. *Museums Journal* 58(10):223–224.

Hatchfield, P. B., and J. M. Carpenter. 1986. The problem of formaldehyde in museum collections. *The International Journal of Museum Management and Curatorship* 5:183–188.

Hearne, M., and D. Catania. 1983. Formaldehyde vapor masks. *Curation Newsletter* 6:7.

Hewitt, A. D. 1995. Evaluation of methanol and $NaHS_{O4}$ for preservation of volatile organic compounds in soil samples. *American Environmental Laboratory* 7(6):16–18.

Hicks, A. J., and P. M. Hicks. 1978. A selected bibliography of plant collection and herbarium curation. *Taxon* 27(1):63–99.

Hochberg, F. G. 1992. Colored dots for coding fluid collections. Pages 267–268 in Rose, C. L., and A. R. de Torres (editors). *Storage of Natural History Collections: Ideas and Practical Solutions.* New York: Society for the Preservation of Natural History Collections.

Hollis, D., A. C. Jeremy, and R. J. Lincoln. 1977. Biological collecting for the small expedition. *Geographical Journal* 143(2):249–265.

Hollister, G. 1932. A new use of ultra-violet radiation. *Bulletin of the New York Zoological Society* 35:49–50.

Hood, R. C. W. S., and W. Neill. A modification of alizarin red S technic for demonstrating bone formation. *Stain Technology* 23:209–218.

Hopwood, D. 1965. Some aspects of gluteraldehyde fixation. *Journal of Anatomy* 99:928.

Horie, C. V. (editor). 1989. *Conservation of Natural History Specimens: Spirit Collections.* Manchester: Manchester Museum and Department of Environmental Biology, University of Manchester.

Howie, F. 1989. Health and safety considerations. Pages 39–45 in Horie, C. V. (editor). *Conservation of Natural History Specimens: Spirit Collections.* Manchester: Manchester Museum and Department of Environmental Biology, University of Manchester.

Humason, G. L. 1962. *Animal Tissue Techniques.* San Francisco: W. H. Freeman.

Inouye, M. 1976. Differential staining of cartilage and bone in fetal mouse skeleton by alcian blue and alizarin red S. *Congenital Anomalies* 16:171–173.

Israel, M. S., and L. F. Young. 1978. Use of liquid paraffin in the preservation of pathological specimens. *Journal of Clinical Pathology* 31(5):499–500.

Jackson, H. H. T. 1926. The care of museum specimens of Recent mammals. *Journal of Mammalogy* 7:113–118.

Jackson, M. K. 1971. A new "syringe" for injecting herptiles. *Herpetological Review* 3(4):75.

Jannett, F. J. 1989. Some tests of synthetic paper and polyethylene sacks for specimens preserved in fluids. *Curator* 32(1):24–25.

Jaques, H. E. 1939. *How to Know the Insects*. Pictured-Key Nature Series, Mt. Pleasant, 140.

Jewett, S. L. 1995. Laser printer labels: potential disaster while specimens are on loan. *Curation Newsletter* 11:6–7.

Johnson, J. A., and B. J. Parker. 1980. An effective method of preserving large reptiles. *Herpetological Review* 11(3):77–78.

Kannemeyer, S. 1973. The storage of a wet collection. *South African Museum Association Bulletin* 10(7):274–277.

Kellog, V. L., and E. E. Slosson. 1889–1890. A comparison of preservative fluids for museum use. *Transactions of the Kansas Academy of Science* 12:83–84.

Kimmel, C. A., and C. Trammell. 1981. A rapid procedure for routine double staining of cartilage and bone in fetal and adult animals. *Stain Technology* 56:271–273.

Koller, J., Y. Kaup, H. Etspuler, and U. Weser. 1998. Embalming was used in Old Kingdom. *Nature* 391:343–344.

Krishna, D. 1948. Shrinkage of animal cells due to fixatives. *Nature* 161(4084):202–203.

Kyaw, A., K. A. Tha, H. Pe, K. P. P. Kyaw, and S. Kun. 1992. Reversion of formaldehyde linkage in Russell's viper venom toxoid on storage. *The Snake* 24(2):147–150.

Labella, F. S. 1971. Cross-links in elastin and collagen. Pages 243–301 in Elden, H. R. (editor). *Biophysical Properties of the Skin*. New York: Wiley.

Lake, P. S. 1973. Trialdehyde fixation of crustacean tissue for electron microscopy. *Crustaceana* 24(Part C):244–245.

Lamb, I. M. 1959. A new method for the permanent preservation of algal material for microscopical study. *Revue Algologique* 4(3):198–201.

Leek, F. F. 1976. An ancient Egyptian mummified fish. *Journal of Egyptian Archaeology* 62:131–133.

Legler, J. M. 1981. Inexpensive containers for large herpetological specimens. *Herpetological Review* 12(1):9–10.

Lendrum, A. C. 1941. On the fixation of post-mortem material. *Journal of Pathology and Bacteriology* 52(1):132–137.

Lettsom, J. C. 1772. *The Naturalist's and Traveler's Companion: Containing Instructions for Discovering and Preserving Objects of Natural History*. London: George Peach.

Lettsom, J. C. 1774. *The Naturalist's and Traveler's Companion: Containing Instructions for Discovering and Preserving Objects of Natural History*. London: E. and C. Dilly.

Levine, R. J., R. D. Dal Corso, P. B. Blunden, and M. C. Battigelli. 1983. The effects of occupational exposure on the respiratory health of West Virginia morticians. Pages 212–226 in Gibson, J. E. (editor) *Formaldehyde Toxicity*. Washington, D.C.: Hemisphere Publishing.

Levison, L. A. 1904. A case of fatal formaldehyde poisoning. *Journal of the American Medical Association* 42:1492.

Levy, M., and D. E. Silberman. 1937. The reactions of amino and imino acids with formaldehyde. *Journal of Biological Chemistry* 118:723–734.

Lincoln, R. 1989. Storage containers. Pages 53–63 in Horie, C. V. (editor). *Conservation of Natural History Specimens: Spirit Collections*. Manchester: Manchester Museum and Department of Environmental Biology, University of Manchester, Manchester.

Lipman, H. J. 1935. Staining the skeleton of cleared embryos with alizarin red S. *Stain Technology* 10:61–63.

Lyons, W. 1993. Methods used in the preparation of specimens for teaching anatomy and pathology in the 1900s. Pages 332–345 in Palacios, F., C. Martinez, and B. Thomas (editors). *Contributions on the Status, Preservation, and Conservation of Natural History Collections.* Vol. 2. Madrid: International Symposium and First World Congress on the Preservation and Conservation of Natural History Collections, Dirección General de Bellas Artes y Archivos, Ministero de Cultura.

Mall, F. P. 1906. On ossification centers in human embryos less than one hundred days old. *American Journal of Anatomy* 5:433–458.

Masner, L. 1994. Effect of low temperature on preservation and quality of insect specimens stored in alcohol. *Insect Collection News* 9:14–15.

Matheus-Pozo, J. C. 1983. Introduccion a la herpetologia. Publicaciones del Museo Ecuatoriano de Ciencias Naturales, serie Miscelaneas. *Manual Para Museos* 4(2):69–75.

Maynard, C. J. 1873. *The Naturalist's Guide in Collecting and Preserving Objects of Natural History, with a Complete Catalogue of the Birds of Eastern Massachusetts.* Boston: James R. Osgood.

McAllister, D. E. 1965. The collecting and preserving of fishes. Pages 152–170 in Anderson, R. M. (editor). *Methods of Collecting and Preserving Vertebrate Animals.* National Museum of Canada Bulletin.

McElhaney, J., J. Fogle, E. Byars, and G. Weaver. 1963. Effect of embalming on the mechanical properties of beef bone. *Journal of Applied Physiology* 19(6):1234–1236.

Medawar, P. B. 1941. III. The rate of penetration of fixatives. *Journal of the Royal Microscopical Society* 61:46–57.

Michalski, S. 1994. Leakage prediction for buildings, cases, bags and bottles. *Studies in Conservation* 39:169–186.

Miller, R. R. 1952. Treated formalin as a permanent preservative. *Turtox News* 30(10):178–179.

Moore, J. A. 1993. *Science as a Way of Knowing.* Cambridge, MA: Harvard University Press, Cambridge.

Moore, S. 1989. Conservation of spirit specimens. Pages 65–90 in Horie, C. V. (editor). *Conservation of Natural History Specimens: Spirit Collections.* Manchester: Manchester Museum and Department of Environmental Biology, University of Manchester.

Moore, S. J. 1989. Narcotising sea anemones. *Journal of the Marine Biology Association of the United Kingdom* 69:803–811.

Moore, S. J. 1990. Investigation into the state of preservation of the E. T. Browne collection of hydromedusae *Journal of the Marine Biology Association of the United Kingdom* 70:477–480.

Moore, S. J. 1994. Liquid preservation—how little we know. *Biology Curator's Group Newsletter* 6(4):45–47.

Moore, S. J. 2002a. Histological effects of fixation and long-term preservation. Are preservatives beneficial or not? *Natural Sciences Conservation Group Newsletter* 18:24–26.

Moore, S. 2002b. Fixation: An overview. *Natural Sciences Conservation Group Newsletter* 18:27–31.

Myers, G. S. 1956. Manual of tropical herpetological collecting. *Natural History Museum of Stanford University Circular* 4:1–13.

Nelson, D. W. 1990. Further comments on laser printer generated jar labels. *Curation Newsletter* 10:3–4.

Nelson, D. W., P. W. Willink, and B. A. Sheilds. 1995. Bis-carbonyl fixatives for museum specimens? *Curation Newsletter* 11:1–3.

Nelson, D. W. 1999. Multi-lot jars: a space saver. *Curation Newsletter* 12:11.

Newell, G. W. 1983. Overview of formaldehyde. Pages 2–12 in Gibson, J. E. (editor). *Formaldehyde Toxicity*. Washington, D.C.: Hemisphere Publishing.

Newman, S. M., T. S. Dugan, and J. N. Dumont. 1983. A rapid and simple procedure for visualization of amphibian skeletons for teratological studies. *Transactions of the American Microscopical Society* 102:300–305.

NIC. 1993. *Summary Report, American Society of Ichthyologists and Herpetologists Discussion Group*. National Institute for the Conservation of Cultural Property.

NIC. 1993. *Summary Report, Materials Science Panel on Fluid-Preserved Collections*. National Institute for the Conservation of Cultural Property.

Nicol, A. 1994. A short note on preservatives. The identification problem—a possible solution. *Biology Curator's Group Newsletter* 6(4):45.

NIOSH/OSHA. 1980. Formaldehyde: evidence of carcinogenicity. *NIOSH/OSHA Current Intelligence Bulletin* 34:1–15.

Nizinskii, M. 1987. Discoloration of alcohol by rubber gaskets. *Curation Newsletter* 9:6.

Noble, G. K., and M. E. Jaeckle. 1926. Mounting by paraffin infiltration. A method for the permanent preservation of whole specimens or dissections. *American Museum Novitates* 233:1–7.

Ojeda, J. L., E. Barbosa, and P. G. Bosque. 1970. Selective skeletal staining in whole chicken embryos. A rapid alcian blue technique. *Stain Technology* 45:137–139.

Ortenburger, A. I. 1923. A method of preparing reptile penes. *Copeia* 1923(119):71–73.

OSHA. 1992. Occupational exposure to formaldehyde: final rule. *Federal Register*, 29 CFR Part 1910:22290–22328.

Palacios, F., and J. Gisbert. 1990. An indelible printing system for permanent records in natural history collections. *Collection Forum* 6(1):38–39.

Palmer, W. H. 1974. Inexpensive jars for museum specimens. *Curator* 17(4):321–324.

Pantin, C. F. A. 1959. *Notes on Microscopical Technique for Zoologists*. Cambridge: Cambridge University Press.

Parks, H. B. 1939. Color preservation. *Turtox News* 17(8):113.

Paulduro, E. 1988. Fachgerechte konservierung und praparation von amphibien und reptilien. Teil I. Konservierung. *Sauria* 10(3):13–16.

Peake, J. 1989. Spirit collections—costs and benefits. Pages 47–51 in Horie, C. V. (editor). *Conservation of Natural History Specimens: Spirit Collections*. Manchester: Manchester Museum and Department of Environmental Biology, University of Manchester.

Peden, A. E. 1976. Collecting and preserving fishes. *Museum Methods Manual Number 3*. Victoria: British Columbia Provincial Museum.

Peden, A. E. 1980. Preservative evaporation in bottled collections. *Association of Systematics Collections Newsletter* 8(6):84.

Piechocki, R. 1985. *Makroskopische Preparationstechnik, Teil II. Wirbeltiere*. Stuggart, Germany: Gustav Fisher Verlag.

Piechocki, R. 1986. *Makroskopische Preparaationstechnik, Teil I. Wirbellose*. Stuggart, Germany: Gustav Fisher Verlag.

Pisani, G. R., and J. Villa. 1974. *Guia de Tecnicas de Preservacion de Anfibios y Reptiles*. Society for the Study of Amphibians and Reptiles Herpetological Circular 2:1–24.

Pitter, P., and J. Chuboda. 1990. *Biodegradablity of Organic Substances in the Aquatic Environment*. Boca Raton, FL: CRC Press.

Prum, R. F. 1993. Las colecciones anatomicas de los museos de historia natural y la investigación de la filogenia y la evolución de las aves. Pages 31–40 in Escalante-Pliego, P. (editor). *Curación Moderna de Colecciones Ornitológicas.* Washington, D.C.: American Ornithologists Union.

Radcliffe, B. C., and C. Messenger. 1992. Vial support system for fluid collections. Pages 213–214 in Rose, C. L., and A. R. de Torres (editors). *Storage of Natural History Collections: Ideas and Practical Solutions.* New York: Society for the Preservation of Natural History Collections.

Raloff, J. 1985. Unwanted formaldehyde "breakthrough." *Science News* 127(19):297.

Rees, N. E., W. M. Boyce, I. A. Gardner and D. M. Nelson. 1996. Fixation affects morphometric characters of *Psoroptes cuniculi* mites (Acari: Psoroptidae). *Journal of Medical Entomology* 33(5):835–838.

Reilly, M. 1989. Curation of a small collection. Pages 95–109 in Horie, C. V. (editor). *Conservation of Natural History Specimens: Spirit Collections.* Manchester: Manchester Museum and Department of Environmental Biology, University of Manchester.

Richmond, G. W., and L. Bennett. 1938. Clearing and staining of embryos for demonstrating ossification. *Stain Technology* 13:77–79.

Rogers, S. P., M. A. Schmidt, and T. Gutebier. 1989. *An Annotated Bibliography on Preparation, Taxidermy, and Collection Management of Vertebrates with Emphasis on Birds.* Carnegie Museum of Natural History Special Publication No. 15.

Romero-Sierra, C., P. Lane, W. Lyons, and J. C. Webb. 1983. An overview of formalin techniques for the fixation of cadavera. Pages 45–52 in Faber, D. J. (editor). Proceedings of the 1981 Workshop on Care and Maintenance of Natural History Collections. *Syllogeus* 44:1–196.

Roonwall, M. L. 1968. *A Handbook for Zoological Collectors.* Calcutta: Zoological Survey of India.

Rose, A., and E. Rose. 1951. Theory. Pages 1–230 in Weissberger, A. (editor). *Technique of Organic Chemistry: Distillation.* Vol. 4. New York: Interscience Publishers.

Rose, A., and E. Rose. 1965. Theory. Pages 1–231 in Weissberger, A. (editor). *Technique of Organic Chemistry: Distillation.* Vol. 4. New York: Interscience Publishers.

Russell, H. D. 1963. Notes on methods for narcotization, killing, fixation, and preservation of marine organisms. *Marine Biology Lab* (Woods Hole, Massachusetts) 59(26):579.

Sanchiz, B. 1994. *Manual de Catalogacion y Gestion de las Colecciones Cientificas de Historia Natural.* Madrid: Manuales Tecnicos de Museologia, Museo Nacional de Ciencias Naturales.

Sanders, A. E. 1997. Alexander Garden (1730–1791), pioneer naturalist in colonial America. Pages 409–437 in Pietsch, T. W., and W. D. Anderson (editors). *Collection Building in Ichthyology and Herpetology.* American Society of Ichthyologists and Herpetologists Special Publication No. 3.

Saul, W. G. 1981. Paraformaldehyde problems. *Curation Newsletter* 3:1.

Saul, W. G. 1982. Field preservation of large fishes. *Curation Newsletter* 4:1.

Schleifer, C. 1982. Alcohol recycling and topping off. *Curation Newsletter* 4:9.

Schmidt, K. P. 1947. Zoological collecting. *Encyclopaedia Britannica.*

Schmidt, R. 1981. Polyethylene bags for permanently sealing jars of pickled collections. *Taxon* 30(1):369–370.

Schrock, J. R. 1992. OSHA further reduces exposure limits to formaldehyde with implications for biology teachers and dissection. *Kansas Biology Teacher* 31(1):47.

Schueler, F. W., and J. Aniskowicz. 1982. A brittle bucket caution. *Herpetological Review* 13(1):5.

Schwaner, T. D., and C. B. Banks. 1993. Collection and preservation of the reptilia. Pages 98–103 in Glasby, C. J., G. J. B. Ross, and P. L. Beesley (editors). *Fauna of Australia. Volume 2A. Amphibia & Reptilia*. Canberra: Australian Government Publishing Service.

Seal, K. J., and H. O. W. Eggins. 1981. The biodeterioration of materials. Pages 8/1–8/30 in Norris, J. R., and M. H. Richmond (editors). *Essays in Applied Microbiology*. New York: Wiley.

Sedra, S. N. 1950. Decreasing the time required for making an alizarin skeleton preparation. *Stain Technology* 25:223–224.

Sibley, C. G., and J. E. Ahlquist. 1981. Instructions for specimen preservation for DNA extraction: a valuable source of data for systematics. *Association of Systematic Collections Newsletter* 9(3):44–45.

Simmons, J. E. 1987. *Herpetological Collecting and Collections Management*. Society for the Study of Amphibians and Reptiles, Herpetological Circular 16:1–70.

Simmons, J. E. 1991. Conservation problems of fluid-preserved collections. Pages 69–89 in Cato, P. S., and C. Jones (editors). *Natural History Museums: Directions for Growth*. Lubbock: Texas Tech University Press.

Simmons, J. E. 1992. Vial supports for cleared and stained specimens. Pages 215–216 in Rose, C. L., and A. R. de Torres (editors). *Storage of Natural History Collections: Ideas and Practical Solutions*. New York: Society for the Preservation of Natural History Collections.

Simmons, J. E. 1993. Blithe spirits: problems and potentials of fluid preservation. Pages 12–27 in Snyder, A. M. (editor). *1992 ASIH Workshop on Collection Care and Management Issues in Herpetology and Ichthyology*. Albuquerque: American Society of Ichthyologists and Herpetologists.

Simmons, J. E., and R. R. Waller. 1994. Assessment of a fluid preserved herpetological collection. Pages 11–15 in Snyder, A. M. (editor). *1993 ASIH Workshop on Collection Care and Management Issues in Herpetology and Ichthyology*. Albuquerque: American Society of Ichthyologists and Herpetologists.

Simmons, J. E. 1995. Storage in fluid preservatives. Pages 161–186 in Rose, C. L., C. A. Hawks, and H. H. Genoways (editors). *Storage of Natural History Collections: A Preventive Conservation Approach*. New York: Society for the Preservation of Natural History Collections.

Simmons, J. E. 1998a. Almacenamiento de colecciones y datos a largo plaza. *Mesoamerica* 3(4):38–43.

Simmons, J. E. 1998b. Long-term storage of collections and data. *Mesoamerica* 3(4):44–49.

Simons, E. V., and J. R. Van Horn. 1971. A new procedure for whole-mount alcian blue staining of the cartilaginous skeletons of chicken embryos, adapted to the clearing procedure in potassium hydroxide. *Acta Morphologica Neerlando. -Scandinavica* 8(1970):281–292.

Sims, L. 1990. How permanent are they? *Curation Newsletter* 10:2.

Sims, L. 1990. An analysis of the suitability of various laser and impact printer generated labels for wet label storage. *Curation Newsletter* 10:2–3.

Sims, L. 1990. Further comments on laser printer generated jar labels. *Curation Newsletter* 10:3–4.

Sloane, T. O. 1935. *Henley's Twentieth Century Book of Formulas, Processes and Trade Secrets*. London: G. D. Hizcox Scientific and Commercial Publishing.

Slosson, E. E. 1890. The relative sweetness of different alcohols. *Transactions of the Kansas Academy of Science* 12:104–105.

Smith, J. B. L. 1958. Three useful laboratory techniques for the biologist. *South African Journal of Science* 54(3):71–73.

Smith, C. L. 1965. Maintaining inactive fish collections. *Curator* 8(3):248–255.

Snyder, R. W., and E. L. Cheatle. 1965. Alkaline gluteraldehyde as an effective disinfectant. *American Journal of Hospital Pharmacy* 22:321–327.

Snyder, A. M. (editor). 1993. *1992 ASIH Workshop on Collection Care and Management Issues in Herpetology and Ichthyology.* Champaign, IL: American Society of Ichthyologists and Herpetologists.

Snyder, A. M. 1996. New source for specimen tags. *Society for the Preservation of Natural History Collections Newsletter* 11(1):6.

Song, J., and L. R. Parenti. 1995. Clearing and staining whole fish specimens for simultaneous demonstration of bone, cartilage, and nerves. *Copeia* 1995:114–118.

SPNHC. 1994. Guidelines for the care of natural history collections. *Collection Forum* 10(1):32–40.

St. Amand, G. S., and W. St. Amand. 1950. Shortening maceration time for alizarin red S preparations. *Stain Technology* 26:271.

Stansfield, G. 1989. Bibliography on the liquid preservation of biological material. Pages 111–113 in Horie, C. V. (editor). *Conservation of Natural History Specimens: Spirit Collections.* Manchester: Manchester Museum and Department of Environmental Biology, University of Manchester, Manchester.

Staples, R. E., and V. L. Schnell. 1964. Refinements in rapid clearing technic in the KOH-alizarin red S method for fetal bone. *Stain Technology* 39:61–63.

Steedman, H. F. 1950. Alcian blue 8GS: a new stain for mucin. *Quarterly Journal of Microscopical Science* 91:477–479.

Steedman, H. F. 1976a. Narcotizing agents and methods. Pages 87–94 in Steedman, H. F. (editor). *Zooplankton Fixation and Preservation.* Monographs on Oceanographic Methodology 4. Paris: Unesco Press.

Steedman, H. F. 1976b. Miscellaneous preservation techniques. Pages 175–181 in Steedman, H. F. (editor). *Zooplankton Fixation and Preservation.* Monographs on Oceanographic Methodology 4. Paris: Unesco Press.

Steedman, H. F. 1976c. Examination, sorting and observation fluids. Pages 182–183 in Steedman, H. F. (editor). *Zooplankton Fixation and Preservation.* Monographs on Oceanographic Methodology 4. Paris: Unesco Press.

Steedman, H. F. 1976d. Permanent mounting media. Pages 189–195 31 in Steedman, H. F. (editor). *Zooplankton Fixation and Preservation.* Monographs on Oceanographic Methodology 4. Paris: Unesco Press.

Steedman, H. F. 1976e. Some technical aspects of leptocephalus fixation and preservation. Pages 321–322 in Steedman, H. F. (editor). *Zooplankton Fixation and Preservation.* Monographs on Oceanographic Methodology 4. Paris: Unesco Press.

Steigerwald, M., and S. Laframboise. 1994. A method for reducing ethanol evaporation in wet collections. Pages 1–10 in Snyder, A. M. (editor). *The 1993 American Society of Ichthyologists and Herpetologists Workshop on Collections Care and Management Issues.* Albuquerque: American Society of Ichthyologists and Herpetologists.

Storey, M., and N. J. Wilimovsky. 1955. Preliminary report on containers and closures for storing specimens preserved in fluid. *Curatorial Practices in Zoological Research Collections Circulars* 3:1–23.

Subrahayan, V., N. L. Lahiry, M. N. Moorjanji, R. Ralanthrisham, and M. N. Krishnaswami. 1963. Ammonia: possible use for preserving fish. *Science* 142:233–234.

Suzumoto, A. Y. 1989. [Review] Conservation of Natural History Specimens: Spirit Collections, 1989, C. V. Horie, Ed. *Collection Forum* 5(2):80–81.

Suzumoto, A. Y. 1992. Storage containers for fluid-preserved specimens. Pages 217–220 in Rose, C. L., and A. R. de Torres (editors). *Storage of Natural History Collections: Ideas and Practical Solutions*. New York: Society for the Preservation of Natural History Collections.

Swainson, W. 1822. *The Naturalist's Guide for Collecting and Preserving Subjects of Natural History and Botany: Both in Temperate and Tropical Countries, Particularly Shells. . .* London: W. Wood.

Tang, P. Y. 2006. *Path to Effective Recovering of DNA from Formalin-fixed Biological Samples in Natural History Collections. Workshop Summary*. Washington, D.C.: National Academies Press.

Tarek, M., D. J. Tobias, and M. L. Klein. 1996. Molecular dynamics investigation of an ethanol-water solution. *Physica A* 231(1/3):117–122.

Tarr, H. L. A., and B. E. Bailey. 1940. Effectiveness of benzoic acid for fish preservation. *Journal of the Fisheries Research Board of Canada* 3:327–336.

Taylor, W. R. 1981. On preservation of color and color patterns. *Curation Newsletter* 3:2–10.

Taylor, W. R. 1984. Comments on "Acetone in Isopropanol" by George Burgess (Curation Newsletter 6:1, May 1983). *Curation Newsletter* 7:2–4.

Thompson, G. 1986. *The Museum Environment*. Stoneham: Butterworth-Heinemann.

Torres Nuñez, M. C., A. Salama, and J. J. Castano Caastrillon. 1998. Uso de las crisalidas del gusano de seda *Bombyx mori linn* en cosmetologia. *Sericultura* 5(24):12–15.

Toyama, K., and G. Miyoshi. 1963. Prevention from color fading of aquatic animals under preservation. *Journal of Tokyo University Fisheries* 50:43–48.

Tucker, A. O., M. J. Marciarello, and S. S. Tucker. 1991. A survey of color charts for biological descriptions. *Taxon* 40(2):201–214.

Turoski, V. 1985. *Formaldehyde: Analytical Chemistry and Toxicology*. Advances in Chemistry Series. Washington, D.C.: American Chemical Society.

Tuross, N. n.d. Time in a bottle. *The impact of formaldehyde fixation and alcohol storage on DNA retrieval, PCR amplification, protein preservation and isotopic stability in ichthyology collections*. Conservation Analytical Laboratory, Smithsonian Institution research proposal.

Viidik, A. 1980. Interdependence between structure and function in collagenous tissues. Pages 257–280 in Viidik, A., and J. Vuust (editors). *Biology of Collagen*. London: Academic Press.

von Lundvall, H. 1905. Weiteres über Demonstration embryonaler Skelette. *Anatomischer Anzelger* 27:520–525.

Wake, D. B., R. G. Zweifel, H. C. Dessauer, G. W. Nace, et al. 1975. Recommendations for the management of herpetological museum collections. *Herpetological Review* 6(2):34–36.

Walker, H. J. 1999. Combination jars. *Curation Newsletter* 12:11.

Weiss, J. B., K. Sedowofia, and C. Jones. 1980. Collagen degradation: a defended multi-enzyme system. Pages 113–134 in Viidik, A., and J. Vuust (editors). *Biology of Collagen*. London: Academic Press.

Wentworth, J. E. 1938. A new method of preserving museum specimens in their natural colours. *Journal of Technical Methods* 18:53.

Whitaker, J., and K. M. Dix. 1979. Double staining technique for rat foetus skeletons in teratological studies. *Laboratory Animals* 13:309–310.

Whitehead, P. J. P. 1970. Museums in the history of zoology. *Museums Journal* 70(2):50–57.

Whitehead, P. J. P. 1971. Museums in the history of zoology. *Museums Journal* 70(4):155–160.

Wilke, H.-J., S. Krischak, and L. E. Claes. 1996. Formalin fixation strongly influences biomechanical properties of the spine. *Journal of Biomechanics* 29(12):1629–1631.

Williams, J. T. 1990. Printer ribbons with non-bleeding ethanol resistant ink—finally a wet label that bypasses presoaking! *Curation Newsletter* 10:4.

Williams, S. L., R. Laubach, and H. H. Genoways. 1977. A guide to the management of Recent mammal collections. *Carnegie Museum of Natural History Special Publications* 4:1–105.

Williams, S. L., and C. A. Hawks. 1986. Inks for documentation in vertebrate research collections. *Curator* 29(2):93–108.

Williams, S. L., and C. A. Hawks. 1988. A note on "Inks. . ." *Society for the Preservation of Natural History Collections Newsletter* 2(1):1.

Williams, T. W. 1941. Alizarin red S and toluidine blue for differentiating adult and embryonic bone and cartilage. *Stain Technology* 16:23–25.

Windsor, A. S. 1938. Maintenance measures for the teaching museum. *Turtox News* 16(2):40–41.

Winokur, R. M., and S. Hillyard. 1992. Pelvic cutaneous musculture in toads of the genus *Bufo*. *Copeia* 1992(3):760–769.

Wood, R. M., and S. L. Williams. 1993. An evaluation of disposable pens for permanent museum records. *Curator* 36(3):189–200.

Woodburne, R. T., and C. A. Lawrence. 1952. An improved embalming fluid formula. *Anatomical Record* 114(3):507–514.

Wright, V. 1971. Elasticity and deformation of skin. Pages 437–449 in Elden, H. R. (editor). *Biophysical Properties of the Skin*. New York: Wiley.

Yoshida, Y. 1962. A way of making fish specimens with their original body-colours kept. *Bulletin of the Misaki Marine Zoology Institute, Kyoto University* 3:67–68.

Zakhari, S., M. Leibowitz, P. Levy, and D. M. Aviado. 1977. *Isopropyl and Ketones in the Environment*. Cleveland, OH: CRC Press.

Zirkle, C. 1933. Aldehydes as cytological fixatives. *Protoplasma* 20:169–170.

III

REFERENCE TABLES

Table 1. Fluid-Preservation Techniques

Reference	*Organisms*	*Comments*
Adams et al. 1976	Ctenophora (comb jellyfish)	Fix for 30 minutes in a solution of 1 g tri-chloracetic acid and 99 mL of seawater or a solution of 1 g p-toluenesulfonic acid and 99 mL of seawater. Transfer to a solution of 1 mL stock (5 mL propylene phenoxetol, 45 mL propylene glycol, and 50 mL formaldehyde) and 99 mL seawater for 5–7 days, then 5 mL stock with 95 mL seawater.
Ahlstrom 1976	Fish eggs and larvae collected during plankton hauls	Place specimens in a quart jar 3/4 full of seawater, add 50 mL commercial grade formaldehyde and 20 mL of a saturated borax solution, fill to the top with seawater.
Aiyappan and Satyamurti 1960	Amphibians	Preserve in 50% alcohol for a few days; change alcohol 3 times; preserve in 60–65% alcohol. Formaldehyde not recommended for adult amphibians, but is preferable for eggs and larvae.
	Reptiles	Inject or make small incisions in the body. Begin with 50% alcohol, then 65% followed by 70% for a few hours each. Preserve permanently in 95% alcohol, changed after first 24 hr. If alcohol is not available, use 5% formaldehyde.
Anderson 1975	Invertebrates	Preserve in alcohol; formalin buffered with 10% hexamine; 1% propylene phenoxetol; or a solution of 1% propylene phenoxetol, 10% propylene glycol, 2% formaldehyde, 5% sodium acetate, and 82% distilled water or seawater.
Anderson 1932, 1948	Amphibians	Preserve in 1:12 formaldehyde and water, 60% ethyl alcohol, or strong rum.
	Reptiles	Preserve in 1:12 formaldehyde and water, 70% ethyl alcohol, or strong rum; make abdominal slits for penetration of preservative in specimens more than 4 inches in length or inject preservative.
	Fishes	Preserve in 75% alcohol; once cured maintain in 65% alcohol or solution of 1:9 formaldehyde and water.
Anderson 1965	Vertebrates	Preserve in alcohol of at least 85% (replace alcohol as necessary) or in solution of 1 pt formalin and 9 pts water; formalin may be neutralized. Alternative preservatives include denatured alcohol, methanol, a strong brine solution; or injection of small organisms with a weak solution of carbolic acid or Lysol.

(continued)

Table 1. ***(Continued)***

Reference	*Organisms*	*Comments*
	Fishes	Preserve in 70% ETOH or 10% formalin (buffered with 1 tsp borax per half gallon; 2 oz hexamine per pint; or 3 oz concentrated ammonia per pint). Store in 45% isopropyl alcohol or 70% ETOH.
Anon. 1793	Small animals	"Small beasts may be put into a cask of rum, or any other spirits."
Anon. 1831	All natural history specimens	Preferred fluid preservative is alcohol 18–20° Baumé [94–94.5%] in sealed vessel. Saturate specimen for 1–2 d, wipe off mucous, and place in fresh preservative. Alternative preservatives include Nichols's solution (2 pints pure water, 1 pint alcohol, 6 oz sulfate of alumina); Graves solution (8 oz alum, 1 pint water, 1/3 pint alcohol; heat water to boiling, add alum, cool, filter, add alcohol); solution of 1 pint water, 1 pint alcohol, and 12 oz alum, mixed cold; or Abbé Manesse solution (1 lb alum, 1 lb nitre, 1 lb sea salt, 4 pints distilled water, and 1 pint alcohol, mixed cold or boiled before alcohol is added).
Anon. 1905 (Bulletin of the College of Charleston Museum)	Earthworms	95% alcohol
	Slugs	95% alcohol
	Crawfishes	3% formaldehyde
	Grasshoppers and cockroaches	Strong alcohol or 3% formaldehyde
	Myriapods (millipedes, centipedes)	70% alcohol
	Frogs	Strong alcohol or 4% formaldehyde; make incisions in large specimens
	Lizards	Strong alcohol or 3% formaldehyde; make incisions in large specimens
	Salamanders, newts, mud puppies, and all other tailed amphibians	6% formaldehyde or strong alcohol, renewed after a few days; open abdominal cavity; make incisions in large specimens.

	Half-grown kittens	7% formaldehyde
Anon. 1919 (American Museum of Natural History; n.d. on publication, but ca. 1919 according to Myers 2000)	Amphibians and reptiles	Preserve in a solution of 1 pt commercial formaldehyde with 12 pts water; 70% alcohol, "pure Aguardiente, strong rum, or wood alcohol" if formalin is not available.
Anon. 1921 (British Museum)	Reptiles	Slit body open and preserve in alcohol; if not available use beverage alcohol strong enough to catch fire. Formaldehyde not recommended.
	Amphibians	Preserve in 50% alcohol for a few days, then 70% alcohol. Formaldehyde not recommended.
	Fish	Slit body open and preserve in alcohol; if not available use beverage alcohol strong enough to catch fire. Formaldehyde is not recommended.
	Insects	Preserve in mixture of 3 pts 90% alcohol, 2 pts distilled water, and 1 pt glycerin.
	Dipteran larvae	Immerse briefly in boiling water; preserve in mixture of 2 pts strong alcohol and 1 pt water; transfer to stronger alcohol after hardening.
	Mosquito larvae	Kill and preserve in 70% alcohol or 4% formalin.
	Centipedes and millipedes	Preserve in alcohol, "Experiments with *formalin* as a preservative for these animals hitherto met with failure."
	Peripatus	Drown in water with a few drops of ammonia; preserve in 70% alcohol.
	Soft-bodied invertebrates	Most can be preserved in 70% alcohol or mixture of 1 pt formaldehyde with 9 pts water; transfer to 5% or less formalin for storage.
	Porifera	Preserve in alcohol (formalin not recommended).
	Hydrozoa	Preserve in 70% alcohol or 5–7% formalin.
	Ctenophora	Stage up to 70% alcohol.
	Anthozoa or Actinozoa	Stage up to 70% alcohol.
	Stony Corals	Inject 35% alcohol into the mouth of each polyp; preserve in 70% alcohol.
	Rotifera	Preserve in 6% formalin.
	Polyzoa	Preserve in 70% alcohol or 5% formalin.
	Mollusca	Fix in Bouin's solution (picro-formol-acetic); preserve in 70% alcohol.

(continued)

Table 1. ***(Continued)***

Reference	*Organisms*	*Comments*
	Crustacea	Kill and fix in 30% alcohol; preserve in strong spirit.
	Worms (parasites)	Preserve in 70% alcohol.
	Pulpy fruits	Preserve in alcohol or mixture of 1 pt formaldehyde with 5 pts water.
Anon. 1935	Plants	Place pressed plants in a ten gal tin box with a tight-fitting lid; pour in 1 pint alcohol and seal box.
Anon. 1943 (Turtox News, credited to "A.H.B.")	Decapods (crayfish, crabs, lobsters, shrimp)	70% alcohol or 10% formaldehyde.
	Marine cladocera (water fleas)	70% alcohol
	Ostracods (seed shrimp)	70% alcohol
	Copepods	70% alcohol
	Barnacles	10% formaldehyde or 70% alcohol
	Acorn barnacles	70% alcohol for 24 hr; store in 70–80% alcohol
	Sacculina and *Peltogaster* (parasitic barnacles)	70% alcohol
	Amphipods (shrimp-like crustaceans; scud)	5% formaldehyde or 70% alcohol
	Larval stages of crabs and lobsters	30% alcohol; then staged up to 70% alcohol
Anon. 1944 (Turtox News)	Freshwater sponges	Fix and preserve in 70% ETOH.
	Hydra	Fix in Bouin's solution (picric acid, formaldehyde, and glacial acetic acid); preserve in 70% alcohol.
	Freshwater planaria	Fix in Gibson's fixative (nitric acid, glacial acetic acid, corrosive sublimate, 60% alcohol, water) or corrosive sublimate; preserve in formaldehyde or alcohol.
	Tapeworms	Bouin's solution (picric acid, formaldehyde, and glacial acetic acid); preserve in alcohol or formaldehyde.

	Ascaris (roundworms)	Fix in 5% formaldehyde or saturated corrosive sublimate; preserve in 5% formaldehyde or alcohol.
	Rotifers	Fix in killing solution of cocaine hydrochloride, alcohol, and water with a few drops of osmic acid; preserve in 10% formaldehyde.
	Pectinatella and plumatella (freshwater bryozoa)	Fix in Bouin's solution (picric acid, formaldehyde, and glacial acetic acid); preserve in 70% alcohol.
	Earthworms	Fix and preserve in 5% formaldehyde.
	Leeches	Fix—inject with 10% formaldehyde; preserve in 8% formaldehyde.
	Crayfish	Fix and preserve in 70% alcohol or 8% formaldehyde.
	Ticks and mites	Fix and preserve in 70% alcohol.
	Centipedes and millipedes	Fix—inject with Carl's solution (alcohol, formaldehyde, glacial acetic acid, water); preserve in Carl's solution.
	Insects	Fix in alcohol, Carl's solution (alcohol, formaldehyde, glacial acetic acid, water), or chloral hydrate; preserve in alcohol, Carl's solution, or by drying.
	Slugs	Fix in alcohol or formaldehyde; preserve in 70% alcohol or 8% formaldehyde.
	Aquatic snails	Fix in 10% formaldehyde; preserve in 8% formaldehyde.
	Clams	Fix in 10% formaldehyde; preserve in 8% formaldehyde.
	Lampreys	Fix in 10% formaldehyde; preserve in 8% formaldehyde.
	Fishes	Fix—inject with 10% formaldehyde; preserve in 8% formaldehyde.
	Grass frogs	Fix—inject with 5% formaldehyde; preserve in 5% formaldehyde.
	Grass frog eggs	Fix in 8% formaldehyde or Tellyesniczky's fixative (1L ETOH, 100 mL formaldehyde and 50 mL glacial acetic acid); preserve in 8% formaldehyde.
	Salamanders	Fix in 5% formaldehyde; preserve in 5% formaldehyde.
	Reptiles	Fix in 10% formaldehyde; preserve in 8% formaldehyde.
	Large mammals	Fix by embalming or injection with 8% formaldehyde; preserve in 8% formaldehyde.
Anon. 1944 (British Museum)	Balanoglossids (acorn worms)	Allow to evacuate sand and mud from gut for 1 hr, then preserve in 1 pt formalin and 10 pts sea or freshwater, or in 50% alcohol for 3–4 hr and then 70% alcohol.
	Mollusks	Preserve in 80% alcohol, with several changes of fluid.
	Spiders	Preserve in alcohol.

(continued)

Table 1. ***(Continued)***

Reference	*Organisms*	*Comments*
	Crustaceans and miscellaneous invertebrates	Kill in 30–40% alcohol, preserve in 70% alcohol.
	Jellyfish	Preserve in 1 pt formalin and 20 pts water.
	Worms and other soft-bodied invertebrates	Kill by slowly adding alcohol or formalin to water; kill marine species in 154 g magnesium sulfate per L water, then transfer to weak alcohol or formalin solution
	Earthworms	Kill in 8–10 percent alcohol, then preserve in solution of 1 pt formalin and 10 pts water.
	Leeches	Kill in narcotic solution, soda water, chloroform, weak nicotine, magnesium sulfate, alcohol, or weak acid solution; preserve in 50% alcohol or 2 percent formaldehyde, then transfer to 85% alcohol or 5% formaldehyde.
	Echinoderms	Preserve in 70% alcohol with at least two changes to prevent dilution
	Brittlestars	Drop into 70% alcohol.
	Crinoids or featherstars	Drop into 75% alcohol.
	Holothurians (sea cucumbers)	Drop small specimens into 70% alcohol; narcotize and inject specimens more than a few inches in length.
	Amphibians	Arrange in a natural position in a pan of 10% formalin for 3–4 hours to stiffen, transfer to 60% alcohol or 2–3% formaldehyde. Make slits in venter of anurans over 5 inches long.
	Small lizards and snakes	Preserve in solution of 3 pts alcohol and 1 pt water or 4–5% formalin. Cut slits in body to allow for penetration of preservative.
	Large lizards and snakes	Remove skin by cutting mid-ventrally from anus to chin, leaving head, 2–3 inches of neck, and tail in skin; place in solution of 3 pts alcohol and 1 pt water or 4–5% formalin. Cut slits in body to allow for penetration of preservative.

	Turtles	With a large hypodermic needle inject pure alcohol or formalin at the base of each leg, or make deep cuts in limbs and skin between limbs and shell to allow for penetration of preservative; prop mouth open with a stick, extend neck fully, and place entire animal in 75 or 90 percent alcohol.
	Crocodilians, large lizards and snakes	Preserve specimens less than two feet long as for turtles; make dry skins of larger animals.
	Small fishes	Formaldehyde preservation: place alive in solution of 1 pt formalin and 9 pts water; make abdominal slit in specimens longer than 3 inches, or inject with hypodermic needle; after minimum of 5–7 days, transfer to water, then to 75% alcohol. For long-term storage in formaldehyde mix 1 pt formalin with 15–18 pts water with 2 tsp borax per gallon of preservative. Alcohol preservation: place alive in 35% alcohol for six hr, transfer to 75% alcohol.
Anon. 1953 (British Museum)	Reptiles, amphibians, and fishes	Inject 70% alcohol or 5% formalin into the alimentary canal and body cavity, or make short slits to one side of the mid-ventral line, immerse specimen in 50% alcohol or 2–3% formalin for a few hours up to 48 hr; transfer specimens to 60–70% alcohol or 2–5% formalin for 7–14 days; transfer amphibians and geckos to 60% alcohol or 3% formalin, transfer other reptiles and most fishes to 70–75% alcohol or 4–5% formalin. Buffer formalin solutions with 200 g of hexamethyltetramine (hexamine) per L of full-strength formalin (formaldehyde) solution.
Anon. 1942–1959 (American Museum of Natural History, n.d.; ca 1942–1959 according to Myers 2000)	Amphibians and reptiles	Fix in 1 pt formaldehyde with 12 pts water; 70% ETOH; or "pure Aguardiente, strong rum, or wood alcohol." Make slits in body cavity or inject fluid into body. Allow to harden for 12–24 hr.
Anon. 1957 (Armed Forces Institute of Pathology Medical Museum Laboratory)	Anatomical specimens (to preserve color)	Fix specimen in a solution of 89.1 g sodium phosphate monobasic, 112.5 g sodium phosphate dibasic, 950 cc formaldehyde, and 19,000 cc distilled water. After fixation, preserve in a solution of 89.1 g sodium phosphate monobasic, 112.5 g sodium phosphate dibasic, 950 cc formaldehyde, 95 g sodium hydrosulfite, and 19,000 cc distilled water.

(continued)

Table 1. *(Continued)*

Reference	*Organisms*	*Comments*
Anon. 1963 (American Museum of Natural History, "supplement to pamphlet," n.d.; 1963 according to Myers 2000)	Lizards	Inject body cavity with 1 solution of pt formaldehyde and 10 pts water; slit tail.
	Salamanders	Inject body cavity with 1 solution of pt formaldehyde and 13 pts water if body length is greater than 3 inches.
	Frogs	
	Turtles	Inject body cavity with 1 solution of pt formaldehyde and 10 pts water.
	Snakes	Inject body cavity with 1 solution of pt formaldehyde and 10 pts water.
	Reptiles and amphibians	Inject body cavity with 1 solution of pt formaldehyde and 10 pts water.
		If formaldehyde is not available, preserve specimens in brine ("saturated solution of ordinary table salt"); or use 1 pt isopropyl alcohol and 2 pts water.
Anon., n.d. (Field Museum of Natural History)	Reptiles and amphibians	Fix in 1 pt formaldehyde with 12 pts water; or 70% ETOH; use pure Aguardiente or strong rum if formaldehyde is not available. Do not use alcohol stronger than 60% ("ordinary rum"). Make slits in body cavity or inject fluid into body.
Anon. 1944 (Smithsonian Institution)	Amphibians and small reptiles	Arrange in pan of 10% formaldehyde for 3–4 hr, then move to 60% alcohol, or place directly in 60% alcohol. Make incisions in larger specimens.
	Fishes	Place live in solution of 1 pt formaldehyde and 9 pts water; make slits in specimens more than 3 inches long.
	Small lizards and snakes	Make small cuts in body; use solution of 3 pts alcohol and 1 pt water, or 4–5% formaldehyde.
	Turtles	Inject with "pure alcohol or formaldehyde at the base of each leg," then submerge in 75–90% alcohol.
	Crocodiles	Inject with 75–90% alcohol.
	Balanoglossids (acorn worms)	Fix in 1 pt formaldehyde and 10 pts fresh or seawater, or preserve in 50% alcohol for 3–4 hr then 70% alcohol.

	Mollusks	Preserve in 80% alcohol.
	Spiders	Preserve in alcohol.
	Crustaceans and miscellaneous invertebrates	Preserve in 60% alcohol overnight, then move to 70% alcohol; or use 1 pt formaldehyde with 20 pts water.
	Earthworms	Preserve in 1 pt formaldehyde with 10 pts water.
	Leeches	Preserve in 50% alcohol or 2% formaldehyde; after hardening move to 85% alcohol or 5% formaldehyde.
Ayre et al. 2008	Amphibods, annelids, copepods, isopods, and sea urchins	Narcotize and preserve in 70% IMS (magnesium sulfate and menthol crystals were poor narcotizing agents).
	Crustacea	Narcotize by adding Alka-Seltzer tablets to seawater; preserve in 70% IMS.
	Fish	Narcotize by adding Alka-Seltzer tablets to seawater; preserve in 70% IMS.
Baird 1852	General	Preserve in ETOH; if not available preserve in (in descending order of recommendation) rum or whiskey; Goadby's solution (either 4 oz rock salt, 2 oz alum, 4 grs corrosive sublimate, and 2 qts boiling water; or 8 oz rock salt, 2 grs corrosive sublimate, and 1 qt boiling water); or a strong brine solution. To preserve in Goadby's solution, macerate the specimen for a few hrs in freshwater, transfer to fresh solution after a few days.
Bancroft 1769	Birds	Cover with "High Wines, or the first Running of the Distillation of Rum" for 24 hr or longer, then dry in a moderate oven.
	Snakes	Wash, coil in container, and submerge in rum.
Barker and Grigg 1977	Frogs	Adults: 10% formaldehyde or 70% alcohol with 10% glycerin to preserve softness Larvae: 5–7% formaldehyde or 50% alcohol; ideal solution is Tyler's fixative (10 mL formaldehyde, 4 mL 10% aqueous calcium chloride, and 0.2 g cobalt nitrate in 90 mL water).
Bé and Anderson 1976	Foraminifera and other calcareous plankton	Fix within 5 min of retrieval in a solution of 5 mL 40% formaldehyde (buffered with 30g sodium tetraborate per 1000 mL), 4.5 mL propylene glycol, and 0.1 mL propylene phenoxetol per 90 mL plankton in seawater.

(continued)

Table 1. *(Continued)*

Reference	*Organisms*	*Comments*
Bean 1895	Fish	Inject alcohol through the mouth and vent and make incisions in body; keep specimens in 50% alcohol for 1–2 days, then 75% alcohol.
Beers 1976a	Actinopods	Use strontium to stabilize actinopod skeletons.
Beirne 1955	Insects	For soft bodied insects and larvae and nymphs use 75% ETOH or 75% methyl alcohol with 2% glycerin in case of evaporation. For others, use AGA (8 pts ETOH, 5 pts distilled water, 1 pt glycerin and 1 pt glacial acetic acid) or Pempel's fluid (4 pts glacial acetic acid, 6 pts formaldehyde, 30 pts distilled water, 15 pts ETOH). Cautions that dilution of preservative occurs after initial preservation.
Berger 1955, 1956	Birds	Preserve in alcohol or 10% formaldehyde buffered with sodium phosphates. Make ventral abdominal slit in small birds; use intravenous injection or injection into breast, arm, thigh, crus, neck, thoracic cavity, and abdomen for large birds.
Berland 1984	Helminths	Platyhelminths and acanthocephalans except nematodes: flatten between glass slides; kill with hot 70% ETOH, Looss's fluid (1 pt glycerol and 9 pts 70% ETOH), or Berland's solution (1 pt formaldehyde and 19 pts glacial acetic acid). Transfer to 70–80% ETOH or solution of 1 pt glycerol and 9 pts 70% ETOH.
Blake 1949	Birds	5% formaldehyde with two tablespoons of salt per quart and 85% alcohol, changed at intervals as it becomes diluted.
Blaufuss 1933 *Note*: Preservation should be done immediately after specimen is killed; store specimens in same strength solution as used for injecting; cavity injections are preferred.	Fish	Inject and store in 10% formalin.

	Frogs	Drop small frogs into 5% or 10% formalin or inject body cavity through thigh muscles to prevent loss of preservative; inject larger frogs.
	Lizards and snakes	Cavity injections of 5% or 10% formalin
	Turtles	Cavity injections of 5% or 10% formalin; replace storage fluid after 3 months.
	Alligators	Cavity and muscle injections of 5% or 10% formalin
	Birds	Cavity injections of 5% or 10% formalin for pigeon-sized birds; inject domestic fowl though femoral artery.
	Mammals	Inject 5% or 10% formalin through femoral or carotid artery.
Borodin 1930	Fish	30 pts alcohol, 2 pts formaldehyde, 1 1/2 pts wood tar, and 66 1/2 pts water saturated with common salt
Borror and DeLong 1964; Borror, DeLong, and Triplehorn 1976	Insects	Preserve in ETOH; XA mixture (1:1 alcohol and xylene); Hood's solution (alcohol with glycerin); KAAD (kerosene, alcohol, glacial acetic acid, and dioxane); Kahle's solution (alcohol, water, formaldehyde and glacial acetic acid); or alcoholic Bouin's solution (alcohol, formaldehyde, glacial acetic acid, picric acid).
Bory de Saint-Vincent 1828	Fishes and some other soft animals	Preserve in oils.
	Fishes and skins of large animals	Brine (water saturated with muriate of soda)
	Soft animals	2 pts pure water, 1 pt alcohol, and 2 oz of aluminum sulfate for every L of solution prepared.
	Natural history objects	Vinegar, particularly if saturated with common salt; red wine with nitrous mercury; pyrolignous acid and sulfurous acid; alcohol including arrack, tafia, rum, whisky, gin, etc. (the more transparent the better), preferably 16–22° Baumé Guyot's liquor: distill 5 pints of spirit from 20 pints of Cognac, add 5 pts well water and 1 lb of flowers, leaves, or green lavender, then distill again to dryness. Mix 11 pts of the spirit distilled from the Cognac with 69 pts well water, and 69 pts of the second distilled solution.
Boulenger 1891	Anuran larvae	Preserve in weak alcohol in test tubes, changing to 40% alcohol within a few hrs, change alcohol after 24 hr. Recommends against preservation with chromic acid "as rendering the specimens too brittle for ordinary study."

(continued)

Table 1. *(Continued)*

Reference	*Organisms*	*Comments*
Boyle 1666	Fetuses and birds removed from eggs	Rinse in alcohol, then preserve "in a distinct Glass (which is to be carefully stopt) in *Spirit of Wine*." Add "a little Spirit of Sal *Armoniack* . . . made with the help of *Quick-lime*."
Bragg 1949	Amphibian eggs and larvae	Recommends (in order), 2–10% (generally 5%) formaldehyde; Bouin's picro-formol fixing solution; or FAA (formaldehyde, acetic acid, and alcohol).
Bridson and Forman 1998 *Note*: IMS = Industrial Methylated Spirit (ETOH, 9% water, 2–4% methyl alcohol).	Algae	Fix and preserve in 1 pt formaldehyde, 9 pts water, neutralized with borax or calcium carbonate (use seawater for seaweeds); add a little glycerol to prevent dehydration.
	Charophytes	Fix and preserve in 1 pt formaldehyde, 9 pts water, neutralized with borax or calcium carbonate (use seawater for seaweeds) with a little glycerol to prevent dehydration; or in 70% ETOH (3 pts 95% IMS and 1 pt water).
	Flowers, soft fruits, some succulent taxa	Fix in Kew mixture (10 pts IMS, 1 pt formalin, 1 pt glycerol, 8 pts water); or 50–70% alcohol (either ETOH or IMS) and 30–50% water. Preserve in Copenhagen mixture (10 pts IMS, 1 pt glycerol, 8 pts water).
	Plant organs intended for anatomical study	Fix in 70% ETOH; Carnoy's fluid (3 pts 96% alcohol and 1 pt glacial acetic acid) for 24–38 hr; or AA (1 pt glacial acetic acid and 8 pts 70% ETOH) for 2–3 days; transfer to Copenhagen mixture (10 pts IMS, 1 pt glycerol, 8 pts water) or 70% ETOH for storage.
		Fix delicate or soft material in FAA (1 pt 40% formaldehyde, 1 pt glacial acetic acid, and 18 pts 70% ETOH) for 2–3 days; may also be fixed in Kew mixture (10 pts IMS, 1 pt formalin, 1 pt glycerol, 8 pts water) although this is "not ideal." After fixation, store in Copenhagen mixture (10 pts IMS, 1 pt glycerol, 8 pts water) or 70% ETOH.
	Very delicate anatomical material (e.g., ovules)	Fix in FPA (1 pt 40% formaldehyde, 1 pt glacial propionic acid; 18 pts 70% ETOH); transfer to Copenhagen mixture (10 pts IMS, 1 pt glycerol, 8 pts water) or 70% ETOH for storage.
	Sporangia, flower buds, root-tips for cytological study	Fix in Carnoy's fluid (3 pts 96% alcohol and 1 pt glacial acetic acid) or 6:3:1 mixture (6 pts chloroform, 3 pts 95% ETOH, 1 pt glacial acetic acid) for 24–48 hr; transfer to Copenhagen mixture (10 pts IMS, 1 pt glycerol, 8 pts water) or 70% ETOH for storage.

	Bark or wood, small samples	Preserve in 70% IMS.
British Museum (Natural History) 1953	Small amphibians and reptiles	Preserve initially in 60% alcohol or 3% formaldehyde, then move to 70–75% alcohol or 5% formaldehyde. Neutralize formaldehyde solutions with hexamine (200 g per L of solution), or use borax or chalk.
	Fishes	Slit open abdomen and place in 2% formaldehyde for a few hours to overnight, then wash in water and move to 50% alcohol for 3 d, then 60–65% alcohol for 3 d, then 70–75% alcohol. Neutralize formaldehyde with 200 g hexamine per L of solution, or use borax or chalk.
Brooks 1932	Hydra (freshwater)	Relax in water from collecting source; flood with hot saturated solution of corrosive sublimate (corrosive sublimate crystals dissolved in 16 pts cold distilled water and 3 pts boiling distilled water) or corrosive sublimate crystals dissolved in 3 pts alcohol to kill and fix; after a few minutes draw off fluid and add iodized alcohol (100 pts water, 6 pts potassium iodide, 4 pts iodine crystals added to 50–70% alcohol until a wine color), then wash and preserve in 70% alcohol.
	Flatworms and roundworms from mammals	Drop freshly skinned mammal specimen into solution of 1 pt formaldehyde and 9 pts water. Preserve cysts containing live roundworms in hot 70% alcohol; preserve cysts containing dead roundworms in solution of 1 pt formaldehyde and 9 pts water, then transfer to alcohol. Preserve live tapeworms by dipping in a warm solution of corrosive sublimate (mercuric chloride crystals dissolved in 16 pts cold distilled water and 3 pts boiling distilled water), raising specimen and repeating until translucent tissue becomes opaque; wash in iodized alcohol (100 pts water, 6 pts potassium iodide, 4 pts iodine crystals added to 50–70% alcohol until a wine color) and preserve in 70% alcohol. Place live flukes in a solution of 0.75 g NaCl and 100 cc distilled water, or in tepid water to relax, then add an equal amount of corrosive sublimate, wash in iodized alcohol, and store in 70% alcohol.
	Echinoderms	Kill in solution of 100 cc seawater and 5 cc absolute alcohol; preserve in alcohol or formalin. Kill brittle-stars in freshwater or solution of 0.5% chromic acid, preserve in alcohol or formalin. Kill crinoids in 90% alcohol and preserve in 70% alcohol.

(continued)

Table 1. ***(Continued)***

Reference	*Organisms*	*Comments*
	Leeches and other annelids	Anesthetize in fresh or salt water, alcohol, chloral hydrate, or chloroform; preserve in alcohol.
Browne 1884	Fish and reptiles	Preserve in "Rectified Spirits of Wine," 56% ETOH or methylated spirits (ETOH with a small amount of impure gum or methyl alcohol); alcoholic solution No. 1 (1.5 pints methylated spirit, 1/2 pint distilled water, 2 oz burnt alum, and 4 oz saltpeter); alcoholic solution No. 2 (3 pts methylated spirits, 1 pt glycerin, 1 pt distilled water). Also recommended Goadby's solution No. 1 (4 oz salt, 2 oz alum, 4 grs corrosive sublimate, 2 quarts boiling water); Goadby's solution No. 2 (0.5 lb salt, 20 grs arsenious acid, 2 grs corrosive sublimate, 1 qt boiling water); Browne's solution (4 oz saltpeter, 2 oz alum, 0.25 oz corrosive sublimate, 1/8 oz sal ammoniac); saline solution (1 gr bichloride of mercury, 90 grs sodium chloride, 1 pint distilled water); Camphorated fluid (16 pts distilled water, 1 pt rectified spirits of wine, a few drops of creosote, and a small quantity of chalk); and Möller's solution (2 oz bichromate of potash, 1 oz soda sulfate, and 3 pints distilled water).
	Mollusca	Preserve in pure alcohol; chloride of zinc; or calcium chloride solution made by dissolving chalk or white marble in hydrochloric acid.
Camacho and Bedoya 1994	Oligochaetes (earthworms)	Fix in 1 mL Angelier's fluid per 100 mL of water for six wks; preserve in 70% ETOH, 5% unbuffered formaldehyde, Angelier's fluid, or Koenike's solution.
Notes: Angelier's fluid (1% acetic acid, 1% chromic acid, 98% distilled water)	Ostracods (seed shrimp)	Preserve directly in 70% ETOH, or fix in 1 mL of Angelier's fluid per 100 mL of water, or Koenike's solution; wash 24 hr after fixing, preserve in 70% ETOH.
Koenike's solution (3 pts glacial acetic acid, 11 pts glycerol, 6 pts distilled water)	Copepods	Preserve directly in 70% ETOH, or fix in 5% unbuffered formaldehyde, then preserve in 70% ETOH; or fix in 1 mL of Angelier's fluid per 100 mL of water for up to 6 weeks or Koenike's solution for up to 24 hr, then wash and preserve in 70% ETOH.
	Syncarids	Preserve directly in 70% ETOH.
	Mollusks	Anesthetize prior to fixation; fix in Angelier's fluid or Koenike's solution for no more than 24 hr; preserve in 70% ETOH.
	Isopods	Preserve directly in 70% ETOH, or fix in 5% unbuffered formaldehyde and preserve in 70% ETOH; or fix in Koenike's solution for up to four wks, then preserve in 70% ETOH or 5% unbuffered formaldehyde.

	Amphipods	Preserve directly in 70% ETOH, or fix in 5% unbuffered formaldehyde and preserve in 70% ETOH; or fix in either Angelier's fluid for up to 24 hr or Koenike's solution for up to 4 wks, then preserve in 70% ETOH.
	Water mites	Fix and preserve in either Angelier's fluid or Koenike's solution.
Cannell 1988	Avian syringes	Preserve fresh syringes in buffered 10% formalin for several days, blot excess fluid, and transfer to 70–75% ETOH for storage; place syringes from fluid-preserved birds directly in 70–75% ETOH for storage.
Casas-Andreu et al. 1991	Amphibians	Inject larger individuals with freshly prepared 10% formaldehyde buffered with 4 g monobasic sodium phosphate monohydrate and 6.5 g dibasic sodium phosphate anhydrate per L (smaller individuals do not have to be injected); preserve in 70–75% ETOH. Preserve amphibian larvae in 10% buffered formaldehyde.
	Reptiles	Inject with freshly prepared 10% formaldehyde buffered with sodium phosphates; preserve in 70–75% ETOH.
Chamberlin 1925	Arachnids and small insects	Preserve specimens in 95% ETOH; change to fresh alcohol solution; transfer to solution of 25% carbolic acid crystals in 75% xylene; transfer to pure xylene; transfer to a highly refined, colorless and highly viscous mineral oil (e.g., Squibbs, Nujol, or Oronite Crystal Oil) for long-term storage.
Clemens 1951	Fish otoliths	Store in tri-sodium phosphate.
Conant 1924, 1926	Plants	Fix in solution of 100 cc of 50% alcohol, 6 1/2 cc formaldehyde, and 2 1/2 cc glacial acetic acid; preserve in two changes of 70% alcohol followed by 80% and then 95% alcohol.
Cook 1965	Amphibians	Inject (except very small frogs) with a solution of 10% formalin (1 pt formalin and 9 pts water), or inject with a 5% formalin solution; after 24 hr transfer to 3% formalin solution. If formalin is not available, preserve in 70% ETOH or 40% isopropyl alcohol; transfer to 70% ETOH after 24–48 hr. Preserve larvae in a 3% formalin solution.
	Reptiles other than turtles	Inject with a solution of 10% formalin (1 pt formalin and 9 pts water), or inject with a 5% formalin solution; after 24 hr transfer to 5% formalin solution. If formalin is not available, preserve in 95% ETOH or 40% isopropyl alcohol; transfer to 70% ETOH after 24–48 hr.

(continued)

Table 1. ***(Continued)***

Reference	*Organisms*	*Comments*
	Turtles	Inject with a solution of 10% formalin (1 pt formalin and 9 pts water). If formalin is not available, preserve in 95% ETOH or 40% isopropyl alcohol; transfer to 70% ETOH after 24–48 hr.
Cooke 1969	Spiders	Kill in 1–2% aqueous propylene phenoxetol; fix in 70% alcohol overnight; store in 1–2% aqueous propylene phenoxetol.
Cristobo et al. 1992	Porifera	Fix in 4% formaldehyde buffered with borax or hexamine; preserve in 70% ETOH.
Cross 1962	Fish	Fix in solution of 1 pt formaldehyde and 9 pts water. Slit open the right side of specimens more than six inches long. After 3–7 days, wash for 2–4 days and transfer to 70% ETOH.
Dall 1892 (based on method of John A. Eyder)	Mollusc soft parts	Euthanize by drowning in airtight vessel completely filled with water for 24 hr. Harden in solution of 1 pt 95–97% alcohol and 2 pts water for 24 hr, then transfer to equal pts water and alcohol, finish hardening in solution of 2–4 pts alcohol (95–97%) and 1 pt water.
		Harden in Müller's Fluid [1 gal water, 3 oz pulverized bichromate of potash, and 1¼ oz sulfate of soda (Gaber's salts)]. Use several volumes of solution to volume of specimens, keep in a cool, dark place, change solution daily for 3 d, then weekly for 3–6 wks. Once hardened, wash with water and store in 70% alcohol.
Davidson 1932	Anuran eggs	Fix in Smith's fixative (0.5 g potassium bichromate, 2.5 cc glacial acetic acid, 10 cc formaldehyde, and 87.5 cc water) for 24 hr, wash, preserve in 3% formaldehyde.
	Anuran larvae	Fix larvae > 10 mm in Smith's fixative; for others use Bouin's solution (75 cc saturated picric acid, 25 cc formaldehyde, and 5 cc glacial acetic acid) for 24 hr and store in 35% alcohol; fix in acetic-alcohol (90 cc absolute alcohol and 10 cc glacial acetic acid) for 8 hr, then wash in absolute alcohol; fix in corrosive-acetic (5 g corrosive sublimate, 10 cc glacial acetic acid, 90 cc water) for 24 hr, wash, and store in 35% alcohol.
De Villiers 1929	Anuran larvae	Fix in corrosive sublimate, formaldehyde, or carbolic acid; do not use alcohol.

DeWolf 1968	Plants	Use "standard botanical preserving fluid" (5 pts formaldehyde, 5 pts glacial acetic acid, and 90 pts 50% alcohol). "For field work, 70% alcohol alone will do. This may be denatured ethyl alcohol or rubbing alcohol (isopropyl alcohol)."
Edmunds et al. 1976	Mayfly adults	80% alcohol with 1% Ionol (2, 6-di-tert-butyl 4-methylphenol), an antioxidant to lesson bleaching
	Mayfly larvae	Fix in Carnoy's fluid [10% glacial acetic acid, 60% ETOH (95%), 30% chloroform]; preserve in 80% ETOH. Kahle's solution: 11% formalin, 28% ETOH (95%), 2% glacial acetic acid, 59% water.
	Mayfly nymphs	Ethyl alcohol, 95%
Eekhout 2010	Amphibians and reptiles	Fix in 10% formalin or 70% ETOH
Emerson and Ross 1965	Invertebrates	Fix with chemical appropriate for the species and purpose; store in 70% ETOH for best results; alternate preservatives include 80% ETOH, isopropyl alcohol, formaldehyde, and formol-alcohol (95 mL of 70% alcohol and 5 mL formaldehyde).
Fenaux 1976	Appendicularians	Fix in 2% formaldehyde prepared with seawater.
Fisk and Addoms 1935	Plants	Color-fixing solution: 500 cc acetic acid (50%) and sufficient cupric acetate to form a saturated solution (about 15 g); Keefe's color-fixing solution: 90 cc alcohol (50%), 5 cc formaldehyde, 2.5 cc glycerin, 2.5 cc glacial acetic acid, 10 g cupric chloride, and 1.5 g uranium nitrate; Formalin-alcohol: 960 cc alcohol (60 or 70%), 40 cc formaldehyde; Rawlin's solution: 100 cc alcohol (50%), 6 cc formaldehyde, 2.5 cc glacial acetic acid
Forman and Bridson 1989	Flowers for morphological study	Preserve in Kew mixture (70% ETOH, 29% water, and 1% glycerol).
Notes: IMS = Industrial Methylated Spirit (ETOH, 9% water, 2–4% methyl alcohol)	Fruit, leaf, axils of leaf nerves (for domatia), stem, wood, root, bulbs, fleshy plants (e.g., orchids or cacti)	Fix in 70% alcohol or Carnoy's fluid (3 pts 96% ETOH and 1 pt glacial acetic acid); store in Kew mixture (70% ETOH, 29% water, and 1% glycerol).

(continued)

Table 1. *(Continued)*

Reference	*Organisms*	*Comments*
The original Kew mixture consisted of 53% IMS, 37% water, 5% formalin (dilute formaldehyde), and 5% glycerol (also described as 10 pts IMS, 1 pt formalin, 1 pt glycerol, and 8 pts water). Beginning with the second edition of *The Herbarium Handbook* (1989), the formula for Kew mixture was changed to 70% ETOH, 29% water, and 1% glycerol.	Very young sporangia or flower buds (for meiotic study)	Fix in Carnoy's fluid (3 pts 96% ETOH and 1 pt glacial acetic acid) or 6:3:1 mixture (6 pts chloroform, 3 pts 95% ETOH, 1 pt glacial acetic acid); store in Kew mixture (70% ETOH, 29% water, and 1% glycerol).
Fraser 1961	Phytoplankton	Po yethylene glycol 30%
Furnestin 1976	Chaetognatha (arrow worms)	Fix in 4–5% formaldehyde mixed with seawater; store in 2% formaldehyde.
Gerrick 1968	Fishes (*Etheostoma caeruleum*, *Etheostoma blennoides*, *Notropis cornutus*, and *Perca flavescens*)	After testing fifteen commercial antioxidants in 10% formalin and 40% isopropyl for color preservation, selected 1% erythorbic acid in 10% formalin as best; Ionol CP-40 preserved reds. Isopropyl alcohol was ineffective.
Geiger et al. 2007	Molluscs	Crack shells slightly or place specimens in hot water to facilitate entrance of fixative; fix marine samples in 5–10% formaldehyde and seawater buffered with 1 g borax per L for pH 7.5–8.5. Store specimens in 70–80% ETOH (borax, powdered aragonite or powdered calcite may be added to safeguard against shell damage).

Gibb and Oseto 2000	Parasitic Hymenoptera	Kill and preserve in 95% alcohol.
	Orthoptera	Preserve in a solution of 5% formalin and alcohol
	Grylloblattodea	Preserve in mixture of 80% alcohol and glycerol.
	Coleoptera grubs and larvae; Hemiptera, Mecoptera, Neuroptera, Strepsiptera, Trichoptera, Zoraptera	Preserve in 70–80% alcohol.
	Diptera larvae, Odonata nymphs, Siphonaptera	Preserve in alcohol
	Thrips and mites	Preserve in AGA (8 pts ETOH, 5 pts water, 1 pt glycerin, 1 pt glacial acetic acid.
	Most arthropod larvae	Preserve in KAA (10 pts ETOH, 2 pts glacial acetic acid, 1 pt kerosene).
	Most insects	Preserve in glycerin.
Gibbs 1976	Planktonic polychaeta	Fix in 2% formaldehyde mixed with seawater; rinse in water; store in fixative or 70–80% alcohol.
Gloyd 1938 *Notes*: 10% formalin = 1 pt formaldehyde and 9 pts water 5% formalin = 1 pt formaldehyde and 19 pts water 1-3-7 mixture = 1 pt formaldehyde, 3 pts 95% alcohol, and 6–7 pts water	Salamanders	Harden in 5–8% formalin or 65% alcohol; rinse in water; store in 60% alcohol or 5% formalin.
	Frogs and toads	Harden in 5–8% formalin, 65% alcohol, or 1-3-7 mixture. Cut into body cavity of large specimens or inject preservative; rinse in water; store in 60% alcohol or 5% formalin.
	Lizards	Harden in 85% alcohol; harden geckos and other thin-skinned species in 65% alcohol. Cut into body cavity of large specimens, or inject with 95% alcohol; rinse in water; store in 70–75% alcohol or 5% formalin.

(continued)

Table 1. ***(Continued)***

Reference	*Organisms*	*Comments*
	Snakes	Harden in alcohol, 10% formalin, or 1-3-6 mixture; slit body cavity open or inject preservative; rinse in water; store in 70–75% alcohol or 5% formalin.
	Turtles	Harden in alcohol, 10% formalin, or 1-3-6 mixture; slit body cavity open or inject preservative; rinse in water; store in 70–75% alcohol or 5% formalin.
Graves 1817	Small mammals	Preserve in mixture of 1 pt spirit of wine, arrack, rum, or other spirit with 2 pts burnt alum (made by pouring 1 quart of boiling water on 8 ounces of alum that has been heated over a fire until losing its transparency).
	Amphibians, reptiles, and fishes	Preserve in a mixture of 1 pt pure spirit with 1 pt burnt alum (made by pouring 3 pints of boiling water on 1 pound of alum that has been heated over a fire until losing its transparency).
Griffiths et al. 1976	Marine plankton	Fix within 5 minutes of collection with solution of 5 or 10 mL formaldehyde, 90 mL seawater, and 2 g borax to raise pH to 8.0–8.2. Alternate solution is 50 mL propylene phenoxetol, 450 mL propylene glycol, and 500 mL borax buffered formaldehyde.
Haly 1892	Vertebrates and invertebrates	Preserve color and fine features in previously preserved specimens in a mixture of coconut oil and carbolic acid with a specific gravity of 10° to 20° below proof.
Hammel 1987	Cyclanthaceae inflorescences	Preserve in 70% ETOH.
Hangay and Dingley 1985a	Birds	Inject with 75–80% alcohol. Store typical specimens in 75–80% alcohol, store larger specimens in 3% buffered formalin. Store nestlings and hatchlings in 1.5–2% buffered formalin.
	Birds for anatomical studies	Inject with 1 pt formaldehyde and 15 pts water; store in same solution.
	Fishes	Fix in 10% formaldehyde for 4 d (slit fish longer than 15 cm between pectoral and pelvic fin base); wash in tap water 3–5 hr; store in 75% ETOH.
	Amphibians	Fix in 10% neutral formaldehyde (buffered with hexamine, calcium carbonate, or magnesium carbonate) injected in abdominal cavity and muscle masses; preserve in 65% alcohol or 7–10% formalin. Store eggs and larvae in 5% formaldehyde.
	Geckos	Fix by injecting 10% formalin in abdominal cavity, thoracic cavity, tail, and brain, and limbs; preserve in 70% alcohol.
	Reptiles except geckos	Fix by injecting 10% formaldehyde in abdominal cavity, thoracic cavity, tail, and brain, and limbs; preserve in 75% alcohol.

	Mammals	Inject 75% alcohol in thoracic and abdominal cavities; store in 75% alcohol.
Hangay and Dingley 1985b	Plants	Preserve small plants in 4% formaldehyde; fleshy plants in 6% formaldehyde; plants with waxy coating in 4% formaldehyde with 25% volume of 95% alcohol.
Notes: Bouin's fluid = 75 mL picric acid, 25 mL formalin, and 5 mL glacial acetic acid	Algae	Microscopic algae: preserve in 0.5 g iodine, 1 g potassium iodide, 4 mL glacial acetic acid, 24 mL formaldehyde, 400 mL water.
		Microscopic algae: preserve in 0.5 g iodine, 1 g potassium iodide, 4 mL glacial acetic acid, 24 mL formaldehyde, 400 mL water;
		Macroscopic algae: preserve in 10 g potassium chrome alum, 5 mL 4% formaldehyde, 500 mL water
FAA = 90 mL absolute alcohol, 5 mL formaldehyde, and 5 mL glacial acetic acid except as noted		Preserve with one of three FAA formulas: A: 100 mL alcohol (50%), 6.5 mL formaldehyde, and 2.5 mL glacial acetic acid. B: 90 mL absolute alcohol, 5 mL formaldehyde, and 5 mL glacial acetic acid. C: 85 mL alcohol (70%), 10 mL formaldehyde, and 5 mL glacial acetic acid.
		Preserve in 72 mL water, 5 mL formaldehyde, 20 mL glycerin, 3 mL glacial acetic acid
Formol-alcohol = 50 mL 70% alcohol, 45 mL water, 5 mL formalin		Transeau's algal preservative: 12 pts water, 6 pts alcohol (95%), 2 pts formaldehyde, 1 pt glycerin. Add copper sulfate crystals to make a saturated solution to retain the green color of chlorophyll.
Helly's fluid = Stock solution of 49.2 g mercuric chloride, 29 g potassium dichromate, 19 g sodium sulfate, 960 mL distilled water; just before using mix 48 mL of stock solution with 2.5 mL formaldehyde		Preserve in 1 g chromic acid, 3 mL glacial acetic acid, 100 mL water.
		Fix in 0.5–1% copper sulfate in 2% formalin for 24 hr; store in 5% formalin.
		Acid Lugol's solution (stock solution of 10 g iodine, 20 g potassium iodide, 20 mL glacial acetic acid, 200 mL distilled water; dilute 1:100 for preservation)
		Preserve in 5 pts formaldehyde, 5 pts propionic acid, 90 pts alcohol (50%).
		Preserve in 1 g copper acetate, 227 mL camphor water, 5 mL glacial acetic acid, 180 g glycerin, 1 mL corrosive sublimate.
		10 g cupric sulfate, 90 mL boiling water
Pampl's fluid = n4 mL glacial acetic acid, 30 mL distilled water, 6 mL formalin, 15 mL alcohol 95%		Preserve in Montagu Brown's solution (14 g borax, 14 g alum, 2 g potassium bichromate, 56 g sodium chloride, 9 L warm water, 4 drops saturated solution of bichloride of mercury).
		Preserve in 56 g potassium nitrate, 1137 mL water, 1 drop saturated solution of bichloride of mercury.
		Marine algae: 25 mL formaldehyde, 25 mL saturated solution of cupric acetate, 350 mL seawater.
		Preserve in 70 mL ETOH, 30 mL seawater.

(continued)

Table 1. *(Continued)*

Reference	*Organisms*	*Comments*
		Preserve in 3 mL formaldehyde, 97 mL seawater.
		Kelp: preserve in 5 mL formaldehyde, 95 mL seawater.
Macroscopic algae: preserve in 10 g potassium chrome alum, 5 mL 4% formaldehyde, 500 mL water	Ferns	Preserve in FAA (dissolve a crystal of copper sulfate to maintain green color of chlorophyll)
	Seed plants	Preserve in FAA.
	Green plants	Preserve in 200 g phenol crystals, 200 g lactic acid, 400 g glycerin, 2 g cupric chloride, 2 g cupric acetate, and 2 L distilled water.
A: 100 mL alcohol (50%), 6.5 mL formaldehyde, and 2.5 mL glacial acetic acid		Remove air from intercellular spaces with vacuum pump or immersion in 90–95% alcohol, preserve in 5% glycerin with enough copper sulfate or copper acetate to give a bluish tint.
B: 90 mL absolute alcohol, 5 mL formaldehyde, and 5 mL glacial acetic acid		Fix fresh plants in 10 mL sodium silicate and 990 mL water for 24 hr; rinse in water; preserve in 50 g zinc sulfate, 5 g copper sulfate, 12 mL hydrochloric acid, 16 g sodium sulfite, and 950 mL water.
C: 85 mL alcohol (70%), 10 mL formaldehyde, and 5 mL glacial acetic acid		Knudsen (1972) method for preserving color in plants: fix in 5% copper sulfate for 24 hr; wash several times; preserve in 16 mL sulfuric acid, 21 g sodium sulfite, and 1 L water.
		Red and green apples: 40 mL distilled water, 2 g zinc chloride, 2 mL formaldehyde, and 2 mL glycerin.
	Fruits	Yellow and red apples: fix in 750 mL white paraffin oil and 10 mL formaldehyde for 2 wks; store in sulfur dioxide solution.
	Fungi	Preserve in FAA.
		Preserve in 10 mL glacial acetic acid, 1 g mercuric acetate, 10 g neutral lead acetate, and 1 L 90% alcohol.
		Preserve in 5 mL glacial acetic acid, 10 g mercuric acetate, 1L distilled water.
		Preserve in 5 mL formaldehyde, 90 mL ethyl or isopropyl alcohol, 5 mL glacial acetic acid.

Lichens	Preserve in FAA.
Sponges	Preserve in 70% alcohol.
Coelenterates	Jellyfish: kill by adding formaldehyde to seawater to make a 5% solution; store in 70% alcohol.
	Sea anemones—narcotize by dripping clove oil or magnesium chloride in seawater; fix in neutral formaldehyde added to seawater to make a 5% solution; store in 70% alcohol.
	Corals and sea pens: kill by adding two times volume of 100 mL chromic acid 1% and 5 mL concentrated acetic acid to seawater; preserve in 70% alcohol.
Echinoderms	In general, kill and preserve in 70% alcohol or 10% formaldehyde; transfer to 70% alcohol for storage.
	Starfish: place in 20% alcohol until ambulacral feet are extended, then transfer to 70% alcohol.
	Sea cucumbers: sprinkle 2.5 g of magnesium sulfate per 4 L of seawater on the surface every half hour for 4–5 hr until specimen is relaxed, add sufficient neutral formaldehyde to seawater to make 10% formalin solution; transfer to 70% alcohol for storage. May also be killed by plunging into hot water, soda water, 1% 2–phenoxyethanol, or nicotine water (4–5 used cigarette butts in 500 mL water).
Acanthocephala	Squeeze specimen between two glass plates until proboscis is extended, then kill in hot 3–5% formalin, hot 70–90% alcohol, or hot corrosive acetate (followed by rinse in iodized alcohol, made from 3 g iodine, 6 g potassium iodide, and 300 mL 70% alcohol). Preserve in 70% alcohol.
Annelida	Oligochaeta (earthworms): relax in water, adding absolute alcohol drop-by-drop to make a 10% solution, or by adding drops of chloroform, or by immersing specimens in chlorobutanol or 2-phenoxyethanol. Kill in 5% formalin or FAA; wash in tap water, store in 70% alcohol.

(continued)

Table 1. *(Continued)*

Reference	*Organisms*	*Comments*
		Polychaeta: relax by adding drops of 80% alcohol to seawater until 10% solution is achieved. Kill by immersion in FAA; preserve in 5% formalin or 70% alcohol; store in 70% alcohol or 50–70% isopropyl alcohol.
		Hirudinea: relax in water to which 5% ETOH is added drop-by-drop; fix and preserve in 5% formalin. Alternatively, drop 95% alcohol into shallow dish, fix specimens in 5% formalin for 24 hr, store in 5% formalin or 75–85% ETOH.
	Aschelminthes	Nematodes: wash parasitic round worms in several changes of isotonic saline solution; fix in hot 70% alcohol or hot 3–5% formalin. After cooling, transfer to fresh fixative for storage.
		Free living round worms may be fixed in 10% formalin, Bouin's fluid, Helly's fluid, or saturated aqueous solution of mercuric chloride (followed by washing in tap water for 5 hr, several changes of 70% alcohol, or iodized alcohol overnight) for 24–48 hr. Alternatively, free living round worms may be narcotized in 1% magnesium chloride solution, fixed in 7% neutral formalin, and stored in 70% alcohol with a trace of glycerin.
		Nematomorpha—kill and fix in 3–5% formalin; store in 5% formalin.
		Nemertines: narcotize by adding magnesium sulfate or chloral hydrate crystals to water, or by immersion in 7% magnesium chloride. Position in glass tubes and fix with 10% formalin or 70–90% alcohol; store in 3–5% formalin or 70% alcohol.
		Pentastomides: kill, fix, and store in 70% alcohol, 3–5% formalin, or formol-alcohol.
		Phoronides: narcotize in 7% magnesium chloride or by adding alcohol slowly to water; fix in either 5% formalin or formol-alcohol; store in fresh fixative.

	Platyhelminthes: place cestodes in 1% magnesium chloride solution; fix in 7% formalin; store in 70% alcohol with a trace of glycerin. Place trematodes in 10% formalin, Bouin's fluid, Helly's fluid, or saturated aqueous solution of mercuric chloride (followed by washing in tap water for 5 hr, several changes of 70% alcohol, or iodozed alcohol overnight) for 24–48 hr. Store in 3–5% formalin.
	Relax turbellarians in seawater for a few minutes; decant most of the water and add hot 70% alcohol; store in cold 70% alcohol.
	Pogonophora: narcotize in 7% magnesium chloride; kill in 70% alcohol or 2–3% formalin; store in 70% alcohol.
	Priapulida: narcotize in 7% magnesium chloride solution, fix in 5% formalin in seawater 48–72 hr; store in 70% alcohol.
	Sipuncula: narcotize by adding magnesium chloride crystals or alcohol to seawater; fix in 3–5% formalin or 70–90% alcohol for 12–24 hr; store in 70% alcohol or formol-alcohol.
Molluscs	Snails and slugs: drown in sealed container of freshwater or boil; soak in water with a few crystals of chloral hydrate for 24+ hr; fix in 95% ETOH and preserve in 70% alcohol.
	Shelled aquatic molluscs: kill in 10% neutral formalin made with salt water, relax specimen and fix in 10% formalin; store in 5% formalin.
	Cephalopods: fix in 10% neutral formaldehyde and preserve in 5% formalin.
Crustaceans	Fix in 10% neutral formaldehyde for 24 hr; wash in tap water; store in 70% alcohol
Cirripedia	Kill in 35% alcohol, fix in 10% neutral formaldehyde for 24 hr; wash in tap water; store in 70% alcohol.

(continued)

Table 1. ***(Continued)***

Reference	*Organisms*	*Comments*
	Chordates	Urochordates: kill and fix in 5% buffered neutral formalin for 48 hr; store in 5% buffered formalin or 70–90% alcohol.
		Cephalochordates: kill and fix in 5% formalin; transfer to fresh 5% neutral formalin or wash in tap water and transfer to 70–90% alcohol.
	Insects	Kill in 30% alcohol, transfer to 50%, then store in 50–80% alcohol. Alternatively, kill and preserve in Kahle's fluid (100 mL alcohol 95%, 7 mL glacial acetic acid, 40 mL formalin); store in 70% alcohol.
		Hemiptera: kill and store in lacto alcohol (80 mL lactic acid 87.5%, 74 mL 95% alcohol, 46 mL water).
		Insect larvae: kill and fix in Carnoy's fluid (60 mL 95% alcohol, 30 mL chloroform, 10 mL glacial acetic acid).
		Odontata nymphs: kill, preserve, and store in Oudeman's fluid (88 mL 70% alcohol, 4 mL glycerin, 6 mL glacial acetic acid).
		Diptera larvae: kill and fix in Peterson's KAA solution (10 mL kerosene, 100 mL 95% alcohol, 20 mL glacial acetic acid) for 30 min to a few hours; store in 95% alcohol.
		Lepidoptera larvae: kill in Huffaker's XA mixture (100 mL xylene and 100 mL alcohol); store in 95% alcohol.
		Ants: kill and preserve in Dornisthorpe's fluid (5 mL glacial acetic acid, 37 mL 95% alcohol, 58 mL distilled water, 0.12 g corrosive sublimate).
		Trichoptera larvae: kill and fix in Pampl's fluid; store in 70% alcohol.
		Beetle larvae: kill and fix in Van Emden's fluid (2 mL glacial acetic acid, 30 mL distilled water, 5 mL formaldehyde, 15 mL 95% alcohol); store in 70% alcohol. Alternately, kill and fix in Bles' fluid (3 mL glacial acetic acid, 7 mL formaldehyde, 90 mL alcohol 70%).
		Arachnids: kill and preserve in 70% alcohol, or Pampl's fluid followed by storage in 70% alcohol.
		Diplopoda and chilopoda: kill and preserve in 70% alcohol.

		Symphyla and pauropoda: kill and preserve in 70% alcohol.
		Onychophora: kill and preserve in 70% alcohol, or Pampl's fixative followed by storage in 70% alcohol.
Haynes 1984	Delicate leaves of aquatic and marsh plants	Preserve in 50% methyl alcohol or 50% ETOH.
Henderson 1963	Fish gonads	Preserve in Davidson's Solution A (30 cc of 95% ETOH, 20 cc neutralized formaldehyde, 10 cc glycerin, 10 cc glacial acetic acid, 30 cc distilled water) for 24–48 hr; store in Davidson's Solution B (30 cc of 95% ETOH, 20 cc neutralized formaldehyde, 10 cc glycerin, 30 cc distilled water).
Hillcourt 1970	Reptiles	Make a series of 0.5-inch slits in body, then preserve in 70% rubbing alcohol (isopropyl alcohol) or solution of 1 pt formaldehyde and 2 pts water; add 10% sugar to formaldehyde solutions to preserve color.
	Aquatic animals	Preserve in alcohol or formalin.
	Fleshy fruits	Preserve in 10% formalin.
Holst 2003	Flowers and entire small plants	Preserve in a mixture of 70 pts denatured alcohol, 27 pts water, and 3 pts glycerin.
Hiscox 1935	Museum specimens	Preserve in 6 pts formaldehyde, 12 pts glycerin, 3 pts alcohol, and 100 pts water. Filter solution through charcoal to remove color. For larger specimens use 80–100 pts formaldehyde.
	Zoological and anatomical specimens	Place in a solution of 5 pts sodium fluoride, 2 pts formaldehyde, and 100 pts water; then place in solution of 5 pts glycerin (28° Bé), 10 pts water, 1 pt magnesium chloride, and 0.2 pts sodium fluoride.
Hiscox and Sloan 1939	Flowers	20 grs salicylic acid; 10 min formaldehyde; 2 oz alcohol; 1 qt distilled water.
	Museum specimens	Preserve in a solution of 6 pts formaldehyde, 12 pts glycerin, 3 pts alcohol, and 100 pts water (the addition of glycerin becomes necessary only if it is desired to keep the pieces in a soft state). Filter through animal charcoal to render liquid perfectly colorless. For dense objects, such as lungs and liver, make incisions to facilitate penetration of the fluid. For very thick pieces substitute 80–100 pts of formaldehyde solution for above quantities.

(continued)

Table 1. *(Continued)*

Reference	*Organisms*	*Comments*
	Zoological and anatomical specimens	Place first in a mixture of 5 pts sodium fluoride, 2 pts formaldehyde (40%), and 100 pts water. Later, transfer to preservative containing 5 pts glycerin (28° Bé), 10 pts water, 1 pt magnesium chloride, and 0.2 pts sodium fluoride.
Hoffmann et al. 2010	Mammals	Preserve directly in 75% ETOH without fixation.
Hoogmoed 1973	Amphibians	Preserve adults in 70% alcohol, larvae and eggs in 4% formalin.
Hornaday 1905	Fish	Wash mucous from fish with alum before preserving, open abdominal cavity, preserve in 2 pts 50% alcohol and 1 pt water; change to fresh solution after a few days.
	Reptiles	Mount on plaster of Paris plate (tied on with thread) in jar of alcohol.
Hornell 1895	Protozoa, porifera, actinozoan, discomedusae, lucernariidae, echinodermata, polyzoa, mollusca, and tunicata	Fix and preserve in 3% or stronger formaldehyde.
	Crustacea	Fix and preserve in 8% formaldehyde.
	Fish	Fix and preserve in 4% formaldehyde.
Hornell 1900	Plankton	Fix in 5% formaldehyde in seawater for 12 hr, then transfer crustaceans, larval echinoderms, worms and shell-bearing mollusks to a solution of 35 pts alcohol, 6 pts formaldehyde, and 60 pts water; transfer noctiluca and coelenterates into 8% formaldehyde; later transfer miscellaneous plankton to 70% alcohol or formalin-alcohol (5 pts formaldehyde, 50 pts alcohol, and 45 pts water).
	Porifera (sponges)	Immerse calcareous sponges in 70% alcohol; preserve other sponges in formalin-alcohol (50 pts alcohol, 5 pts formaldehyde, and 45 pts water).
	Hydrozoa	Immerse in 8% formaldehyde in seawater.
	Actinozoa	Preserve in formalin-alcohol (50 pts alcohol, 5 pts formaldehyde, and 45 pts water), then treat with 1% chromic acid, and return to formalin-alcohol.

	Siphonactinia (marine worms)	Harden in 1% chromic acid, then preserve in 70% alcohol.
	Alcyonaria (corals, sea pens, gorgonians)	Preserve in 70% alcohol.
	Echinodermata	Preserve in 70% alcohol.
	Enteropneusta (acorn worms)	Harden in a fixative, then preserve in 70% alcohol.
	Turbellaria (flatworms)	Preserve in 70% alcohol.
	Nemertines (ribbon worms)	Preserve in formalin-alcohol (50 pts alcohol, 5 pts formaldehyde, and 45 pts water).
	Annelida (segmented worms)	Immerse for 12 hrs in 5% formaldehyde, then preserve in formalin-alcohol (50 pts alcohol, 5 pts formaldehyde, and 45 pts water).
	Copepods and amphipods	Kill and fix in 8% formaldehyde; preserve in 8% formaldehyde and 35% alcohol.
	Large crustacea	Fix in 6% formaldehyde for 24 hr; grade into 70% alcohol with 4% formaldehyde.
	Mollusca	Preserve in 6% formaldehyde; transfer shelled mollusca to 70% alcohol after a few hours to prevent decalcification.
	Polyzoa	Preserve in 6% formaldehyde.
	Tunicata	Preserve in 8% formaldehyde.
	Fishes	Preserve in 5% formaldehyde.
Jackson 1919	Bryozoa (freshwater)	Narcotize with cocaine or chlorotone; preserve a solution of 10 pts cane sugar, 2 pts formalin, and 100 pts distilled water.
Jaques 1939	Mayflies and stoneflies	Euthanize larvae in boiling water; preserve in 80% ETOH with 5–8% glycerin added; or preserve in 4% formaldehyde.
Jordan 1905, 1907	Fishes	Wash specimens and make an incision on the right side; preserve in 1 pt formaldehyde with 20 pts water. Wrap in cloth and pack in a can for shipment.
Jørgensen et al. 1984	Passionflowers	Preserve flowering material, pieces of stems, and leaves in FAA (no recipe provided).
Karns 1986	Amphibians and reptiles	Fix in solution of 1 pt formaldehyde to 9 pts water or preserve reptiles in full strength ETOH, amphibians in 75% ETOH.

(continued)

Table 1. ***(Continued)***

Reference	*Organisms*	*Comments*
Kellicott 1896	Fish	Fix and preserve in 2% formaldehyde.
	Salamanders	Fix and preserve in 4% formaldehyde
	Organs (including brains)	Fix and preserve in 4% formaldehyde.
Kendrick 1969	Fleshy fungi	Preserve in alcohol, formaldehyde, FAA (mixture of formaldehyde, alcohol, and acetic acid), or lactic acid.
Kimor 1976	Marine protozoa	Fix in buffered formalin solution of less than 2%. Fix free-living protozoans in 1 mL Lugol solution (2 g KI, 1 g iodine, 200 mL distilled water; may add 20 mL glacial acetic acid) per 100 mL sample.
King and Porter 2004	Ants	Short-term tests showed specimens preserved in either 95% ETOH or isopropyl were easier to later dry mount; specimens preserved in 95% ETOH yielded the best DNA.
Klauber 1935	Amphibians and reptiles	Preserve amphibians in 65% alcohol, with abdomen cut open.
		Preserve reptiles in 85% alcohol or 10% formaldehyde after opening abdomen.
Klauber 1972	Rattlesnakes	Preserve in 1 pt formaldehyde with 10 pts water or 75–80% ETOH; slit abdominal cavity or inject.
Knudsen 1966	Algae	Marine algae: preserve in 5% formalin solution. Use FAA (93 pts formaldehyde, 3 pts glacial acetic acid, 50 pts alcohol 95%; 40 pts water) for specimens intended for histological or slide preparations.
	Fungi	Preserve freshwater algae in 3–5% formalin or FAA.
	Mosses	Preserve in FAA or 10% formalin.
	Liverworts and hornworts	Preserve in FAA.
	Marine protozoans	Preserve in 4% formalin (with a small quantity of copper sulfate added if color preservation is desired); or FAA with copper sulfate.
	Sponges (Porifera)	Kill in 95% isopropyl alcohol; preserve in 70% ETOH.
	Freshwater hydra (Coelenterata)	Wash specimen thoroughly and preserve in 75–95% alcohol; change after 24 hr
		Place specimens in 3 mm of water to relax, then pour hot (50°C) Bouin's fixative over them, allow to soak for a few minutes; decant and replace with fresh Bouin's fixative for 30 min. Wash in several changes of 30% alcohol, then stage to 50% and on to 70% alcohol.
	Colonial hydroids (Coelenterata)	Relax in clean seawater in semi-dark area, then add small quantity of Epsom salts every 20–30 min for several hours; add enough formaldehyde to make a 5% solution. Store in 5% formalin or 70% alcohol.

Notes:
Bouin's fixative = 75 pts picric acid, 25 pts formalin, 5 pts glacial acetic acid
Chrom-acetic fixative = 100 cc chromic acid, 5 cc glacial acetic acid)
FAA = 2 pts formaldehyde, 50 pts alcohol 95%, 2 pts acetic acid, 40 pts water unless otherwise noted

Medusae and jellyfish	Narcotize with urethane, Chloretone crystals, or Epson salts; kill and fix in 5% formalin; stage to 70% alcohol for storage.
Siphonophora	Preserve in 5% formalin; store in 70% alcohol.
Anemones	Narcotize with drops of clove oil or Chloretone overnight, preserve in 5% formalin.
Horny corals, sea pens, sea whips, stony corals	Narcotize and preserve in 5% formalin or 70% alcohol.
Turbellaria	Narcotize larger specimens; fix in FAA, Gibson's fixative (5 gm mercuric chloride, 5 cc nitric acid 80%, 1 cc glacial acetic acid, 25 cc alcohol 70%, 220 cc water; mix and filter after 3 days) for 24 hr, or 5% formalin in seawater. Preserve in 5% formalin or 70% alcohol.
Cestoidea	Fix in FAA; preserve in 5% formalin or 70% alcohol.
Nemertea	Narcotize in seawater with small quantities of Chloretone, menthol, or Epsom salts. Siphon off water and add 10% formalin or hot (60°C) FAA. Store in 5% formalin or wash and store in 60% isopropyl or 70% ETOH.
Rotifera	Narcotize in a watch glass with a few crystals of Chloretone. When celia almost stop moving add drops of 1% osmic acid. Wash 3x with water for 15 min to remove acid. Preserve in 2–5% formalin or stage through 30% and 50% to 70% alcohol.
Gastrotricha	Narcotize with a few crystals of Chloretone followed by a few drops of 1% osmic acid. Wash 3x with water for 15 min to remove acid. Fix in 2% osmic acid or FAA. Preserve in 5% formalin or stage through 30% and 50% to 70% alcohol.
Nematoda	Fix small specimens in FAA. Fix medium to large specimens in 5–10% formalin or 75% alcohol. Preserve in 5% formalin or stage to 70% alcohol.
Nematomorpha	Fix in 5% formaldehyde or 70–80% alcohol. For histological use, fix and preserve in hot saturated mercuric chloride containing 5–10% acetic acid.
Acanthocephala	Evert proboscis before killing by chilling specimens, and then immersing them in hot FAA. Fix and preserve in FAA or 5% formalin.

(continued)

Table 1. ***(Continued)***

Reference	*Organisms*	*Comments*
	Polychaeta	Narcotize by adding 75% alcohol to seawater drop-by-drop to make a 10% solution, stretch on a board and add FAA; or narcotize by sprinkling menthol crystals on surface and waiting 12–24 hr. Fix in 10% formalin. Preserve in 5% formalin, 70% ETOH, or 50–70% isopropyl alcohol.
	Oligochaeta	Narcotize and preserve aquatic species on a wet plate with hot (60C) FAA or Bouin's fixative.
	Hirudinea	Narcotize with Chloretone, chloroform vapor, or chloral hydrate, then fix in FAA, 5% formalin, 70% alcohol, or Bouin's fixative.
	Sipunculoidea, Echiuroidea, Priapuloidea	Narcotize with menthol crystals in seawater, Chlorotone, or Epsom salts; fix in 5–10% formalin; store in 80% alcohol or 5% formalin.
	Gastropoda	Narcotize by adding alcohol drop-by-drop into water until reaching a 10% concentration over several hours. Fix in 10% formalin or FAA, or 5–10% formalin. Transfer to 35% to 50% to 70% alcohol for storage.
	Pelecypoda	Narcotize marine forms in freshwater, then stage through 35% to 50% to 75% alcohol for preservation.
	Amphineura	Narcotize or bind feet flat, then kill in 10% formalin or 35% alcohol. Wash specimens, then stage through 35% and 50% to 70% alcohol for storage.
	Cephalopoda	Kill specimen by adding formalin to water to make a less than 1% solution; after death, fix in fresh 5% formalin. Store small specimens in 5% formalin and larger specimens in 8% or 10% formalin, or stage through 50% alcohol to 70% alcohol.
	Brachiopoda	Relax in seawater until shells open, then insert wooden plug and preserve in 35% alcohol; transfer to 50% and then 70% for storage.
	Bryzoans	Narcotize marine species with menthol, Chloretone, Epsom salts, or clove oil for 6 hr, then fix by adding sufficient formaldehyde to make a 5% formalin solution. Stage through 50% to 70% alcohol to preserve.
		Narcotize freshwater species with menthol crystals overnight, then add formaldehyde to make a 4% solution. Preserve in 70% alcohol or 5% formalin.

Crustaceans	Add 4–5 drops of 10% formalin to water to euthanize specimens, then replace water with 35% alcohol for 30 min and 50% alcohol for 30 min. Store in 70% alcohol.
Cirripedia	Place in finger bowls to relax, then narcotize with Chloretone or menthol for 6 hr. Fix in 5% formalin, then wash and transfer to 35%, then 50%, and then 70% alcohol.
Arameae	Kill and preserve in 70 to 80% alcohol; change alcohol after 24 hr.
Ticks	Kill and preserve in 70% alcohol.
Mites and water mites	Kill and preserve in 70% alcohol.
Centipedes and millipedes	Kill in insect killing jar or by dropping into 70% alcohol. Arrange legs, straighten body, and preserve with 70% alcohol.
Asteroidea and Ophiuroidea	Narcotize and euthanize with Epsom salts overnight. Place in water and arrange arms, then add formaldehyde to make a 5–10% solution; fix for 24 hr. Wash, then transfer to 50% and on to 70% alcohol for storage.
Echinoderm larvae	Narcotize with Chloretone crystals in water for 4–6 hr; stage through 35%, 50%, 75%, and leave in 95% alcohol for 24 hr to harden. Store in 70% alcohol.
Echinoidea	Kill and preserve in 70% alcohol or 5% formalin (followed by wash and transfer to 70% alcohol). Change alcohol after 24 hr. Isopropyl can be used instead of ETOH.
Crinoidea	Kill in 90% alcohol, store in 70% alcohol.
Holothuroidea	Relax in seawater until expanded, then add small quantities of Epsom salts every 30 min for several hr. Kill in 10% formalin, wash, and preserve in 70% alcohol; or kill and preserve directly in 70% alcohol, injecting body cavity with alcohol.
Hemichordata	Narcotize in seawater; add alcohol drop-by-drop over 1 hr to make a 10% solution; leave for 4–6 hr. Kill in 5% formalin, Bouin's fixative, or Kleinenberg's solution, or 50% alcohol. Store in 70% ethyl or isopropyl alcohol or 5% formalin.
Urochordata	Narcotize in seawater with chloral hydrate 12–24 hr; kill in chrom-acetic fixative, 5% formalin, or FAA. Alternatively, add chrom-acetic fixative or formaldehyde to water; fix and store in chrom-acetic fixative or 5% formalin; or place directly in 50% alcohol, then transfer to 70% alcohol.
Cephalochordata	Kill and store in 4% formalin; or kill in 9 pts seawater and 1 pt alcohol, then transfer to 50% and to 70% alcohol.

(continued)

Table 1. ***(Continued)***

Reference	*Organisms*	*Comments*
	Fish	Kill in air or in 5% formalin; open abdomen of fishes larger than 5–6 inches; harden n 10% formalin. Wash slime from body prior to preservation in 70% ETOH or 40–50% isopropyl alcohol. Change alcohol after 24 hr. Preserve color with solution of 20 cc of 1% phenol, 12 cc formaldehyde, 8 cc glycerin, and 60 cc water.
		To preserve color in fish, use Kotthaus method of a solution of 20 cc 1% phenol, 12 cc formaldehyde, 8 cc glycerin, and 60 cc water.
	Amphibians	Kill in Chloretone (1 tsp per gal water) or 2–3 drops clove oil in 2/3 quart water. Position in pan and fix in 8–10% formaldehyde for 2–7 days, injecting body cavity of large specimens. Transfer to 5% formaldehyde, 70% ETOH, or 40–50% isopropyl for preservation.
	Reptiles	Euthanize by freezing or injection of 10% Nembutal. Inject or open body cavity to permit entrance of 10% formalin fixative; preserve in 6% (small specimens) or 10% (large specimens) formalin, 70% ETOH, or 40–50% isopropyl.
	Mammals	Inject body cavity and large muscle areas with 10% formalin; wash with detergent; immerse in 10% formalin.
Köhler 2001	Amphibians and reptiles	Inject and submerge adults in a mixture of 10 mL of formaldehyde mixed with 1 L of 95% ETOH.
		Preserve amphibian larvae in 4–8% formaldehyde.
Koeppen 1958	Amphibians and reptiles	To preserve color, fix and store in 10–20g of 40% formol-alcohol in 1 L distilled water .
Koeppen 1961	Anura and small salientia	To preserve color, kill by immersion in solution of 2 pts formaldehyde, 40 pts alcohol 96%, and 60 pts water; store in solution of 550 mL glycerin, 150 mL water, and 20 mL tanning solution.
	Salientia (large), reptiles, and fish	To preserve color, kill by immersion in solution of 2 pts formaldehyde, 40 pts alcohol 96%, and 60 pts water; store in solution of 100 mL glycerin, 250 mL water, and 100 mL alcohol 95%, and 40 mL tanning solution.

Krogmann and Holstein 2010	Insects	For morphological analysis euthanize and store in a fixative such as Kahle's solution (30 mL 95% ETOH, 10 mL 35–40% formalin, 2 mL glacial acetic acid, and 60 mL water) or Bouin's solution (70 pts picric acid. 25 pts formalin, 5 pts glacial acetic acid). Euthanize and store in 70–80% ETOH with a small amount of glycerin for natural history collections. Do not use glycerin for micro-Hymenoptera.
	Molluscs	Narcotize and preserve in 80% ETOH with 5% glycerin, or formalin
	Plathelminthes (flatworms)	Preserve in FAA (10 pts formaldehyde, 50 pts 95% ETOH, 2 pts acetic acid, 40 pts water) or 5% formalin.
	Annelida (segmented worms)	Relax and euthanize in 1–2% formalin; preserve in 5% formalin or 70–80% ETOH.
	Arthropods	Preserve in 70–80% ETOH. Coleoptera, Heteroptera, and ants may be preserved in 2% acetic acid; preserve mites (Acari) and thrips (Thysanoptera) in AGA (1 pt acetic acid, 1 pt glycerol, 6 pts 95% ETOH, and 4 pts water); preserve aphids and scale insects (Coccoidea) in lactic alcohol (2 pts 95% ETOH and 1 pt 75% lactic acid).
Lazell 1972	Lizards	Fix in 85–95% ETOH, methanol, or isopropyl; recommends against the use of formaldehyde.
Leach 1995	Spiny euphorbias and aloes	Preserve plant specimens in one of three formulations of FAG. FAG-A (5 mL alcohol, 1 mL formalin, 1 mL glycerol, and 3 mL water); FAG-B (8 mL alcohol, 2 mL formalin, 2 mL glycerol, and 10 mL water); or Bally's Mix (2 mL formalin, 1 mL glycerol, and 20 mL water).
Lendrum 1941	Human postmortem tissue samples	Fix in formol-saline (90 cc saturated aqueous corrosive sublimate) and 10cc formaldehyde; preserve in 60% alcohol.
Ligame 1964	Amphibian eggs	Dissolve gelatinous membrane and fix eggs in 1.3% sulfuric acid (0.25 M) in a 10% Ringer solution.

(continued)

Table 1. *(Continued)*

Reference	*Organisms*	*Comments*
Lillie 1965	Human tissue fixatives	10% formalin: 100 cc formaldehyde, 900 cc tap water; Formol saline: 100 cc formaldehyde, 8.5 g sodium chloride, 900 cc tap or distilled water; Neutral 10% formalin: 100 cc formaldehyde, 900 cc water, calcium or magnesium carbonate to excess; Neutral buffered formaldehyde solution (pH 7.0): 100 cc formaldehyde, 900 cc water, 4 g acid sodium phosphate monohydrate, 6.5 g anhydrous disodium phosphate; Calcium acetate formalin: 100 cc formaldehyde, 900 cc distilled water, 20 g calcium acetate (monohydrate); Ramón y Cajal's formol ammonium bromide, FAB: 140 cc formaldehyde, 20 g ammonium bromide, 1000 cc distilled water; Conn and Darrow's formol ammonium bromide, FAB: 150 cc formaldehyde, 20 g ammonium bromide, 850 cc distilled water; Acetic formalin: 100 cc formaldehyde, 900 cc alcohol (95%), with optional 0.5 g calcium acetate; Acetic alcohol formalin (Tellysnicsky): 5 cc formaldehyde, 5 cc glacial acetic acid, 100 cc alcohol (70%); Acetic alcohol formalin (Fekete): 10 cc formaldehyde, 5 cc glacial acetic acid, 100 cc alcohol (70%); Acetic alcohol formalin (Opie and Lavin): 5 cc formaldehyde, 5 cc glacial acetic acid, 90 cc alcohol (80%); Acetic alcohol formalin (Lillie's AAF): 10 cc formaldehyde, 5 cc glacial acetic acid, 85 cc alcohol (100%)
Lincoln and Sheals 1979 *Notes*: Bouin's fluid = 75 mL picric acid, 5 mL acetic acid, and 25 mL formaldehyde Buffered formalin = to a formaldehyde and water solution, add drops of 4% sodium hydroxide until neutral with phenol red indicator; or add 200 g hexamine (hexamethylenetetramine) per L	Protozoa	Preserve in 50% alcohol followed by 70–90% alcohol, or in a 3–5% neutral formalin solution.
	Porifera (sponges)	Immerse in 50% alcohol for 12 hr; transfer to fresh 50% alcohol for 12 hr, preserve in 70% alcohol. For histological studies, fix specimens in Bouin's fluid.
	Coelenterata	Hydrozoa: anaesthetize by sprinkling menthol crystals, MS222, magnesium sulfate, magnesium chloride, propylene phenoxetol, or Stovaine (amyl chlorohydrin) on surface of water. Fix in 20% buffered formalin. Preserve in 10% formalin or 70% alcohol. Scyphozoa: fix in 20% buffered formalin; preserve in 10% buffered formalin. If relaxation is necessary, use menthol or MS222. Stauromedusae: fix in neutral 10% formalin; preserve in 70% alcohol.

Dowicil 100 - 1-(3-chlorallyl)5,7-triaza-1-azoniaadamantane chloride		Anthozoa: preserve forms with calcareous skeleton in 70% alcohol. Fix other forms in 20% buffered formalin and preserve in 10% formalin or 70% alcohol. If relaxation is necessary, use menthol, MS222, magnesium chloride, or magnesium sulfate.
Eucaine = ß-eucaine hydrochloride, usually administered as 1 g eucaine with 10 mL alcohol 90% and 10 mL distilled water	Ctenophora (comb jellies, sea gooseberries)	Anaesthetize with chloral hydrate crystals; fix in chromic/osmic acid mixture (100 mL chromic acid 1%, 2 mL osmic acid 1%) for 1 hr or Flemming's solution (150 mL chromic acid 1%, 40 mL osmic acid 2%, 10 mL glacial acetic acid). After fixing, stage through 30%, 40%, 50% and 60% to 70% alcohol.
MS222 = ethyl-m-aminobenzoate	Turbellaria (planarians)	Preserve in 70–90% alcohol, preferably by using hot alcohol for a rapid fixative.
Phenoxetol BPC = 1–2% solution of B-phenoxyethyalcohol	Trematoda (flukes)	Clean in 1% salt solution; fix in 10% formalin; store in 3–5% formalin.
Propylene phenoxetol = 1–2% solution	Cestoda (tapeworms)	Clean in 1% salt solution; spread on glass plate; fix in 5–10% formalin; store in 3–5% formalin. Alternatively, fix in Schaudinn's solution (2 pts mercuric chloride and 1 pt alcohol 90%) followed by washing for several hours in water or several changes of 70% alcohol or iodized alcohol (3 g iodine, 6 g potassium iodide, 300 mL alcohol 70%). Preserve in 70–90% alcohol or 3–5% formalin. May also fix in Bouin's fluid (75 mL picric acid, 25 mL formaldehyde, and 5 mL glacial acetic acid) or Zenker's fluid (mercuric chloride, glacial acetic acid, potassium dichromate, sodium sulfate, water).
Stovaine = amyl chlorohydrin	Nemertinea	Anaesthetize with chloral hydrate crystals or magnesium sulfate for 6–12 hr; fix in 10% formalin or 30–50% alcohol; store in 3–5% formalin or 70–90% alcohol.
	Rotifera (wheel-animalcules)	Kill and fix in 10% formalin. Illoricate species must be anaesthetized with benzamine hydrochloride [anilinium chloride]. Fix in 10% formalin and store in 2–5% formalin.
	Gastrotricha	Fix in 10% formalin. Freshwater forms may be fixed in 2% osmic acid. Store in 70% alcohol or 5% formalin.
	Kinorhyncha or echiodera	Fix in 10% formalin and store in 2–5% formalin.
	Nematoda	Immerse large, free-living specimens in hot or cold 3–5% formalin or hot (70–80°C) 70–90% alcohol, then transfer to 3–5% formalin. Kill small specimens in a double boiler at 60–65°C, then fix in 20% formalin or double strength TAF (14 mL formaldehyde, 4 mL triethylamine, 82 mL water).
	Nematomorpha (hair worms)	Fix and store in cold 3–5% formalin.

(continued)

Table 1. ***(Continued)***

Reference	*Organisms*	*Comments*
	Acanthocephala (thorny headed worms)	Clean with a moistened paintbrush, fix in hot 3–5% formalin or hot 70–90% alcohol. Apply pressure with two glass plates to extrude proboscis. Store in 70% alcohol.
	Priapulida	Anaesthetize in 7% magnesium chloride, fix and store in 5% formalin; or fix in Bouin's fluid, then place in 50% alcohol for 1 hr and transfer to 70% alcohol.
	Entoprocta	Anaesthetize with 1% Stovaine, eucaine, a similar agent, or menthol or Epsom salts added drop-by-drop at 10–15 min intervals. Fix in Bouin's fluid, wash, and preserve in 70–90% alcohol.
	Phoronidae	Kill and fix in 5% buffered formalin or 10% buffered Dowicil in seawater, then wash and transfer to 70–90% alcohol or 1% propylene phenoxetol. May be anaesthetized with MS222 or magnesium sulfate prior to fixation.
	Brachiopoda	Anaesthetize gradually by adding alcohol to seawater (not to exceed a 10% solution), or with MS222 or propylene phenoxetol, then fix and preserve in 70–90% alcohol.
	Sipuncula (peanut worms)	Anaesthetize by adding small amounts of alcohol, propylene phenoxetol or magnesium chloride to seawater until introvert is everted, then fix in 70–90% alcohol or 4% formalin.
	Echiura (spoonworms)	Anaesthetize by adding a few crystals of menthol or dropping 90% alcohol in water, or by immersion in 7% magnesium chloride or 1% propylene phenoxetol. Fix in 5% neutral formalin and preserve in 70% alcohol.
	Polychaeta (bristleworms)	Fix immediately in 5% buffered formalin or 10% buffered Dowicil and seawater. After 48 hr, wash nonpelagic forms in clean seawater and transfer to 70–90% alcohol or 1% propylene phenoxetol.
	Oligochaeta (earthworms, potworms, freshwater ringed worms)	Anaesthetize in 5–10% alcohol or 1% propylene phenoxetol for 10–15 min, then lay out in flat dish and fix in 4% formalin or 10% Dowicil for 24 hr, then transfer to 70–90% alcohol or 1% propylene phenoxetol for storage.
	Hirudinea (leeches)	Anaesthetize in 5–10% alcohol, 1% propylene phenoxetol, lemon juice, or soda water; flatten and fix in 4% formalin; store in 4% formalin or 10% Dowicil
	Pogonophora	Preserve animals still in tubes in 70% alcohol, changing solution after 24 hr. Fix specimens in soft tubes in 2–3% formalin.
	Onychophora	Euthanize with chloroform or ethyl acetate vapors; dip in alcohol then fix in formalin. Preserve in 5% formalin or 70–90% alcohol.

Myriapoda	Kill and preserve in 70–90% alcohol. Larger specimens may be first anaesthetized with ethyl acetate vapor. Transfer to fresh preservative after 12–24 hr.
Crustacea	Euthanize small specimens by adding a small amount of formaldehyde to water; fix in 5% buffered formalin made with seawater or freshwater as appropriate. Fix amphipods in 5% formalin or 10% Dowicil for 24 hr; transfer to 70–90% alcohol for storage.
	Fix pelagic marine species in Steedman's solution (0.5 mL propylene phenoxetol, 4.5 mL propylene glycol, 5 mL formaldehyde; 90 mL seawater or distilled water). Kill and fix woodlice and other small isopods in 70–90% alcohol. Anaesthetize large crabs, lobsters, shrimp and prawns with 1–2% formalin, chloral hydrate, or MS222. May be euthanized by elevating temperature higher than ambient. Fix in 5% buffered formalin with a small amount of glycerol added for min. 3–5 days; wash in freshwater and store in 70% alcohol.
Xiphosura (king and horse shoe crabs)	Inject formalin into heart and ventral tissues, flush digestive track with formalin; preserve in 70–90% alcohol.
Arachnida	Kill and preserve in 70–90% alcohol. Larger species should be first anaesthetized with ethyl acetate vapor.
Hydracarina (water mites)	Kill and store in Viet's solution (3 pts glacial acetic acid, 11 pts glycerin, 6 pts distilled water).
Terrestrial mites	Kill and store in Oudeman's fluid (8 pts glacial acetic acid, 5 pts glycerin, 87 pts alcohol 70%).
Pycnogonida (sea spiders)	Kill and preserve in 70–90% alcohol.
Tardigrada (bear-animalcules or water-bears)	Kill, fix, and preserve in 5% formalin or 70–80% alcohol.
Pentastomida (tongue worms)	Fix and store in 3–5% formalin or 70–90% alcohol.

(continued)

Table 1. ***(Continued)***

Reference	*Organisms*	*Comments*
	Mollusca	Euthanize terrestrial species by immersion in cold, boiled water for 24 hr, or add urethane or menthol. Anaesthetize aquatic species with magnesium sulfate, magnesium chloride, urethane, or menthol.
		Anesthetize bivalves with propylene phenoxetol or phenoxetol BPC. Fix and preserve marine species in 5% formalin. Fix and preserve terrestrial species in 70% alcohol. Some species may be preserved in propylene phenoxetol.
	Chaetognatha (arrow-worms)	Fix and store in 5% formalin or 70–90% alcohol.
	Echinodermata	Fix hard bodied forms in 10–12% buffered formalin made with seawater or 95–100% alcohol; store in 70–90% alcohol. Anaesthetize soft bodied forms with a few drops of 1–2% propylene phenoxetol, magnesium sulfate crystals, or menthol added to seawater; fix in 10–12% buffered formalin made with seawater or 95–100% alcohol; store in 70–90% alcohol.
	Hemichordata	Anaesthetize by gradual addition of 95% alcohol to seawater to make a 1 pt alcohol to 95 pts water. Enteropneusts: anaesthetize in 7% aqueous magnesium chloride. Fix in 5% formalin in seawater for 24 hr; transfer to fresh 5% formalin in seawater for storage. For histological specimens, fix in Bouin's fluid and preserve in 70% alcohol.
	Urochordata	Anaesthetize with a few drops of 1–2% propylene phenoxetol or crystals of magnesium chloride or menthol. Kill and fix by adding buffered formaldehyde to seawater. Preserve sessile tunicates in 70–90% alcohol or buffered 5% formalin. Kill and preserve pelagic tunicates in 5% formalin or anaesthetize and fix in Bouin's fluid or Schaudinn's fluid (2 pts mercuric chloride and 1 pt alcohol 90%), followed by washing and preservation in 5% formalin or 70–90% alcohol.
	Cephalochordata	Fix and preserve in 5% buffered formalin.
Lo Bianco 1899	Soft or gelatinous marine animals	Preserve initially in 35–50% alcohol; transfer to 60%; transfer to 70% alcohol. May also preserve in 1–4% formaldehyde, but not recommended for long-term storage. May be fixed in 5% bichromate of potassium.
Notes: Alcoholized seawater = 100 cc of seawater and 5 cc absolute alcohol	Marine animals (general)	Fix in alcohol, chromic acid, acetic acid, osmic acid, or corrosive sublimate; preserve in 70% alcohol. May preserve in 1–6% formaldehyde but not recommended for long-term storage.

- Chrom-acetic mixture #1 = 100 cc chromic acid (1%) and 5 cc concentrated acetic acid
- Chrom-acetic mixture #2 = 10 cc chromic acid (1%) and 100 cc concentrated acetic acid
- Chrom-osmic mixture = 100 cc chromic acid (1%) and 2 cc osmic acid (1%)
- Iodized alcohol = 100 cc of 35% or 70% alcohol and 2.5 cc tincture of iodine
- Kleinenberg's solution = 100 cc saturated picric acid and 2 cc concentrated sulfuric acid, filtered and mixed with 3 volumes distilled water
- Müller's solution = 2 gm potassium bichromate, 1 gm sodium sulfate, 100 gm distilled water
- Perenyi's solution = 40 cc 10% nitric acid, 30 cc 0.5% chromic acid, and 30 cc 90% alcohol

Radiolaria: *Thalassicolla*	Fix in chromic acid (0.5%) for 1 hr; transfer to 70% alcohol.
Radiolaria: Aulacanthidae and Acanthometrae	Place in 50% alcohol for a few hours; transfer to 70%. Alternatively, add a few drops of 1% osmic acid to seawater with animals, then wash and transfer to 70% alcohol.
Radiolaria: Spherozoidae	Fix in iodized alcohol (35%) for 15 minutes to 1 hr; transfer to 35% alcohol; transfer to 50% alcohol; transfer to 70% alcohol.
Porifera	For exhibit specimens, immerse in 70% alcohol; replace preservative when it becomes discolored. For research specimens, immerse <10 cm specimens in 90% alcohol, replacing preservative after 3 hr and again after 24–48 hr.
Anthozoa	Preserve in chrom-acetic mixture #2 for a short time; transfer to 70% alcohol.
Anthozoa: *Cornularia*, *Clavularia*, *Rhizoxenia*, *Sympodium*	Allow to expand; carefully siphon off water to just leave specimens covered, then quickly pour in double the volume of chrom-acetic mixture #2; transfer to 35%; transfer to 50% alcohol.
Anthozoa: *Pennatula phosphorea* and *Kophobelemnon*	Allow to distend, then immerse in chrom-acetic mixture #2; after a few seconds transfer to 50% alcohol, inject; transfer to 70% alcohol. Suspend in the container with a glass float.
Anthozoa: *Pennatula rubra*, *Pteroides spinulosus*, *Veretillum*, *Funiculina*	Immerse in chrom-acetic mixture #2; after a few seconds transfer to 50% alcohol; transfer to weak alcohol; transfer to 70% alcohol.
Anthozoa: *Gorgonia*, *Gorgonella*, *Primnoa*, *Muricea*, *Isis*, etc.	Allow to distend; immerse in chrom-acetic mixture #2; after a few seconds transfer to 50% alcohol, inject; transfer to 70% alcohol.
Anthozoa: *Corallium rubrum*	Allow to expand in running seawater; kill with boiled saturated sublimate solution; transfer to weak alcohol.
Zoantharia	Fix with saturated sublimate solution; store in 70% alcohol.
Actiniaria: *Anemonia sulcata*	Distend in running water; kill with solution of 50 cc chromic acid (1%) and 50 cc Kleinenberg's solution. Alternatively, use a solution of 1 pt chromic acid (1%), 1 pt saturated picric acid, and 1 pt formalin (4%). Transfer to 70% alcohol.

(continued)

Table 1. *(Continued)*

Reference	*Organisms*	*Comments*
	Actiniaria: *Eloactis, Sagartia dohrni, Paranthus, Corynactis,* small specimens of *Aiptasia*	Kill in boiling saturated sublimate; transfer to chromic acid (0.5%) for a few minutes; transfer to 70% alcohol.
	Actiniaria: *Heliactis bellis, Bunodes gemmaceus, Bunodes rigidus*	Allow to distend; replace 2/3 of seawater with chloral hydrate solution (1%); pour off and cover animals with cold saturated sublimate to kill; preserve in 70% alcohol.
	Actiniaria: *Adamsia rondeleti, Adamsia palliata*	Narcotize with tobacco smoke for several hours; kill in chrom-acetic mixture #2; transfer to weak alcohol; transfer to 70% alcohol.
	Actiniaria: *Cladactis, Cereactis,* small *Bunodeopsis strumosa*	Kill in chrom-acetic mixture #2; harden immediately in solution of 1% chromic acid.
	Actiniaria: *Actinia equina, Actinia cari*	Immerse in boiling sublimate and acetic acid; transfer to 0.5% chromic acid for hardening.
	Actiniaria: *Edwardsia*	Narcotize by dropping 70% alcohol in seawater; kill with hot saturated sublimate.
	Actiniaria: actinian larvae	Kill with saturated sublimate or chrom-acetic mixture #2.
	Madreporaria: *Astroides calycularis*	Allow to distend in seawater; add 1 volume of boiling sublimate and acetic acid to kill; immediately transfer to 35% alcohol and inject each polyp; transfer to 70% alcohol and inject each polyp.
	Madreporaria: *Caryophpyllia, Dendrophyllia, Cladocora*	Fix in boiling saturated sublimate.
	Hydromedusae: Hydroidea	Kill in hot saturated sublimate; cool in freshwater; wash; transfer to 50% alcohol.
	Hydromedusae: *Tubularia, Pennaria*	Kill with 1 volume solution of sublimate and chromic acid; wash; transfer to 50% alcohol and then 70% alcohol.

Turularidae: medusa forms (*Eleutheria*, *Cladonema*, *Podocoryne*, etc.)	Kill in chrom-acetic mixture; transfer to 70% alcohol.
Turularidae: medusa forms (*Eleutheria*)	Kill in Kleinenberg's solution.
Turularidae: medusa forms (*Lizzia koellikeri*, *Oceania pileata*}	Allow tentacles to extend; kill with concentrated acetic acid; immediately transfer to a mixture of alcohol and chromic acid and agitate for 15 min. Transfer to 35% alcohol; gradually transfer to70% alcohol. Alternatively, distend and then add 1 volume acetic acid; transfer at once to alcohol and chromic acid; after a few minutes add more chromic acid for 15 min; wash in freshwater and transfer to 35% alcohol.
Turularidae: medusa forms (*Oceania conica*, *Tiara pileata*)	Narcotize in 3% solution of 100 cc seawater with 5 cc absolute alcohol.
Campanularidae: medusa forms (*Eucope*, *Gastroblasta*, *Obelia*)	Fix in a mixture of 100 cc copper sulfate (10%) and 10 cc corrosive sublimate; rinse in freshwater; transfer through weak alcohol to 70% alcohol.
Campanularidae: medusa forms (*Mitocoma*, *Aequorea*)	Kill with acetic acid; immediately transfer to chrom-osmic mixture for 15 to 30 min.
Campanularidae: medusa forms (*Tima flavilabris*)	Kill in 4% formalin; after 5–20 days transfer gradually through weak alcohol to 70% alcohol.
Campanularidae: medusa forms (*Olindias mülleri*)	Relax in seawater; quickly add 1 volume of 6% formalin; after 1 week, transfer to 35% alcohol; gradually transfer to 70% alcohol.
Campanularidae: medusa forms—Trachymedusae (*Rhopalonema*, *Cunina*, *Aegineta*, *Aeginopsis*, *Liriope*, *Carrmarina*)	Fix in chrom-osmic mixture for 5–20 min; wash in freshwater; gradually transfer to 70% alcohol.

(continued)

Table 1. *(Continued)*

Reference	*Organisms*	*Comments*
	Acalephae: *Charybdoea*	Kill in chrom-acetic mixture #2; immediately transfer to 0.5% chromic acid for 0.5 hr; transfer to alcohol.
	Acalephae: *Nausithoe*, *Pelagia* ephyra, *Rhizostoma*	Kill by adding 3% solution of osmic acid (1%) to seawater; when brown tint appears, wash in freshwater; transfer to 35% alcohol or 4% formalin.
	Acalephae: *Pelagia noctiluca*	Kill in chrom-osmic mixture; after 1 hr wash and transfer to weak alcohol.
	Acalephae: *Cotylorhiza tuberculata*	Kill in 2–3% formalin. Alternatively, kill by adding 3% solution of osmic acid (1%) to seawater; when brown tint appears, transfer to 5% bichromate of potassium for a few days; transfer to 35% alcohol.
	Acalephae: larval forms	Kill with hot saturated sublimate.
	Acalephae: *Strobila* larvae	Fix in 9 pts concentrated acetic acid and 1 pt osmic acid (1%); transfer quickly to few water to wash; transfer to alcohol.
	Siphonophora: *Athorybia rosacea*	Kill with mixture of 100 cc copper sulfate (10%) and 10 cc corrosive sublimate; wash in freshwater; transfer to alcohol.
	Siphonophora: delicate species (*Physophora, Agalma, Halistemma*)	Kill by adding 1–2 volumes of a mixture of 100 cc copper sulfate (10%) and 10 cc corrosive sublimate to water; transfer to 35% alcohol for a few hours; transfer to 70% alcohol.
	Siphonophora: *Forskalia*	Kill by adding 1–2 volumes of a mixture of 100 cc copper sulfate (10%) and 10 cc corrosive sublimate to water; transfer to Flemming's solution (25 cc chromic acid 1%, 10 cc osmic acid 1%, 5 cc glacial acetic acid, 60 cc distilled water) for 2–6 hr; wash in freshwater and transfer to weak alcohol; gradually transfer to 70% alcohol
	Siphonophora: *Hippopodius, Galeolaria, Abyla*	Kill with mixture of 100 cc copper sulfate (10%) and 10 cc corrosive sublimate; transfer immediately to weak alcohol.
	Siphonophora: *Prya*	Kill with mixture of 100 cc copper sulfate (10%) and 10 cc corrosive sublimate; harden in solution of 100 cc potassium bichromate (5%) and 2 cc osmic acid (1%) for 1–2 days.

Siphonophora: *Diphyes*	Kill with hot sublimate.
Siphonophora: *Velella*	Kill with solution of 100 cc chromic acid (1%) and 10 cc Kleinenberg's solution; after a few minutes transfer to weak alcohol.
Siphonophora: *Porpita*	Distend and kill by adding drops of Kleinenberg's solution to water; when blue colony color begins to turn red, remove to Kleinenberg's solution for 15 min; transfer to weak alcohol.
Ctenophora (*Beroe ovate, Hormiphora, Callianira, Lampetia, Euchlora,* young specimens of *Cestus, Eucharis, Bolina*)	To kill place in chrom-acetic mixture #2 for 15–60 min; gradually transfer to 70% alcohol.
Ctenophora (*Callianira*)	Kill in a solution of 1 pt concentrated pyroligneous acid, 2 pts saturated sublimate, and 1 pt chromic acid (0.5%).
Ctenophora (*Cestus veneris* and *Vexillum*)	Kill by rapidly adding chrom-acetic mixture #1; arrange specimen in a coil; after 10 min decant and replace with chrom-osmic mixture for 15 min; wash in freshwater; place in 35% alcohol; gradually transfer to 70% alcohol.
Echinoderma (Crinoidea)	Kill in 90% alcohol; preserve in 70% alcohol. Narcotize larval forms with ½% chloral hydrate solution for 2–4 hr, then harden in alcohol.
Echinoderma (Asteroidea)	Kill in 20–30% alcohol.
Echinoderma (Luidia)	Relax and euthanize with chrom-acetic mixture #2; immediately transfer to 50% alcohol for 2 hr; remove to 70% alcohol.
Echinoderma (Brisinga)	Quickly immerse in absolute alcohol to prevent arms from breaking.
Echinoderma (Bipinnaria)	Kill with chrom-acetic mixture or chrom-osmic mixture.
Echinoderma (Echinoidea)	Relax in seawater; add 1 volume of chrom-acetic mixture; transfer at once to alcohol.
Echinoderma (Holothurioidea)	Relax in clear seawater; place immediately in concentrated acetic acid and inject through anal aperture with 90% alcohol; immerse in 70% alcohol to preserve.
Echinoderma (*Thyone, Thyonidium, Phyllophorus*)	Immerse in acetic acid; remove to alcohol to preserve.

(continued)

Table 1. ***(Continued)***

Reference	*Organisms*	*Comments*
	Echinoderma (*Cucumaria plancii*)	Relax in clear seawater; place immediately in concentrated acetic acid and inject through mouth opening with 90% alcohol; immerse in 70% alcohol to preserve
	Echinoderma (*Synapta*)	Kill in a 1:1 mixture of seawater and ether or chloroform; wash in freshwater; transfer gradually to 70% alcohol. Alternatively, harden in 2–3 cc of chromic acid (1%) in water.
	Enternopneusta (*Balanoglossus*)	Kill with Kleinenberg's solution; transfer to 70% alcohol.
	Enternopneusta (*Tornaria*)	Kill in a mixture of 100 cc copper sulfate (10%) and 10 cc corrosive sublimate; preserve in 70% alcohol
	Vermes (Cestodes)	Fix in cold saturated sublimate; preserve in 70% alcohol.
	Vermes (Trenatides)	Fix in hot saturated sublimate; preserve in 70% alcohol.
	Vermes (Rhabdocoelia, Dendrocoelia)	Distend in water; kill with boiling saturated sublimate; transfer immediately to freshwater for a few minutes; transfer to alcohol.
	Vermes (Nemertinae)	Plunge specimen rapidly head first into a mixture of 7 pts Müller's solution and 1 pt concentrated hydrochloric acid for a few minutes; preserve in alcohol. Alternatively, narcotize in a solution of 0.1% chloral hydrate in seawater for 6–12 hr; harden in alcohol.
	Vermes (Chaetognatua)	Euthanize with a mixture of 100 cc copper sulfate (10%) and 10 cc corrosive sublimate or chrom-osmic mixture; preserve in 70% alcohol.
	Vermes (*Gephyrea*)	Kill with a solution of 0.1% chromic acid (or weaker solution), or narcotize with 1% solution of chloral hydrate (0.1%) in water; preserve in 70% alcohol.
	Vermes (*Phascolosoma*)	Place in a solution of 100 cc seawater and 5 cc absolute alcohol for 3–6 hr; transfer to 70% alcohol.
	Vermes (*Phoronis*)	Place in a solution of 100 cc seawater and 5 cc absolute alcohol for 0.5 hr; transfer to boiling saturated sublimate; transfer to 70% alcohol.
	Vermes (Bonellia)	Extend proboscis and submerge in Kleinenberg's solution; transfer to 70% alcohol to preserve.
	Vermes (Hirundinei, Pontobdella, Branchellion)	Kill in 0.1% chromic acid; transfer to 70% alcohol.

Vermes (Chaetopoda)	Place in shallow dish of seawater and slowly add a mixture of 1 pt glycerin, 2 pts 70% alcohol, and 2 pts seawater; after a few hours, transfer to 70% alcohol. Alternatively, place in seawater with 5% absolute alcohol for 2–12 hrs; harden in 70% alcohol.
Vermes (Annelids: Polygordiidae, Opheliadae, Capitellidae, Telethusidae, Maldanidae, Ariciidae, Cirratulidae, Spionidae, Terebellidae)	Place in seawater with 5% absolute alcohol for 2–12 hrs; harden in 70% alcohol.
Vermes (Annelids: *Polymnia*, *Lanice*)	Kill in solution of 100 cc saturated corrosive sublimate and 50 cc concentrated acetic acid; transfer to 70% alcohol.
Vermes (Chlorbaemidae: *Stylaroides*, *Trophonia*	Narcotize in a mixture of 100 cc seawater and 5 cc absolute alcohol; harden in 1% chromic acid; transfer to 70% alcohol.
Vermes (Chlorbaemidae: *Siphonostomum diplochaitos*	Kill in 5% chloral hydrate; harden in 1% chromic acid for 15 min; transfer to 70% alcohol.
Vermes (Hermionidae)	Immerse directly in 70% alcohol.
Vermes (Chaetopteridae, Sternaspidae, *Spirographis*, *Protula*)	Place in 1% chromic acid for 0.5 hr or longer to kill; wash in freshwater; transfer to 70% alcohol; transfer to 90% alcohol for storage.
Vermes (Amphictenidae, Hermellidae, Serpulidae, Aphroditidae, Eunicidae)	Place in cold saturated sublimate for less than 15 min to kill; preserve in 70% alcohol.
Vermes (Alciopidae)	Place in a mixture of 100 cc copper sulfate (10%) and 10 cc corrosive sublimate for 5 min or less; wash thoroughly in freshwater; transfer to 70% alcohol.

(continued)

Table 1. ***(Continued)***

Reference	*Organisms*	*Comments*
	Vermes (Tomopteridae)	Kill in a mixture of 100 cc copper sulfate (10%) and 10 cc corrosive sublimate or cold saturated sublimate; rinse in iodine solution; transfer to 70% alcohol.
	Crustacea (Cladocera)	Kill in saturated sublimate or with a few drops of 1% osmic acid in seawater; wash; transfer to 70% alcohol.
	Crustacea (Ostracoda, *Balanus*)	Place directly in 70% alcohol.
	Crustacea (Copepoda)	Kill by adding saturated sublimate to seawater for 5–10 min; transfer to 70% alcohol.
	Crustacea (Cirripedia)	Euthanize in 35% alcohol; preserve in 70% alcohol.
	Crustacea (Rhizocephala)	Place in a mixture of 1 pt 90% alcohol and 1 pt sublimate for 15 min; transfer to 70% alcohol.
	Crustacea (Amphipoda)	Place in Perenyi's solution for 15 min; immerse in 70% alcohol.
	Crustacea (Isopoda)	Place directly in 70% alcohol.
	Crustacea (Isopoda: *Bopyrides, Entoniscides*)	Place in a mixture of 1 pt 90% alcohol and 1 pt saturated sublimate; then place in 70% alcohol.
	Crustacea (Cumacea, Stomatopoda, Schizopoda)	Place directly in 70% alcohol.
	Crustacea (Decapoda)	Kill in freshwater; place directly in 70% alcohol. Larvae should be killed in sublimate or with a few drops of 1% osmic acid added to water.
	Pantopoda	Euthanize with 0.5% chromic acid; transfer to 70% alcohol.
	Mollusca (Lamellibranchs)	Narcotize in mixture of 100 cc seawater and 5 cc absolute alcohol for 6–12 hr; transfer to 1% chromic acid (1%) for 0.5 hr; preserve in 70% alcohol.
	Mollusca (Lima)	Kill with 0.25% chromic acid; transfer to 70% alcohol.

Mollusca (Scaphopoda—Dentalium)	Narcotize with 0.2 to 1% chloral hydrate for 10–12 hr; preserve in 70% alcohol.
Mollusca (Gastropoda)	Narcotize with a few drops of a solution of 2 gm cocaine powder dissolved in 100 cc of alcohol (50%); transfer to 70% alcohol.
Mollusca (Placophora, Patellidae, Fissurellidae, Haliotidae)	Narcotize in mixture of 100 cc seawater and 5 cc absolute alcohol; transfer to 70% alcohol.
Mollusca (*Natica josephinia*)	Narcotize by dropping 70% alcohol into seawater; kill by submerging in concentrated acetic acid; transfer to weak alcohol; transfer to 70% alcohol.
Mollusca (*Natica millepuncata, Natica hebrea, Nassa, Columbella, Conus, Trochus*)	Leave several days in a mixture of 1 pt freshwater and 1 pt seawater; kill with acetic acid; transfer to 70% alcohol.
Mollusca (Heteropoda—Atlantidae)	Narcotize in mixture of 100 cc seawater and 5 cc absolute alcohol for 6–12 hr or with a few drops of a solution of 2 gm cocaine powder dissolved in 100 cc of alcohol (50%); transfer to 70% alcohol.
Mollusca (Pterotrachaidae)	Euthanize in chrom-acetic mixture #1 for 10–30 min; wash in freshwater; transfer gradually to 70% alcohol.
Mollusca (Opistobranchiata)	Narcotize in a solution of 1 pt seawater and 1 pt freshwater; or in mixture of 100 cc seawater and 5 cc absolute alcohol; kill with concentrated acetic acid and transfer to 70% alcohol.
Mollusca (*Gastropteron meckeli*)	Kill in Kleinenberg's solution; preserve in 70% alcohol.
Mollusca (Doridium, Scaphander)	Narcotize in a mixture of 100 cc seawater and 5 cc absolute alcohol; kill with concentrated acetic acid and transfer to 70% alcohol.
Mollusca (Philine)	Relax in seawater; kill with concentrated acetic acid or pyroligneous acid; transfer to alcohol.
Mollusca (Pleurophyllidia)	Narcotize with alcoholized seawater; kill with concentrated acetic acid.

(continued)

Table 1. *(Continued)*

Reference	*Organisms*	*Comments*
	Mollusca (*Aplysia limacine, Aplysia punctata*)	Fix in 1% chromic acid for 15–60 min; transfer to 70% alcohol.
	Mollusca (*Aplysia depilans*)	Narcotize in 1% chloral hydrate (for up to 12 hr); kill with concentrated acetic acid; transfer immediately to 1% chromic acid for 30 min; transfer to 50% alcohol and gradually to 70% alcohol.
	Mollusca (*Pleurobranchia meckeli*)	Narcotize with cocaine; put in 1% chromic acid for 1 hr; wash; move to 70% alcohol.
	Mollusca (*Pleurobranchus meckeli, Pleurobranchus testudinarius*)	Kill in 5% chromic acid; transfer to 1% chromic acid for 15–60 min.
	Mollusca (*Umbrella*)	Kill in alcoholized seawater; transfer to weak alcohol.
	Mollusca (Elysiidae, Aeolidiidae; Triopa, Idalia, Polycera)	Relax in seawater; add 1–2 volumes of concentrated acetic acid; transfer to weak alcohol.
	Mollusca (*Phyllirrhoe bucephalum*)	Fix in chrom-osmic mixture for a few minutes or chrom-acetic mixture #2.
	Mollusca (*Doris, Chromodoris*)	Narcotize large specimens by adding 70% alcohol a little at a time; kill with concentrated acetic acid or boiling saturated sublimate. Alternatively, narcotize with cocaine; kill with concentrated acetic acid; place in 1% chromic acid for 10 min; transfer to 50% alcohol and on to 70% alcohol.
	Mollusca (Tritonia)	Kill large specimens in freshwater with a few drops of acetic acid; harden in 0.5% chromic acid. Narcotize small specimens with cocaine, then harden and preserve as for large specimens.
	Mollusca (Marionia)	Narcotize in alcoholized seawater; kill in acetic acid.
	Mollusca (Tethys)	Relax in small volume of seawater; kill in 1 or more volumes of acetic acid; siphon off fluid and substitute 1% chromic acid.

Mollusca (Pteropoda: Hyaleidae)	Relax in a small volume of water; kill in saturated sublimate.
Mollusca (Pteropoda: *Criseis acicula*)	Kill with alcoholized seawater.
Mollusca (Pteropoda: Cymbuliidae)	Place in Perenyi's solution for 15 min; transfer to 50% alcohol.
Mollusca (Gymnosomata)	Place in 1% chloral hydrate for 6–12 hr; quickly kill with acetic acid or sublimate.
Mollusca (Cephalopoda)	Soak dead specimens in seawater to restore shape; then harden in 1% chromic acid for 15–60 min.
Mollusca (small octopods)	Narcotize in 0.2% chloral hydrate; immerse in alcohol to kill; position in 1% chromic acid to harden (0.5–2 hr); rinse in freshwater; transfer to 70% alcohol; change alcohol several times.
Mollusca (*Ocythoe catenulate*)	Immerse directly in 70% alcohol to kill; position in 1% chromic acid to harden (0.5–2 hr); rinse in freshwater; transfer to 70% alcohol; change alcohol several times.
Mollusca (decapods)	Kill in 70% alcohol; position arms and harden in 0.2–1% chloral hydrate or alcoholized seawater; transfer to 70% alcohol.
Mollusca (*Loligopsis, Verania*)	Kill in Kleinenberg's solution for 1 hr; transfer weak alcohol; transfer gradually to 70% alcohol.
Bryzoa (Pedicellina, Loxosoma)	Place in 0.1–1% chloral hydrate for 1 hr; kill with hot or cold saturated sublimate; wash immediately; place in alcohol.
Brachiopoda	Narcotize in alcoholized seawater for several hours; transfer to 70% alcohol.
Tunicata (*Ascidiae simplices*)	Relax in seawater; immerse in 0.1% chloral hydrate; kill with chrom-acetic mixture; transfer to 1% chromic acid and inject each individual; after 0.5 hr transfer to 35% alcohol and inject; transfer to 70% alcohol.
Tunicata (Ascidia: *Phallusia*)	Narcotize in 0.1–1% chloral hydrate; harden in 1% chromic acid for 0.5 hr; wash; transfer to alcohol.
Tunicata (*Ciona intestinalis*)	Kill by adding a few drops of chrom-acetic mixture #2 to water; after 1 hr, harden in 1% chromic acid, injecting the body cavity; transfer to alcohol, injecting the body cavity.

(continued)

Table 1. ***(Continued)***

Reference	*Organisms*	*Comments*
	Tunicata (*Ascidia, Rhopalea*)	Relax in a small volume of seawater; slowly drop in 1% chromic acid and allow it to diffuse through water for 12–24 hr; harden in 1% chromic acid; wash in freshwater; transfer to alcohol.
	Tunicata (*Molgula, Polycarpa, Rhopalea, Chevreulius*)	Place in 0.1% chloral hydrate for 12 hr; kill with chrom-acetic mixture; transfer at once to 1% chromic acid to harden.
	Tunicata (*Cynthia, Styela*)	Narcotize in 0.2% chloral hydrate for 24 hr; kill with chrom-acetic mixture; transfer at once to 1% chromic acid to harden.
	Tunicata (Ascidiae compositae gelatinous forms)	Place in 0.1% chloral hydrate for a few days; kill with hot saturated sublimate; immediately transfer to 0.5% chromic acid for 0.5 hr; wash; transfer to 70% alcohol.
	Tunicata (*Distaplia*)	Narcotize with 0.1% chloral hydrate; kill with chrom-acetic mixture #2; wash; transfer to weak alcohol.
	Tunicata (*Phrysoma*)	Suspend by a thread in a mixture of alcohol and hydrochloric acid for 15 min; transfer to 60% alcohol; gradually increase alcohol concentration.
	Tunicata (Salpidae with hard bodies)	Immerse in 100 cc of freshwater and 10 cc concentrated acetic acid for 15 min; wash in freshwater 10 min; gradually transfer to 70% alcohol.
	Tunicata (Salpidae of medium consistency)	Place in chrom-acetic mixture #1 for 10 min; transfer to weak alcohol.
	Tunicata (Salpidae of soft consistency)	Immerse in chrom-osmic mixture for 0.25–1 hr; wash in freshwater; transfer to weak alcohol.
	Tunicata (Doliolids)	Kill in mixture of 100 cc copper sulfate (10%) and 10 cc saturated corrosive sublimate, or in saturated sublimate alone, or in chrom-osmic mixture; after a few minutes, wash in freshwater; gradually transfer to 70% alcohol.
	Fish	In general, place fish live into the fixing fluid; inject through the anus with 90% alcohol and store in 70% alcohol.
	Fish (*Amphioxus*)	Kill in seawater alcoholized to 10%; transfer to 50% alcohol; gradually increase concentration to 70%. Alternatively, kill in Müller's solution or 1% chromic acid.

	Fish (Cyclosomans, Selachians, Ganoids)	Place small specimens in 70% alcohol; inject large specimens with 90% alcohol through the anus with each change of fluid.
	Fish (of soft consistency such as *Torpedo*)	Fix in 1% chromic acid for 0.5 hr; transfer to alcohol.
	Fish (Selachian embryos)	Fix in saturated sublimate 5–15 min; wash well; transfer to alcohol.
	Fish (Torpedo embryos)	Fix in 1 pt chromic acid (1%) and 1 pt saturated sublimate for 15 min; wash in freshwater; transfer to alcohol.
	Fish (Teleosts)	Place small specimens in 70% alcohol; repeatedly inject large specimens with 90% alcohol through the anus with each change of fluid.
	Fish (Teleosts with silvery skin, e.g., *Trachypterus*)	Place in saturated sublimate for a few minutes; transfer to alcohol.
	Fish (translucent larvae of Teleosts)	Place directly in weak alcohol, or fix first with saturated sublimate.
	Fish (transparent fertilized eggs)	Place in a mixture of 100 cc alcohol (50%) and 5 cc hydrochloric acid (concentrated) for a few minutes; transfer to pure alcohol.
Mandal 1984	Mammals	Preserve in 70% ETOH, 45% isopropyl alcohol, or 10% buffered formaldehyde.
Martin 1977	Arachnids, larval, nymphal insects, and soft-bodied adults	Preserve in 70–80% ETOH or isopropyl with 10% acetic acid as a hardener and 2–3% glycerin to protect from dehydration.
	Mites	Preserve in Oudeman's fluid (87 cc alcohol, 8 cc glacial acetic acid, 5 cc glycerin).
	Insects	Preserve in AGA (8 pts ETOH, 5 pts distilled water, 1 pt glycerin, 1 pt glacial acetic acid), Pempel's fluid (4 pts glacial acetic acid, 6 pts formaldehyde, 30 pts distilled water, 15 pts alcohol); recommends against use of formaldehyde.
Martin and Terrence 1962 (reported in Knudsen 1972)	Fish	To preserve color, preserve in solution of 20 cc 1% phenol, 12 cc formaldehyde, 8 cc glycerin, and 60 cc water.
Maynard 1870	Amphibians, fish, reptiles, and eggs	Preserve in alcohol; inject large specimens with alcohol before submerging.

(continued)

Table 1. ***(Continued)***

Reference	*Organisms*	*Comments*
McDiarmid 1994	Amphibians	Fix in 1 pt formaldehyde with 9 pts water, neutral buffered with 1/2 tsp. powdered magnesium chloride or 4.0 g sodium phosphate monobasic and 6.5 g sodium phosphate dibasic per L.
McDiarmid and Altig 1999	Anuran larvae and eggs	Fix in a mixture of 1800 mL formaldehyde, 16,200 mL distilled water, 72 g sodium phosphate monobasic, and 117 g sodium phosphate dibasic (anhydrous). Check pH before and during use. Increase ratio of fixative to water if processing a large number of specimens to counteract the dilution of the fixative by the water in the specimens.
Michaelis 1981	Freshwater angiosperms and bryophytes	Wash in distilled water, shake dry, submerge in 50% glycerin for 3d; transfer to 100% glycerin for 1 wk. May be removed and dried or maintained in 100% glycerin in vials.
Middleton 2007	Succulents, delicate plant structures, fruits	Preserve in 70% ETOH; may add 1% glycerin to prevent embrittlement.
Miller 1899	Mammals	Preserve in alcohol with water equivalent to 1/5 of the volume of alcohol, or formaldehyde with 15 to 10 times its volume in water.
Moore 2009a	Animals	Fix in Serventy's fluid (90 pts 70% alcohol; 5 pts 40% formaldehyde; 5 pts glycerin) (used in the 1950s).
Moore 2010	Plants	Fix in FAA (48 mL formaldehyde, 160 mL IMS, 8 mL glacial acetic acid, and 320 mL deionized water) or Kew mixture (4–5% formaldehyde, 53% IMS, 5% glycerin, and 37% water; or 1 pt formaldehyde, 10 pts IMS, 1 pt glycerin, and 8 pts deionized water). Preserve in alcohol with 5–10% glycerin added to reduce evaporation, Kew mixture, or Copenhagen mixture (53% IMS, 5% glycerin, and 37% water; or 10 parts IMS, 1 part glycerin, and 9 parts deionized water).
Mori and Prance 1987	Lecythidacae buds	Preserve in Farmer's solution (3 pts ETOH, 1 pt glacial acetic acid) or Jackson's solution (4 pts ETOH, 2 pts methanol, 2 pts chloroform, 1 pt proprionic acid, 1 pt acetone).
	Lecythidacae flowers	70% ETOH or FAA (10 pts ETOH, 7 pts distilled water, 2 pts formaldehyde, 1 pt glacial acetic acid).
	Lecythidaceae twigs and bark	FAA (10 pts ETOH, 7 pts distilled water, 2 pts formaldehyde, 1 pt glacial acetic acid).

Morton 1950	Marine invertebrates	Fix in a solution of 1 pt formaldehyde with 10 pts water or in Bouin's fluid; preserve in 70–80% ETOH with a teaspoon of glycerin added per quart.
Notes: Bouin's fluid = 75 pts picric acid; 25 pts formaldehyde; 5 pts glacial acetic acid	Coelenterates	Preserve in 70% ETOH (preferred) or 5% formalin.
	Platyhelminthes (flatworms) and Nemertea (proboscis worms)	To fix, pipette warm Bouin's fluid over the animal from tail forward to minimize contraction; store for several days in Bouin's fluid.
	Polychaeta (bristle worms)	Preserve in 70% ETOH, brought up to 80%, with a drop of glycerin added.
	Echinodermata	Preserve in 70–80% ETOH with glycerin.
	Ascidians and Enteropneusta	Preserve in formalin. Fix compound ascidians in glacial acetic acid, then transfer after 5 minutes to 95% ETOH.
	Crustacea	Preserve in 70–80% ETOH with glycerin; do not use formalin.
Myers 1956	Amphibians, fish and reptiles	Fix in 1:9 formaldehyde and water; preserve in 70–75% ETOH for reptiles and fishes; 60–65% ETOH for amphibians.
Nagorsen and Peterson 1980	Mammals	Fix in 10% formalin (neutralized with 1 tsp borax or 1 tbs ammonia per gal, or 4 g monobasic sodium phosphate monohydrate and 6.5 g dibasic sodium phosphate anhydrate per L), Bouin's solution, or sodium acetate. Preserve in 65–70% ETOH or 45–60% isopropyl alcohol.
Necker 1940	Amphibians and reptiles	Preserve in ETOH; if ETOH not available, preserve in formaldehyde.
Neill 1950	Reptiles	Cut open body cavity to permit entrance of full-strength ETOH; after 3–4 days move to a solution of 3 pts alcohol to 1 pt water.
	Amphibians	Submerge in or inject with a solution of 2 pts alcohol and 1 pt water, or 1 pt formaldehyde to 19 pts water.
Neumann 2010	Fishes	Fix in 4% (temperate regions) or 10% (tropical regions) formalin buffered with 6 g sodium phosphate dibasic anhydrous and 4 g sodium phosphate monobasic monohydrate per L or a carbonate buffer (inject specimens larger than 30 mm); wash after fixation, then transfer to 20%, 40%, 60% to 70% ETOH.
Nieuwland and Slavin 1928	Monotropa	To preserve color in monotropa, use saturated alcohol solution of sulfur dioxide, obtained by adding sodium sulfite and hydrochloric acid to the alcohol.

(continued)

Table 1. ***(Continued)***

Reference	*Organisms*	*Comments*
Oldroyd 1958	Insects	Best general preservative is 75–80% ETOH. Alternatives include AGA (8 pts 95% alcohol, 5 pts distilled water, 1 pt glacial acetic acid), Bouin's fixative (30 pts saturated picric acid in water, 10 pts formaldehyde, 2 pts glacial acetic acid), Pampel's fluid (4 cc glacial acetic acid, 30 cc distilled water, 6 cc formaldehyde, 15 cc alcohol 95%), Picro-chlor-acetic fixative (12 pts picric acid 1% in 95% alcohol, 12 pts alcohol 95%, 2 pts chloroform, 1 pt glacial acetic acid). Fix for 12 hr, rinse in several changes of 80% ETOH, preserve in 80% ETOH.
Omori and Fleminger 1976	Crustacean zooplankton	Allow stock solution (30–40 g sodium tetraborate per 1,000 mL formaldehyde) to rest for 2 months; filter before using. For fixation and preservation, use 5–10 mL stock to 95 mL seawater; or 5 mL stock, 0.5 mL propylene phenoxytol, 4.5 mL propylene glycol, and 90 mL seawater.
Ouwens 1907	Reptiles	Immediately after shedding, euthanize reptile with chloroform or carbon disulfide, remove gut contents, dip specimen briefly in boiling water, wash out mouth, inject preservative through anus and mouth, then coil in container and add preservative.
Owen and Steedman 1956 (also Hildebrand 1968)	Animals	Fix for a minimal time in formalin or alcohol with 10 g of sodium acetate per L; wash 12–24 hr in running water. Preserve in 1% ethylene glycol monophenyl ether in water; 0.2% para-hydroxybenzoic acid in water; or 1% phenoxetol.
Papparitz 1933	Animals	To preserve color, inject with a mixture of 100 pts water, 5 pts chloral hydrate, 5 pts magnesium sulfate, and 5–10 pts formaldehyde; fix in same solution for 6 hr; rinse; store in solution of 100 pts glycerin, 20 pts potassium chloride, and 299–300 pts water.
Parsons 1831	Birds	Remove liver, stomach, intestines, and oviduct; preserve in spirit.
	Fish and reptiles	Preserve small specimens entire in spirit (ETOH), cut open abdomen. Large specimens must be cut into pieces prior to submersion in alcohol. Use a quantity of proof spirit equal to the weight of the specimen; fish may require more spirit. Watery or pulpy specimens require rectified [distilled] spirit. Fill the thorax and abdomen with spirit. Inject snakes through mouth and anus.
Penrith 1971	Fish	Preserve in N-propyl alcohol (Propylol), 45%.

	General	Preserve in Dowicil 100 powder mixed to 10% aqueous solution by weight (cautions that Dowicil is a formaldehyde preservative); ethylene glycol (50%), Nipa ester 82121 (0.2%, dissolved in alcohol before adding to water); Perminal K.B. (10%), or propylene phenoxetol (1%), dissolved in alcohol before adding to water.
Perry and Thomas 1992	Plants	Preserve in 70% ETOH, 29% water, and 1% glycerol.
Petersen 1976	Planktonic coelenterates	Add 10% formalin to seawater or transfer specimens to 4% formalin; store in 4% formalin.
Phillips 1984	Mammals	For general purposes, fix in Bouin's fixative (25 mL formaldehyde, 75mL saturated aqueous picric acid, 5 mL glacial acetic acid) for 24 hr; store in 70% ETOH.
		For bone, cartilage, mast cells, proteins use Carnoy H (60 mL 100% ETOH, 30 mL chloroform, 10 mL glacial acetic acid) for 6–8 hr, wash 2–3hr in 100% ETOH to remove chloroform.
		For general purposes, muscle, connective tissue, glands, and gastrointestinal track, use neutral buffered formalin (100 cc formaldehyde, 900 cc H_2O, 4 g acid sodium phosphate monohydrate and 6.5 g anhydrous disodium phosphate) for 24–48 hr, wash in water, store in 70% ETOH.
		For general purposes, nuclei, and chromosomes, use Sanfelice fixative (80 mL 1% CrO_2, 40 mL formaldehyde, and 5 mL glacial acetic acid) mixed immediately before use, fix for 24 hr, wash thoroughly in running water.
		For general microanatomy, mitochondria, connective tissue, myelin, and neurosecretry material, fix in Susa (Heidenhain's fixative, 45 g $HgCl_2$, 5 g NaCl, 20 g trichloroacetic acid, 40 mL glacial acetic acid, 200 mL formaldehyde, 1000 mL H_2O) for 24 hr, transfer to 95% ETOH.
		For hemopoietic, hypothalamic, neurosecretary material and bone marrow, use Helly's fixative (5 g $HgCl_2$, 2.5 g $K_2CR_2O_7$, 1 g $Na_2SO_4 10H_2O$, and 100 mL H_2O) for 12–24 hr.

(continued)

Table 1. ***(Continued)***

Reference	*Organisms*	*Comments*
		For ultrastructure, organelles, cell membrane, fix in trialdehyde (5 g paraformaldehyde dissolved in 25 mL H_2O with a drop of NaOH; add 15 mL 50% gluteraldehyde, 6.25 mL DMSO, 0.25 mL of 0.1 m $CaCl_2$, 2.5 mL acrolein, and dilute to 250 mL with 0.05 M cacodylate buffer with 0.1 M sucrose (makes 250 mL, mix fresh daily) for 10–24 hr, replace with 3% gluteraldehyde, store at 4–5°C; in the field store at ambient temperature in 0.05 M cacodylate buffer with 0.01 M sucrose.
		For general microanatomy and cell organelles, fix with Zenker's fixative (stock solution is 2.5 g $K_2Cr_2O_7$, 5 g $HgCl_2$, 1 g Na_2SO_4, in 100 mL H_2O. Add 5 mL glacial acetic acid just before use).
Piechocki and Händel 2007	Hydroidea and Siphonophora	Fix and preserve in 10% buffered formalin.
Note: Agitate specimens in 40–50% ETOH before transferring to final preservative to remove the white precipitates that forms when alcohol is contaminated with formalin mixed with seawater.	Nemertini (ribbon worms)	Relax in water sprinkled with menthol or chloral hydrate crystals; fix in 10% buffered formalin; preserve in 75% ETOH; or fix in 10% formalin; rinse in 40–50% ETOH; preserve in 75% ETOH.
	Heteropoda	Fix and preserve in 10% buffered formalin.
	Cephalopoda	Fix in 10% buffered formalin; preserve in 75% ETOH.
	Polychaeta	Relax in water sprinkled with menthol crystals; fix in 10% buffered formalin; rinse in 40–50% ETOH; preserve in 75% ETOH.
	Crustacea	Fix in 10% buffered formalin; rinse in 40–50% ETOH; preserve in 75% ETOH. Large, transparent amphiopodes (cystosoma) and decopode larvae can be preserved in formalin.
		Fix crayfish in 10% formalin; rinse in 40–50% ETOH; preserve in 75% ETOH.
	Chaetognatha	Fix and preserve in 10% formalin.
	Foraminifera and other small organisms	Fix in 10% buffered formalin; rinse in 96% ETOH, preserve in 96% ETOH.
	Porifera	Fix and preserve in 96% ETOH.

Coelenterates (hydroidea, alcyonaria, corals, sea anemones)	Relax with menthol crystals in water; fix in 10% formalin, rinse in 40–50% ETOH; preserve in 75% ETOH.
Parasitic worms and free-living nematodes	Fix in hot sublimate; rinse in 40–50% ETOH; preserve in 75% ETOH.
Opistobranchia, Tectibranchiata, Nudibranchiata, etc.	Relax with menthol crystals in water; fix in 10% formalin; rinse in 40–50% ETOH; preserve in 75% ETOH.
Other molluscs	Fix in 10% formalin, rinse in 40–50% ETOH; preserve in 75% ETOH.
Sipunculidae	Relax with menthol crystals in water; fix in 10% formalin; rinse in 40–50% ETOH; preserve in 75% ETOH.
Pantopoda	Fix in 10% formalin; rinse in 40–50% ETOH; preserve in 75% ETOH.
Bryozoa	Relax with menthol crystals in water; fix in 10% formalin. Rinse in 40–50% ETOH. Preserve in 75% ETOH.
Pterobranchia	Relax with menthol crystals in water; fix in 10% formalin. Rinse in 40–50% ETOH. Preserve in 75% ETOH.
Crinoidae	Fix, rinse, and preserve in 75% ETOH.
Sea cucumbers (Holothuroidea)	Relax with menthol crystals in water; fix in 10% formalin; rinse in 40–50% ETOH; preserve in 75% ETOH. To prevent ejection of the guts, inject 40% formalin through the anal opening, repeat with 40–50% and 75% ETOH.
Sea urchins (Echinoidea) and starfish (asteroidean)	Fix in 10% formalin, rinse in 40–50% ETOH; preserve in 75% ETOH.
Brittle stars (Ophiuroidea)	Fix in 10% formalin, rinse in 40–50% ETOH; preserve in 75% ETOH.
Sea squirts (Ascidiacea)	Fix in 10% formalin; rinse in 40–50% ETOH; preserve in 75% ETOH.
Soft-bodied small insects and larvae	AGA (6 pts alcohol, 1 pt glycerin, and 1 pt acetic acid)

(continued)

Table 1. *(Continued)*

Reference	*Organisms*	*Comments*
Pisani 1973	Amphibians	Fix in 1 pt formaldehyde and 9 pts water, buffered with baking soda or borax; FAA; or 70% ETOH. Store in 40% ETOH or isopropyl.
Note: FAA = 10 pts formaldehyde, 50 pts alcohol, 2 pts glacial acetic acid	Reptiles	Fix in 1 pt formaldehyde and 9 pts water, buffered with baking soda or borax; FAA; 95% alcohol. Store in 50% ETOH or isopropyl.
Pope 1928	Small animals	40–50% isopropyl alcohol
Post 1967	Bird, fish, mammal tissue samples for bacterial and viral diseases	Preserve "half-fist size piece" in 10% formalin, Bouin's fluid, or Zenker's fluid using 10 times volume of preservative to tissues.
Notes: Bouin's Fluid = 75 pts picric acid, 25 pts formaldehyde, and 5 pts acetic acid.	Ectoparasites	Preserve in mixture of 70% alcohol and 5% glycerin (ethyl or isopropyl preferred; may use methyl if others are not available).
Zenker's Fluid = Prepare a saturated solution of mercuric chloride in 5% potassium dichromate. Immediately before use add 5 pts acetic acid to 95 pts of the stock solution; discard after 24 hr.	Roundworms	Drop into mixture of hot 70% alcohol and 5% glycerin (ethyl or isopropyl preferred; may use methyl if others are not available).
	Tapeworms and flukes	Relax in water, then preserve in mixture of 70% alcohol and 5% glycerin.
	Bladder worms or Cysticercus	Evert by immersion in 1% potassium dichromate; preserve in mixture of 70% alcohol and 5% glycerin.
	Amoeba	Preserve piece of liver with at least one ceca in 10% formalin, Bouin's fluid, or Zenker's fluid.
	Coccidia	Preserve portion of inflamed intestine and ceca in 10% formalin, Bouin's fluid, or Zenker's fluid.
	Sarcocystis	Preserve piece of muscle with one or more cysts in 10% formalin, Bouin's fluid, or Zenker's fluid.
	Throat bots and warbles (insect larvae found beneath the skin)	Preserve larvae in mixture of 70% alcohol and 5% glycerin.
	Bird and mammal specimens suspected of having nutritional disease	Place cube of liver, kidney, spleen, lung and lymph node in 10% formalin, Bouin's fluid, or Zenker's fluid using 10 times volume of preservative to tissues.

	Fish specimens suspected of having nutritional disease	Open abdomen, remove guts, and place fish in 10% formalin, Bouin's fluid, or Zenker's fluid using 10x volume of preservative to tissues.
	Animal tissue samples suspected of having selenium or other toxic poisoning	Place one-inch cubes of liver, kidney, spleen, lung, heart and lymph node in 10% formalin, Bouin's fluid, or Zenker's fluid using 5 times volume of preservative to tissues. Cut a slit in the back of eyeball and preserve in Bouin's fluid, or Zenker's fluid (use 10% formalin if others not available).
	Zooplankton	Add formaldehyde to a jar containing plankton in water to make a 10% formalin solution.
	Phytoplankton	Add formaldehyde to a jar containing plankton in water to make a 5% formalin solution.
	Benthic organisms	Preserve in 10% formalin.
	Stomach samples	Preserve stomach contents of mammals or fish and crop and gizzards of birds in 10% formalin.
	Fish (as museum specimens)	Place specimens 5 inches or shorter directly in 10% formalin or 70% alcohol; open left body wall of larger specimens or inject preservative. Use at least 10 times preservative volume to volume of fish.
Providence Franklin Society 1829	Amphibians and reptiles	Preserve in sulfurous acid (burn "sulphur matches" over agitated water, filter the water); a mixture of one pt molasses with one pt salt and 15–16 pts water; solution of three pounds clean salt and one gallon water, strained, diluted with half ounce clean water per quart of mixture; or chloride of lime (one dram to half an ounce of "common bleaching salts" mixed with a pint of water). Wash specimens thoroughly before preservation, make incisions in body; suspend specimen inside solution in jar.
Quay 1974	Birds and mammals	Perfuse with 10% formaldehyde buffered with two sodium salts.
Réaumur 1748	Birds	Put specimen in a vessel full of spirit of wine or a very strong brandy, taking care to clean off any blood and to smooth out the feathers; recommends removing the intestines as well.
Redenbaugh 1895	Coelenterates (comb jellies and corals)	Preserve in magnesium sulfate (saturated solution), followed by immersion in Perenyi's fluid, picro-sulfuric acid, or 4% formaldehyde.
	Echinoderms	Preserve in saturated solution of Epsom salts followed by immersion in 4% formaldehyde.

(continued)

Table 1. ***(Continued)***

Reference	*Organisms*	*Comments*
	Ctenophores (comb jellies)	Preserve in solution of 1 pt 2% formaldehyde, 1 pt Perenyi's fluid, and NaCl to equal the density of seawater. Transfer after 30 minutes to 4% formaldehyde.
Reid 2000	Mollusks	Narcotize in 7.5% magnesium chloride; fix in 1 pt formaldehyde with 9 pts seawater, buffered with a spoonful of sodium tetraborate per L; preserve in 70–80% ETOH.
Riley 1892	Micro-larvae (Lepidoptera)	Submerge in 90% alcohol, boil to bleach out black color that forms upon submergence (H. Dewitz's method).
	Soft-bodied insects	Preserve in diluted alcohol; store in alcohol of higher concentration.
		Submerge first in 1 quart of a solution of 2.5 lbs common salt and 4 oz niter dissolved in 1 gal water and filtered, and 2 oz sodium arsenite of potash; then preserve in a solution of 2.5 lbs common salt and 4 oz niter dissolved in 1 gal water and filtered (Laboulbene method).
	Caterpillars	Preserve in a solution of 2.35 g common salt, 55 g alum, 18 centigrams corrosive sublimate, and 5 L boiling water that is allowed to cool before adding 50 g carbolic acid and filtering (Verrill method).
	Delicate soft-bodied insects	Preserve in glycerin, either pure or mixed with water or alcohol.
		Submerge for 6–12 days in a solution of 100 g alum, 25 g common salt, 12 g potash, and 10 g arsenious acid in 3,000 g boiling water that is then filtered and cooled, then mixed in a ratio of 10 L to 4 L glycerin and 1 L m ETOH (Wickersheim Preserving Method).
Robins and Backus 1952	Fish otoliths	Store in tri-sodium phosphate; write on labels in pencil as the preservative affects ink.
Roper and Sweeney 1983	Cephalopods	Fix in 4–10% formalin, depending on specimen size and tissue density in a volume of 2:1 to 4:1 of fixative fluid to specimen volume. For greater fixative penetration, use a fixative that is also a lipid solvent (based on Steedman 1976: 1.5 pts propylene phenoxetol, 5.0 pts propylene glycol, 10.0 pts formaldehyde, and 83.5 pts distilled water). For histological preparations, fix in Bouin's fixative (25 pts saturated aqueous picric acid, 5 pts formaldehyde, and 1 pt glacial acetic acid). In an emergency, fix in 70–75% ETOH. Field fixation may also be done with paraformaldehyde or gluteraldehyde. Buffer formaldehyde solutions with calcium carbonate, sodium borate, or hexamine. After fixing preserve in 70–75% ETOH; 50% isopropyl alcohol, or 10% buffered formalin.

Ruthven 1912	Reptiles and amphibians	Inject adults with a 4% formaldehyde solution; put eggs and larval stages in a 2% formaldehyde solution.
Schutlz 1924	Mammal embryological material	Perfuse fixative through an artery with a solution of 1 pt formaldehyde and 9 pts water; add 5 pts glycerin or salt to every 100 pts stock solution. To prepare specimen for studies of blood vessels, after fixing for 24–48 hr and perfuse with starch (the consistency of cream) colored with Prussian blue or cinnabar, or India ink. Perfuse fetuses through the umbilical cord with a 10% formaldehyde solution. Perfuse fetuses less than 2 in long with Bouin's solution (75 pts picric acid, saturated; 20 pts 40% formaldehyde, and 5 pts glacial acetic acid) and later preserve in alcohol.
Scrocchi and Kretzschmar 1996	Amphibians and reptiles	Fix in 10% formaldehyde buffered with 4 g monobasic sodium phosphate monohydrate and 6.5 g dibasic sodium phosphate anhydrate per 250 cc; or preserve in 50–70% ETOH; preserve in 70% ETOH. Preserve amphibian larvae in 10% buffered formaldehyde.
Scully 1937	Plants	To preserve colors in plants, preserve for 24 hr in a solution of 5% copper sulfate, then transfer to a solution of 16 cc commercial sulfuric acid, 21 g sodium sulfite, and 1000 cc water.
Sheim 1951 (also Hildebrand 1968)	Animals	Fix in 50 mL formaldehyde with 5.9 g dibasic sodium phosphate and 4.7 g monobasic sodium phosphate with sufficient tap water to make 1000 mL for 1 wk; preserve in 50 mL formaldehyde with 5.9 g dibasic sodium phosphate, 4.7 g monobasic sodium phosphate, and 5 g sodium hydrosulfite with sufficient tap water to make 1000 mL.
Simmons 2002	Amphibians and reptiles	Fix in 10% formaldehyde buffered with 4 g monobasic sodium phosphate monohydrate and 6.5 g dibasic sodium phosphate anhydrate per L. Introduce fixative into the body cavity by making cuts, perfusion, or injection; preserve in 70% ETOH. Preserve amphibian larvae in 10% buffered formaldehyde.
Slevin 1927	Smaller amphibians and reptiles	Make one or more deep cuts into the body cavity just to one side of the median ventral line; place reptiles in 95% ETOH, geckos and amphibians in 65–75% ETOH. Remove large masses of food and fat deposits from the body cavity. Do not allow alcohol concentration to fall below 55–60%. Do not use formaldehyde.
	Large turtles, crocodilians, and huge snakes	Remove skin and preserve in alcohol, brine, or dry.

(continued)

Table 1. ***(Continued)***

Reference	*Organisms*	*Comments*
Smith 1951	Amphibians	Preserve in 1 pt formaldehyde with 9 pts water; ETOH at 65%; or Viosca's fluid (1 pt formaldehyde, 3 pts ETOH, 6 pts water).
	Reptiles	Preserve in 1 pt formaldehyde with 9 pts water; ETOH at 75%; or Viosca's fluid (1 pt formaldehyde, 3 pts ETOH, 6 pts water).
Soderstrom and Young 1983	Bamboo	For morphological and anatomical studies, preserve in FAA (90 pts 50% ETOH, 5 pts glacial acetic acid, and 5 pts formaldehyde) and store in either FAA or 70% ETOH. For cytological studies, fix in solution of 3 pts absolute ETOH and 1 pt glacial acetic acid.
Steedman 1976d	Invertebrate shells, spicules, and calcium supports	Preservation of calcareous marine zooplankton is affected by low pH, changes in pH during preservation, pH of 8.5 and higher, mechanical pressure, ratio of specimens to preservative, handling, osmotic changes, variability of seawater as a diluents, evaporation, high storage temperatures, and ammonia from amine breakdown.
Steenis 1935	Fruits	Preserve in 20% ammonium sulfate in water to preserve dark color.
Stejneger 1891	Amphibians and reptiles	Open body cavity or inject with alcohol weaker than 75% for 24 hr, then transfer to reptiles to 75% ETOH and amphibians to 50–60% ETOH.
Stejneger 1911	Amphibians and reptiles	Use a 1:20 mixture of formaldehyde and water, tested from time to time for acidity, and neutralized with bicarbonate of soda when necessary. Open body cavity or inject into specimen.
Stephenson and Riley 1995	Marine animals	Fix and preserve in a mixture of 2.5% formaldehyde and 1% glutaraldehyde buffered with 2% sodium glycerophosphate.
Stone 1899	Plants	Fix and preserve plants in a solution of 4 pts 40% formaldehyde and 100 pts water; add nitric acid to retain blue color of starch grains with iodine stain.
Stone 1983	Pandanaceae (*Pandanus*, *Freycinetia*, and *Sararanga*)	FAA (no recipe provided); "If FAA is lacking, strong alcohol or just formalin may suffice."
Tate 1883	Slugs	Allow to extend fully on a glass slide, then plunge into cold saturated corrosive sublimate solution; preserve in a mixture of 1.5 pts water and 1 pt glycerin.

Taylor 1976	Flagella	Thecate dinoflagellates—preserve in 1–2% or 5% formalin made with seawater. Fix in 0.4 mL Lugol's iodine (saturated iodine and potassium iodide with some acetic acid) per 100 mL sample; in 2–5% glutaraldehyde or 1–2% osmium tetroxide; or in 8 mL modified Karnovsky's fixative (50 mL saturated paraformaldehyde 35%, 50 mL gluteraldehyde 25% and 3 g hexamine in 20 mL distilled water) and 92 mL filtered seawater.
Terentjev and Chernov 1949	Amphibians	Kill by submersion in 60% used yellow or reddish alcohol. Anesthesize amplecting amphibians in 2–3 cm water by adding strong boiled tobacco solution drop by drop, then transfer to weak alcohol for 30 min; preserve in 70–75% alcohol. After euthanizing, wash specimen. For display specimens, after euthanizing arrange in desired shape on cork plate using pins and place in 10% solution of commercial formalin; transfer to 70–75% alcohol for preservation.
	Amphibian eggs	Place in weak formalin solution.
	Amphibian larvae	Place in very weak alcohol; gradually increase to 70%.
	Reptile eggs	Place in very weak alcohol; gradually increase to 70%.
	Reptiles	Kill by submersion in 96% alcohol for several hours. After euthanizing, wash specimen and either make superficial cuts to allow penetration of preservative into body cavity, or inject into stomach, intestines, mouth, and cloaca. For display specimens, after euthanizing arrange in desired shape on cork plate using pins and place in 10% solution of commercial formalin; transfer to 70–75% alcohol for preservation.
Tokioka 1976	Thaliacea	Fix in 4% formaldehyde and seawater; preserve in 70% ETOH.
Tomlinson 1965	Plants	Fresh CRAF (5 g chromium trioxide, 35 cc glacial acetic acid, 350 cc 10–20% formaldehyde plus water to make one quart).
Treadwell 2006	Plants	Fix specimens in Kew Mix (5% formaldehyde, 5% glycerin, 53% IMS, and 37% water); preserve in Copenhagen Mix (70% IMS, 28% water, and 2% glycerin).
Turner 1976	Molluscan zooplankton	Fix in 4% formalin prepared with seawater or 90–95% ETOH; preserve in 80% ETOH; buffer alcohol of below 80% with borax.

(continued)

Table 1. ***(Continued)***

Reference	*Organisms*	*Comments*
	Pteropoda and Heterpoda	Relax specimens by adding 7% magnesium chloride; 1% acetone-chloroform; a solution of 1 pt glycerin, 2 pts alcohol 70% and 2 pts seawater; MS222; or urethane in seawater. Fix in 3 times volume of sample with 4% formalin in seawater; Bouin's fixative (15 pts picric acid, 5 pts formaldehyde, 1 pt acetic acid); Newcomer's solution (6 pts isopropyl alcohol, 3 pts propionic acid, 1 pt petroleum ether, 1 pt dioxane, 1 pt acetone); or Perenyi's solution (4 pts nitric acid 10%, 3 pts alcohol 70%, 3 pts chromic acid 0.5%). For best shell preservation, fix in 70% acid-free alcohol or 4% formalin made with seawater and buffered to pH 8. Preserve in 70% acid-free alcohol; preserve naked forms in 3–4% formalin.
Tyler 1962	Anuran larvae	Recommends a mixture of 10 mL of formaldehyde, 90 mL of water, 4 mL calcium chloride (10% aqueous solution), and 0.2 g cobalt nitrate.
Valdecasas and Baltanás 1989	Freshwater invertebrates	Preserve in Angelier's fluid (1% acetic acid, 1% chromic acid, 98% water), recommended as a "near universal fixative" for freshwater invertebrates.
van Dijk 1966	Amphibian larvae	Fix and preserve in 4% formaldehyde; or fix in Bouin's fluid and preserve in 70% alcohol.
van Steenis 1935	Plants	Recommends adding kalium or natrium bisulfite ($NaHSO_2$) to alcohol to prevent plant specimens from turning black.
Viswanathan et al. 1991	Plants	Preserve with cupric acetate, glacial acetic acid, and formaldehyde.
Wagstaffe and Fidler 1955	Small invertebrates (general)	Fix in Flemming's fluid (150 mL chromic acid 1%, 40 mL osmic acid 2%, 10 mL glacial acetic acid) or Hermann's Solution (15 pts platinum chloride 1%, 1 pt glacial acetic acid, 2–4 pts osmic acid 2%). Wash in running water 24 hr, store in 70% alcohol or 5% formalin.
Notes: Alcohol = either ETOH (ethyl alcohol) or IMS (Industrial Methylated Spirit)	Water mites	Fix and preserve in Viet's fluid (5 pts distilled water, 10 pts glycerin, 2 pts glacial acetic acid).

Bouin's fluid = 30 pts picric acid, 10 pts formaldehyde, 2 pts acetic acid	Actinozoa	Alcyonaria: narcotize with menthol, euthanize with hot corrosive sublimate, store in 70% alcohol or euthanize in 5% formalin, rinse in cold saturated solution of corrosive sublimate, and store in 70% alcohol.
Carnoy's fluid = 5 mL glacial acetic acid, 30 mL ETOH, 15 mL chloroform		Zoantharia: when fully extended in seawater, add menthol crystals and leave overnight; reduce fluid volume and add 10% formalin, pumping into mouth with a pipette if necessary; store in 5–10% formalin.
Formalin = commercial formaldehyde (saturated solution)	Annelida	Place specimen in clean water, add 70% alcohol drop by drop until flaccid; spread on glass plate with soft brush; harden and store in formol-alcohol (95 mL 70% alcohol and 5 mL formalin).
Pampl's fluid = 4 mL glacial acetic acid, 30 mL water, 6 mL formaldehyde, 15 mL ETOH	Arachnida	Kill and store in 70% alcohol or Oudeman's fluid (22 pts 70% alcohol, 1 pt glycerin, 5 pts glacial acetic acid).
	Brachiopoda	Relax in seawater for 30 min; add alcohol to make a 5% solution; transfer to 70% alcohol.
	Chaetognatha	Narcotize with menthol or 5% chloral hydrate; position specimen on glass plate and flood with Bouin's fluid for 30–60 min followed by 2–3 changes of 50% alcohol, or Gibson's fluid (5 gm corrosive sublimate, 4 mL nitric acid 80%, 1 mL glacial acetic acid, 15 mL alcohol 95%, 220 mL distilled water) for 30–60 min followed by 2–3 changes of iodized alcohol. Store in 5% formalin.
	Coelenterata	Hydroids: add 10 mL of 10% formalin to 100 mL of seawater containing hydroids, fix for 3–4 hr; store in 5% formalin.
		Medusae: replace seawater with 5% formalin.
	Crustacea	Branchiopoda and Ostracoda: kill in 5–10% formalin; after 10 min transfer to 50% alcohol; store in 70% alcohol.
		Copepoda: kill and store in 5–10% formalin.
	Ctenophora	Place specimens in solution of 100 mL of 1% chromic acid and 2 mL of 1% osmic acid for 15–60 min; stage through 30% and 50% alcohol; store in 70% alcohol.
	Echinodermata	Asteroidea: add magnesium sulfate crystals to water; wash quickly, then place in 70% alcohol.

(continued)

Table 1. *(Continued)*

Reference	*Organisms*	*Comments*
		Ophiuroidea: add weak alcohol to seawater or place in freshwater; preserve in 70% alcohol.
		Echinoidea: drop in warm 70% alcohol or strong acetic acid, rinse, preserve in 70% alcohol.
		Holothuroidea: add magnesium sulfate or menthol crystals to water; fix in 15% formalin, inject with 70% alcohol in body cavity; preserve in 70% alcohol.
		Crinoidea: position in container and add 90% alcohol for 12–24 hr; preserve in 70% alcohol.
	Myriapoda	Chitopoda and Diplopoda: kill in 50% alcohol; store in 70% alcohol.
		Progoneata: kill and store in 70% alcohol.
	Insecta	Aphaniptera: kill and preserve in 70% alcohol with a few drops of glycerin.
		Apoda and Vespoidea larvae: kill in Carnoy's fluid; store in 70% alcohol.
		Coleoptera larvae: kill in Carnoy's fluid; store in 70% alcohol; or fix in Van Emden's fluid (2 mL glacial acetic acid, 30 mL distilled water, 5 mL formaldehyde, 15 mL IMS).
		Coleoptera pupae: kill in Pampl's fluid; store in 70% alcohol.
		Diptera: preserve larvae in 70% alcohol; preserve pupae and puparia in 80% alcohol.
		Dermaptera: kill and store in 80% ETOH with a few drops of glycerin added.
		Ephemeroptera: kill with ammonia or cyanide fumes; rinse in 70% alcohol; store in 2–3% formalin.
		Hymenoptera larvae: kill in Carnoy's fluid; store in 70% alcohol.
		Lepidoptera ova: store in 70% alcohol containing a few drops of glycerin.
		Odonata: kill by depravation of food; fold wings to body and secure with paper band, then submerge in 20% alcohol for 2–3 hr; in 50% alcohol for 2–3 hr; store in 70% alcohol.
		Plecoptera: kill in 70% alcohol; store in 2% formalin.
		Thysanoptera: kill in 50% m ETOH and 0.5% ethyl acetate; macerate max 2–3 days; transfer to 80% alcohol for storage.
		Thysanura, Collembola, and Protura: kill and store in 70% alcohol containing a few drops of glycerin.
		Trichoptera: fix in 2 pts 2% formalin and 1 pt 90% alcohol; store in 2% formalin.

	Trichoptera larvae: kill in Pampl's fluid; store in 2% formalin.
Mollusca	Amphineura and Scaphopoda: kill in seawater with menthol crystals or alcohol; preserve in 50% alcohol for 48 hr, store in 70% alcohol.
	Cephalopoda: to kill inject mantle cavity with chloroform, ether, or ethyl bromide. Begin fixation with injection of 5% formalin; rinse and position on flat surface and submerge in 10% formalin; when hardened, rinse in water and preserve in 10% formalin or 70% alcohol.
	Gastropoda: euthanize terrestrial snails and slugs in tobacco water or 9:1 alcohol and turpentine, slugs in saturated corrosive sublimate, marine snails in seawater with menthol crystals or alcohol, nudibranchiata in glacial acetic acid, tectibranciata in Perényi fluid (4 pts 10% nitric acid, 3 pts alcohol, 3 pts chromic acid) or seawater with glycerin and alcohol. Preserve in 50% alcohol for 48 hr, store in 70% alcohol.
	Lamellibranciata: kill freshwater forms in boiling water or cold water with menthol crystals; kill marine forms in freshwater or water with menthol or chloroform. To preserve, rinse in freshwater, stage through 30% and 50% ETOH, store in 75% alcohol with a small amount of glycerin or 5% formalin.
Polyzoa	Kill and store in 70% alcohol or 1:9 formalin and water.
Scyphomedusae	Place mouth upward in 5% formalin (mixed with seawater); after 10 minutes add enough 5% picric acid to color the formalin; gradually increase concentration to 10% formalin in seawater.
Siphonophora	Kill in saturated solution of picric acid, transfer to 5% formalin; increase to 10% formalin for storage.
Tunicata	Add menthol crystals or chloral hydrate to water; flood with Bouin's fluid for 1–24 hr followed by several changes of 50% alcohol; or saturated corrosive sublimate for 1–24 hr followed by several changes of iodized alcohol. Preserve in 4–5% formalin or 70% alcohol.

(continued)

Table 1. *(Continued)*

Reference	*Organisms*	*Comments*
Wagstaffe and Fidler 1968 *Notes:* Calberla's fluid = 30 mL alcohol 100%, 30 mL glycerin, 30 mL distilled water Formol-alcohol = 1 pt alcohol 95% and 3 pts formalin 5%	Amphibia	Preserve very small adult specimens in 35% alcohol for 1 day, transfer to 50% alcohol for 2 days, and preserve in 65% alcohol.
		Slit small to medium salamanders in ventral body wall or inject with 10% formalin, arrange in tray in desired position, and submerge in 10% formalin for 1–4 days, then wash for 6 hr in running water. Preserve in 65% alcohol.
		Slit small to medium anura in ventral body wall or inject with 65% alcohol, arrange in tray in desired position, and submerge in 65% alcohol for 1–4 days. Preserve in 65% alcohol.
		Amphibian ova and embryos: fix in 10% formalin for 24 hr, preserve in 5% formalin.
		Larvae: fix in 10% formaldehyde, wash in water, preserve in 65% alcohol.
	Aves	Plug mouth with cotton wool, incise abdomen without perforating bowels, immerse in 4–10% formalin or 65% alcohol for 48 hr, then in 75% alcohol for 48 hr; preserve in 75% alcohol.
	Cephalochordata	Fix in 5% formalin (2% formaldehyde) in seawater for several days; preserve in 1 pt alcohol (75%) and 1 pt formalin (5%) made with freshwater.
	Chelonia	Place very small specimens in 50% alcohol; preserve in 70–75% alcohol.
		Slit small to medium specimens through skin of body wall between carapace and plastron. Keep mouth open with piece of wood or cork, arrange in desired position. Cover with 75% alcohol, preserve in 70–75% alcohol.
		Inject small to medium specimens thoroughly with 10% formalin in axilla and groin. Keep mouth open with piece of wood or cork, arrange in desired position. Cover injected specimens with 10% formalin. Once fixed, wash in water 2–3 days, preserve in 70–75% alcohol.
	Chondrichthyes	Fix whole small to medium specimens and skins of large specimens by injection with 10% formalin (9 pts water, 1 pt formaldehyde); submerge in 10% formalin for 2–3 weeks; wash in running water 2–3 days; preserve in 75% alcohol.

Crocodilia	Small specimens: make a slit in the ventral body wall. Position in desired posture and cover with 85% alcohol. Preserve in 75% alcohol.
	Medium specimens: make a slit in the ventral body wall and slits along ventral surface of the tail, thighs, and arms; remove large food masses. Position in desired posture and cover with 85% alcohol. Preserve in 75% alcohol.
	Inject thoroughly in body cavity and tail with 95% alcohol, position in desired posture, cover with 95% alcohol. Preserve in 75% alcohol.
Cyclostomata	Fix ova and larvae in 5% formalin (2% formaldehyde) in seawater for several days; preserve in 1 pt alcohol (75%) and 1 pt formalin (5%) made with freshwater.
Mammalia	Incise body wall and insert a wad of cotton wool to prevent viscera from bulging. Plug mouth with cotton wool, incise abdomen without perforating bowels, immerse in 4–10% formalin or 65% alcohol for 48 hr, then in 75% alcohol for 48 hr; preserve in 75% alcohol.
Osteichthyes	Wrap very small specimens in tissue paper or muslin and fix 1–2 days in 10% formalin. Sponge mucus off small to medium specimens with brine or solution of 4.5 L water and 28 g alum. Inject with 10% formalin (9 pts water, 1 pt formaldehyde); then set fins with pins on cork board and submerge in 10% formalin for 2–3 weeks; wash in running water 2–3 days; preserve in 75% alcohol.
Reptile eggs	Puncture and preserve in 70% alcohol.
Reptile embryos	Harden in 70% alcohol; or fix in Bouin's fluid (30 pts picric acid, 10 pts formaldehyde, 2 pts acetic acid) for 2–3 days, wash in several changes of 70% alcohol, and preserve in 75% alcohol.
Sauria	Small specimens: make a slit in the ventral body wall. Position in desired posture and cover with 85% alcohol. Preserve in 75% alcohol.
	Medium specimens: make a slit in the ventral body wall and slits along ventral surface of the tail, thighs, and arms; remove large food masses. Position in desired posture and cover with 85% alcohol. Preserve in 75% alcohol.

(continued)

Table 1. *(Continued)*

Reference	*Organisms*	*Comments*
		Inject specimens (except geckos) thoroughly in body cavity and tail with 95% alcohol. Position in desired posture and cover with 85% alcohol (some skinks require 95% alcohol). Preserve in 75% alcohol.
		Inject small geckos with 75% alcohol and large geckos with 85% alcohol; store geckos in 65% alcohol. Position in desired posture and cover with 65% alcohol. Preserve in 75% alcohol.
		For small to medium specimens make series of small slits ventrally along entire length of body wall and tail. Remove large food masses (which should be preserved separately in 70% alcohol). Arrange in desired posture and cover with 70–75% alcohol until hardened.
		Inject tail and body at intervals with 10% formalin. Thoroughly inject any large food masses. Arrange in desired posture and cover with 10% formalin to harden. Wash in running water for 18 hr, preserve in 70–75% alcohol.
		Large specimens: if not possible to preserve entire, remove skin, roll flat, and place in 70% alcohol or 10% formalin for at least a week. Wash in running water for 18 hr, preserve in 70–75% alcohol.
	Algae	Chlorophyta: preserve in 3% formalin.
		Cyanophyta: preserve in 3–5% formalin.
		Phaeophyta: preserve in 8–10% formalin.
		Rhodophyta: preserve in 10% formalin.
	Anthophyta	Preserve in 5% formalin or formol-alcohol (1 pt alcohol 95% and 3 pts formalin 5%).
	Bryophyta	Preserve in formol-alcohol or Calberla's fluid.
	Lichens	Preserve in formol-alcohol.
	Mycophyta	Preserve in 70–75% alcohol, 10% formalin, or formol-alcohol.
	Myxomycetes (slime mold)	Preserve in 4% formalin; 70% alcohol, or a mixture of 5 mL formalin, 10 mL glycerin, and 100 mL distilled water.

Pteridophyta	Preserve in formol-alcohol, 5% formalin, or Calberla's fluid.
Mixtures for preserving color in plants	Green pigment preservation: bubble sulfur dioxide through 750 mL distilled water to increase volume to 755 mL, add 15 g copper sulfate and stir until dissolved. Place specimen in fluid 3–7 days; remove and store in 2–5% formalin. Green pigment preservation: to make stock solution, mix a saturated solution of copper acetate in a 50% solution of glacial acetic acid. Place specimen in 1:4 stock solution and water for 3 weeks, or in stock solution heated to near boiling point for 10 minutes. After treatment, wash thoroughly in tap water and preserve in 5% formalin. Red apples, blackberry springs: dissolve 50 g zinc in 1 L warm water, cool and add 25 mL formaldehyde and 25 mL glycerin. Store permanently in this fluid. Yellow apples with red streaks; prepare mixture of 5 mL formaldehyde and 750 mL white paraffin oil. Place specimen in solution for 2 weeks, store in a solution prepared by bubbling sulfur dioxide through 750 mL distilled water to increase volume to 755 mL. Red and yellow flowers (especially bright blue flowers with high pH): preserve in a solution of 100 gm tertiary butyl alcohol, 1 gm thiourea, and 2 gm sodium citrate. Scarlet-red and yellow flowers with low pH: preserve in a solution of 100 gm tertiary butyl alcohol, 1 gm thiourea, and 2 gm citric acid. Green algae: preserve in solution of 0.5% copper acetate in 2% formalin for 1 day; store in 5% formalin. Fungi (Basidiomycetes) with color that does not leech in water: preserve in solution of 1 gm mercury acetate, 5 mL glacial acetic acid, and 1,000 mL water. Fungi (Basidiomycetes) with color that leeches easily in water: preserve in solution of 1 gm mercury acetate, 10 gm neutral lead acetate, 10 mL glacial acetic acid, and 1,000 mL alcohol (90%).

(continued)

Table 1. ***(Continued)***

Reference	*Organisms*	*Comments*
Ward 1890	Birds, fish, and reptiles	Make a moderate incision in the body cavity, then place specimen in large container of spirit for 7–10 days "so that the mucus, water, etc., may be drawn out of them" (Ward 1890, 33). Transfer to stronger, fresh spirit, then to spirit "quite strong enough to be readily inflammable" (Ward 1890, 33) for storage.
Weber et al. 1984	Birds	Open abdominal cavity and remove digestive track; inject with a solution of 2 pts phenoxetol, 2 pts ETOH, 2 pts formaldehyde, 96 pts water, neutralized with phosphate salts; immerse bird in solution. In the laboratory, wash specimen for 24 hr before preparing as dry skin.
Wentworth 1939	Medical museum specimens	Fix in Kaiserling I (30 g sodium acetate, 15 g potassium nitrate, 100 cc formaldehyde, 1000 cc water), then transfer to Kaiserling III for a few days (100 g sodium acetate, 200 cc glycerin, 1000 cc water) to remove formaldehyde, then transfer to fresh Kaislerling III for storage. Before sealing container, add 1 g sodium hydrosulfite per L of fluid.
Wentworth 1942	Museum specimens in wartime (to compensate for shortages of glycerin and alcohol)	Fix for 3 months in a formol-saline solution or a mixture (40 g sodium acetate, 100 cc formaldehyde, and 1000 cc water), then submerge for 2 wk in 10% sodium acetate (to remove excess formalin). Preserve in 10% watery sodium acetate with 2.5 g sodium hydrosulfite per 1000 g of tissue. Mold growth can be prevented with the addition of 1.5 cc 80% phenol or 1 g of camphor per L of preservative.
White and Peters 1969	Aquatic vertebrates and invertebrates (color preservation)	Fix in the field with solution of 1 pt formaldehyde, 9 pts water (seawater for marine species), with 1 pt antioxidant such as butyl hydroxytoluene (BHT; trade name Ionol-40, Shell Company) added to 200 pts fixative solution (shake for 30 sec, and keep chilled until ready for use).
Williamson and Russell 1965	Marine plankton	Kill and preserve in 1 pt seawater and 1 pt ethylene glycol; store in 50% ethylene glycol or pure ethylene glycol.

Windsor 1971	Frogs (color preservation)	Preserve in saturated or 50% solution of ammonium sulfate.
Withering 1801	Fungi	Prepare pickle by pouring 1 tea cup of cold water over 2 oz powdered vitriol of copper [copper sulfate]), stir 1 min, pour off excess water. Add 1 pint boiling water, and allow to dissolve and cool, then add half-pint spirit of wine and filter through paper. Partially dry fungi, then immerse in pickle for 3 hr to 3 d according to bulk. Remove specimen and place in a clean jar. Add stronger liquor (8 pints spring water and 1.5 pint spirit of wine) to large fleshy plants; add weaker liquor (8 pints spring water and 1 pint spirit of wine) for smaller, thinner plants.

Abbreviations, equivalents, and unusual names used in Table 1:
arrack = a southeast Asian distilled alcoholic beverage made from coconut flower sap, sugarcane, rice, or fruit
cc = cubic centimeter
corrosive sublimate = mercuric chloride
d = day
dr - dram (1/8 fluid ounce or 3.6966697109375 mL)
ETOH = ethanol, ethyl alcohol
g = gram
gal = gallon (3.7854 liters)
gr = grain (64.79891 mg)
hr = hour
IMS = Industrial methylated spirits (ethyl alcohol with methyl alcohol)
kg = kilogram
L = liter
lb = pound (454 g)
min = minim (1/480 fluid oz or 1/60 fluidram)
min = minute
mL = milliliter
mg = milligram
oz = fluid ounce (29.573 mL; 28.35 g)
pt = part
pts = parts
qt = quart (0.94635 liters)
sal ammoniac, sal armoniak = ammonium chloride
tafia = an unaged alcoholic beverage made from sugar cane
tbs = tablespoon
tsp = teaspoon
wk = week

Table 2. Timeline of Milestones in Published Fluid-Preservation Techniques

Date	*Reference*	*Organisms*	*Preservation Technique Milestone*
1662	Croone	Dog embryos	Preserve in spirit of wine.
1666	Boyle (1666)	Fetuses and birds	Preserve in spirit of wine with a little sal armoniack made from quick lime.
1681	Grew	Chickens	Preserve in rectified oil of turpentine.
1748	Réaumur	Birds	Preserve in strong brandy.
1769	Bancroft	Birds and snakes	Preserve in high wines or rum.
1801	Withering	Fungi	Fix in vitriol of copper solution.
			Preserve in weak or strong spirits of wine.
1817	Graves	Small animals	Preserve in spirit of wine containing burnt alum.
1828	Bory de Saint-Vincent	Small animals	Preserve in oils.
			Preserve in vinegar.
			Preserve in brine consisting of water saturated with muriate of soda.
			Preserve in solution of aluminum sulfate, water, and alcohol.
			Preserve in red wine containing nitrous mercury.
			Use preservative containing pyrolignous acid and sulfurous acid.
			Preserve in Guyot's liquor (which contains Cognac; water; and flowers, leaves, or green lavender).
1829	Providence Franklin Society	Amphibians and reptiles	Preserve in sulfurous acid solution.
			Preserve in solution of molasses, salt, water, and chloride of lime.
1831	Anonymous	All natural history specimens	Preserve using Nichols method (solution of water, alcohol, and sulfate of alumina).
			Preserve using Graves method (alum, water, and alcohol).
			Preserve using Abbé Manesse method (alum, nitre, sea salt, water, and alcohol).
1852	Baird	General	Preserve in Goadby's solution (rock salt, alum, corrosive sublimate, and water).
1884	Browne	Fish and reptiles	Preserve in spirits of wine containing methyl alcohol, powdered burnt alum, saltpeter, and glycerin.
			Preserve in Browne's solution (bichloride of mercury, sodium chloride and water).

			Preserve in camphorated fluid (distilled water, a few drops of creosote, and chalk).
			Preserve in Möller's solution (bichromate of potash, soda sulfate, and water).
1892	Dall	Mollusc soft parts	Preserve in Müller's Fluid (water, bichromate of potash, sulfate of soda or Gaber's salts).
	Haly	Vertebrates and invertebrates	Preserve color using a mixture of coconut oil and carbolic acid.
	Riley	Soft-bodied insects	Preserve with the Laboulbene method (solution of salt, niter, water, and sodium arsenite of potash).
			Preserve using the Verrill method (solution of salt, alum, corrosive sublimate, and carbolic acid).
			Preserving with the Wickersheim method (solution of alum, salt, potash, and arsenious acid in water).
1895	Redenbaugh	Coelenterates	Preserve in a sequence using a solution of magnesium sulfate followed by Perenyi's fluid, picro-sulfuric acid, and other chemicals.
1899	Lo Bianco	Marine invertebrates	Fix in formaldehyde.
			Fix in chromic acid, acetic acid, or osmic acid.
			Preserve in iodized alcohol (containing tincture of iodine).
			Preserve in Kleinenberg's solution (picric acid, sulfuric acid, and water).
			Preserve in Müller's solution (potassium bichromate, sodium sulfate, and water).
			Preserve in Perenyi's solution (nitric acid, chromic acid, and alcohol).
	Stone	Plants	Fix in formaldehyde; add nitric acid to preserve color.
1911	Stejneger	Amphibians and reptiles	Neutralize formaldehyde with bicarbonate of soda.
1924	Conant	Plants	Preserve in solution of alcohol, formaldehyde, and glacial acetic acid.
1925	Chamberlin	Spiders	Preserve in ETOH; transfer to carbolic acid in xylene; transfer to xylene; transfer to mineral oil.
1928	Nieuwland and Slavin	Monotropa	Preserve color in alcohol with sulfur dioxide.
	Pope	Small animals	Preserve in 40–50% isopropyl alcohol.
1929	De Villiers	Anuran larvae	Fix in corrosive sublimate.
			Fix in carbolic acid.

(continued)

Table 2. ***(Continued)***

Date	*Reference*	*Organisms*	*Preservation Technique Milestone*
1930	Borodin	Fish	Preserve in solution of alcohol, formaldehyde, wood tar, water and salt to preserve color.
	Swingle	Plants	Preserve in solution made with oxyquinoline sulfate powder.
1935	Hiscox	Zoological and anatomical specimens	Preserve in mixture of sodium fluoride, formaldehyde, and water; then glycerin, water, magnesium chloride, and sodium fluoride.
	van Steenis	Plants	Preserve in alcohol containing kalium or natrium bisulphite ($NaHSO_2$).
1937	Scully	Plants	Preserve color in copper sulfate, transfer to sulfuric acid, sodium sulfite, and water.
1938	Gloyd	Reptiles and amphibians	Use mixture of formaldehyde, alcohol, and water to counteract shrinkage and swelling.
1939	Wentworth	Medical museum specimens	Fix in Kaiserling I solution (sodium acetate, potassium nitrate, formaldehyde); transfer to Kaiserling III solution (sodium acetate, glycerin, water) to remove formaldehyde, use fresh Kaislering III for storage.
	Hiscox and Sloan	Flowers	Preserve in salicylic acid, formaldehyde, alcohol, and water.
1944	Anonymous	Invertebrates	Use Carl's solution (alcohol, formaldehyde, glacial acetic acid, and water).
1955	Beirne	Soft bodied insects, larvae and nymphs	Preserve in ethyl alcohol or methyl alcohol.
			Preserve in AGA (alcohol, glycerin, and glacial acetic acid).
			Preserve with Pempel's fluid (acetic acid, formaldehyde, and water).
	Wagstaffe and Fidler	Invertebrates	Preserve in Blès's fluid (glacial acetic acid, formaldehyde, and ethyl alcohol).
			Use Bouin's fluid alcoholic (alcohol, formaldehyde, acetic acid, and picric acid).
			Preserve in Bouin's fluid (picric acid, formaldehyde, and acetic acid).
			Preserve in Carnoy's fluid (glacial acetic acid, ethyl alcohol, and chloroform).
			Preserve in corrosive sublimate with glacial acetic acid.
			Preserve in Donisthorpe's fluid (glacial acetic acid, ethyl alcohol, water, and corrosive sublimate).
			Use Flemming's fluid (chromic acid, osmic acid, and glacial acetic acid).
			Preserve with Hermann's solution (platinum chloride, glacial acetic acid, and osmic acid).
			Use Oudeman's fluid (alcohol, glycerin, and glacial acetic acid).

			Preserve with Pampl's fluid (glacial acetic acid, water, formaldehyde, and ethyl alcohol).
			Use Peterson's KAAD (kerosene, ethyl alcohol, glacial acetic acid, and dioxin).
			Preserve in Zenker's fluid (corrosive sublimate, potassium bichromate, sodium sulfate, and water).
1963	Henderson	Fish gonads	Preserve in Davidson's solution (ethyl alcohol, neutralized formaldehyde, glycerin, glacial acetic acid, and water).
1964	Borror and DeLong	Insects	Preserve in Hood's solution (alcohol with glycerin).
			Preserve with KAAD (kerosene, alcohol, glacial acetic acid, and dioxane).
			Use Kahle's solution (alcohol, water, formaldehyde and glacial acetic acid).
1968	Wagstaffe and Fidler	Plants (color preservation)	Green: use solution of sulfur dioxide, copper sulfate, or copper acetate.
			Red: use solution of zinc, water, formaldehyde, and glycerin.
			Yellow: use formaldehyde and white paraffin oil.
			Red and yellow flowers, bright blue flowers with high pH: preserve in solution of tertiary butyl alcohol, thiourea, and sodium citrate.
			Green algae: use copper acetate in formalin.
			Fungi (Basidiomycetes) with color that does not leech in water: preserve in mercury acetate, glacial acetic acid, and water.
			Fungi (Basidiomycetes) with color that leeches in water: use mercury acetate, neutral lead acetate, glacial acetic acid, and alcohol.
1969	Kendrick	Fleshy fungi	Preserve with lactic acid.
	White and Peters	Aquatic animals (color preservation)	Fix in a formaldehyde antioxidant such as butyl hydroxytoluene (BHT; trade name Ionol-40, Shell Company).
1975	Anderson	Invertebrates	Preserve in propylene glycol, phenoxetol, formaldehyde, sodium acetate, and water.
1976	Adams et al.	Ctenophora (comb jellyfish)	Preserve in tri-chloracetic acid and seawater or p-toluenesulphonic acid and seawater.
	Beers	Actinopods	Use strontium solution to stabilize structure.
	Edmunds et al.	Mayflies	Preserve in alcohol with Ionol (2, 6-di-tert-butyl 4-methylphenol).
			Use Carnoy fluid (glacial acetic acid, ethanol, and chloroform).

(continued)

Table 2. ***(Continued)***

Date	*Reference*	*Organisms*	*Preservation Technique Milestone*
	Taylor	Flagella	Fix in Lugol's solution (iodine, potassium iodide, acetic acid).
			Fix in glutaraldehyde.
			Fix in osmium tetroxide.
			Fix in Karnovsky's fixative (paraformaldehyde, gluteraldehyde, hexamine, and water).
1977	Barker and Grigg	Anuran larvae	Use Tyler's fixative (formaldehyde, aqueous calcium chloride, cobalt nitrate in water).
1979	Lincoln and Sheals	Invertebrates	Preserve in Schaudinn's solution (mercuric chloride and ethanol).
			Preserve in Dowicil 100 [1-(3-chlorallyl)5,7-triaza-1-azoniaadamantane chloride].
			Preserve in propylene phenoxetol.
			Use Zenker's fluid (mercuric chloride, glacial acetic acid, potassium dichromate, sodium sulfate, and water).
			Fix in Heidenhain's Susa (mercuric chloride, sodium chloride, water, trichloracetic acid, glacial acetic acid, and formaldehyde).
1985	Hangay and Dingley (1985b)	Algae	Preserve in mixture of iodine, potassium iodide, glacial acetic acid, formaldehyde, and water.
			Preserve in potassium chrome alum, formaldehyde, and water.
		Green plants	Preserve in phenol, lactic acid, glycerin, cupric chloride, cupric acetate, and water.
		Fungi	Preserve in absolute alcohol, formaldehyde, and glacial acetic acid.
		Hemiptera	Preserve in lacto alcohol (lactic acid, alcohol, water).
		Insect larvae	Carnoy's fluid (alcohol, chloroform, glacial acetic acid)
		Ants	Preserve in Dornisthorpe's fluid (glacial acetic acid, alcohol, water, and corrosive sublimate).
		Beetle larvae	Preserve in Van Emden's fluid (glacial acetic acid, water, formaldehyde, and alcohol).
		Arachnids, diplopoda, chilopoda, onychophora	Preserve in Pampl's fixative (glacial acetic acid, water, formaldehyde, and alcohol).

1987	Mori and Prance	Lecythidacae buds	Use Farmer's solution (ethanol and glacial acetic acid).
			Preserve in Jackson's solution (ethanol, methanol, chloroform, proprionic acid, and acetone).
1989	Forman and Bridsor	Plants	Preserve using Kew preservative (ethanol, water, and methanol).
			Fix in mixture of water, formaldehyde, and glycerol.
	Valdecasas and Baltanás	Freshwater invertebrates	Preserve in Angelier's fluid (acetic acid, chromic acid, and water).
1994	Camacho and Bedoya	Oligochaetes, ostracods, copepods, molluscs, etc.	Preserve in Angelier's fluid (acetic acid, chromic acid, and water). Preserve in Koenike's solution (glacial acetic acid, glycerol, and water).
1995	Stephenson and Riley	Marine animals	Preserve in mixture of formaldehyde, glutaraldehyde, and sodium glycerophosphate.
2000	Reid	Mollusks	Fix in formaldehyde mixed with seawater and buffered with sodium tetraborate.

Table 3. Tissue Matrix Types

Matrix Characteristics	*Example*
Matrix relatively tenuous meshwork	Spleen, lymph nodes
More substantial matrix	Fibrous tissues
Matrix is the bulk of the volume	Connective tissues (e.g., cartilage)
Mineralized matrix	Bone

Table 4. Proprietary Fixatives (Based on Manufacturer's MSDS, Advertisements, and Published Analyses)

Brand Name	*Manufacturer*	*Comments*
BoonFix	Denteck BV	Propylene glycol based
FineFIX	Milestone Microwave Laboratory System	Unknown; website states it is "a patented formalin-free, water-based concentrate," and that when diluted with 98% ethanol it does not cause significant shrinkage
GlycoFix	Thermo Fisher	Glyoxal-based fixative
Histochoice MB	Amresco	A non-cross-linking fixative containing sodium chloride, glyoxal [OCHCHO], and ethanol
HistoFix	Richard-Allen Scientific	Fixative containing pyrrolid-2-one, polyol, urea, and zinc salt
Mirsky's Fixative	National Diagnostics	Clear, colorless petroleum distillate composed mostly of alkanes and cyclic paraffins; "excellent barrier against water vapor"; density 0.8 g/ cm3; Sold as a cross-linking fixative described as a "sugar aldehyde." However, the maximum tissue cube it can penetrate is 1cm3.
Molecular Fixative (UMFIX)	Sakura Finetek UK, Ltd.	Methanol and polyethylene glycol
NoTox Histological Fixative	Earth Safe Industries	Ethanol, bis carbonyl compounds, propylene glycol, and "antiseptic and antifungal agents" (and probably a di-aldehyde glucose)
OmniFix II and OmniFix 2000	AnCon Genetics	Alcohol-based fixative containing glycol and salt
Prefer Prefer Concentrate	Anatech LTD	"Glyoxal, a di-aldehyde" (gluteraldehyde), with ethanol, water, and a proprietary buffer.
RCL2	Alphelys	DNA preservative; according to the data sheet, contains acetic acid, probably in 70% ethyl alcohol, trehalose (a disaccharide)
RNA*later*	Ambion, Inc.	Marketed to preserve RNA and DNA and histology in samples 0.5 cm or thinner; sodium citrate tribasic dehydrate (buffer) and ammonium sulfate (protein precipitator)
SafeFix II	Fisher Scientific	Fixative containing glytoxal
S.T.F.	Nebraska Scientific/ Streck Laboratories	Diazolidinyl urea; 2-bromo-2-nitropropane-1, 3-diol (Bronopol); zinc sulfate; sodium citrate, and some formaldehyde

Table 5. Narcotizing Agents

Chemical Agent	*Administration*	*Notes*
Acid solution (weak)	Add to water.	Anon. 1944 (British Museum)
Alcohol	Add drops of 50–70% solution to water; or use as Lo Bianco's mixture (40 pts 70% alcohol, 20 pts glycerin, 40 pts water).	Anon. 1944 (British Museum); Wagstaffe and Fidler 1955
Bouin's solution	Narcotize aquatic species on wet plate with hot (60°C) Bouin's fixative.	Knudsen 1966
Carbonic acid gas	Add to water in the form of soda water.	Wagstaffe and Fidler 1955
C. F. Rousselet's solution	Use mixture of 3 pts hydrochlorate of cocaine (2%), 1 part methylated spirit, and 6 pts water; Stovaine may be substituted for cocaine.	Wagstaffe and Fidler 1955
Chloral hydrate	Drop crystals into water; place specimens in 5–10% solution; add to water slowly to reach 5–10% concentration.	Wagstaffe and Fidler 1955
Chloretone (acetone chloroform)	Use 1% solution.	Wagstaffe and Fidler 1955
Chloretone	Add to water.	Jackson 1919, Knudsen 1966
Chloroform	Add a small quantity to water every few minutes or use as a vapor.	Anon. 1944 (British Museum); Wagstaffe and Fidler 1955
Clove oil	Drip clove oil in seawater.	Hangay and Dingley 1985b
Cocaine	Add to water.	Jackson 1919
Epson salts	Add to water.	Knudsen 1966
Ether	Add saturated solution (7–8%) to water, or use a 1% solution.	Wagstaffe and Fidler 1955
FAA (formaldehyde-acetic acid-alcohol)	Narcotize aquatic species on wet plate with hot (60°C) FAA.	Knudsen 1966
ß-eucaine hydrochloride	Slowly add 1% solution to water containing specimens; or use as a mixture of 1 gm eucaine, 10 mL alcohol 90%, and 10 mL distilled water.	Wagstaffe and Fidler 1955
Magnesium chloride	Drip magnesium chloride in seawater.	Hangay and Dingley 1985b
Magnesium sulfate	Add crystals to water.	Anon. 1944 (British Museum); Wagstaffe and Fidler 1955
Menthol	Add crystals added to seawater.	Wagstaffe and Fidler 1955
Nicotine	Add weak nicotine to water.	Anon. 1944 (British Museum)
Soda water	Add to water.	Anon. 1944 (British Museum)
Stovaine (amyl chlorohydrins)	Use 1% solution in distilled water; add slowly to water containing specimen.	Wagstaffe and Fidler 1955
Tobacco	Use as smoke or as tobacco water.	Wagstaffe and Fidler 1955
Tobacco smoke	Narcotize with tobacco smoke for several hours.	Lo Bianco 1899
Urethane	Add to water.	Knudsen 1966

Table 6. Disinfectant Mechanisms of Some Preservatives (Based in Part on Volk and Wheeler 1984)

Agent	*Disinfectant Action*
Acids	Stronger acids are more bactericidal than weaker acids; bactericidal action is by hydrolysis and denaturation of cellular proteins (coagulation of proteins); most acids are also destructive to tissues.
Alcohols	Germicidal action is due to protein denaturation (by coagulation of proteins). Germicidal effect increases with the molecular weight of the alcohol. The most effective alcohol concentration depends on the amount of moisture present—for ethanol 70% is most effective for most purposes.
Alkalies	Alkalies are less destructive to bacteria than acids. Alkali disinfectant action is proportional to the concentration of hydroxyl ions (OH^+) in solution, resulting in hydrolysis and denaturation of proteins.
Formaldehyde	Excellent disinfectant; an alkylating agent
Gluteraldehyde	Very effective against most microbes; is an alkylating agent
Mercuric chloride	Disinfectant action comes from oxidation (reaction with sulfhydryl groups of enzymes) and combination with proteins.
Phenols	In high concentration, phenols disrupt the cytoplamic membrane and precipitate cell proteins. At concentrations of 0.1–2.0%, phenol distorts the cytoplasmic membrane, causing leakage of essential metabolites. Phenols are relatively effective in the presence of organic matter.
Soaps and Detergents	Not true germicides, soaps and detergents reduce surface tension so that bacteria become enmeshed in the soap and are removed in the washing process; some fragile bacteria are killed by the chemical action of soaps. Some detergents disrupt cell membranes and denature proteins.

Table 7. Summary of Factors that Affect the Long-Term Usefulness of Fluid-Preserved Specimens (After Simmons 2002)

1. Quality of the Initial Fixation or Preservation	a. Time interval between death and treatment (fixation or preservation)—should be as short as possible.	
	b. The quality of the fixative or preservative—mix and measure carefully, using deionized or distilled water.	
	c. Rate of penetration of the fixative or preservative into the tissues.	
	d. Temperature of fixation or preservation.	
	e. Proportion of fluid preservative to specimen volume (should be at least 2:1).	
2. Postfixation Processing	a. Wash time (as short as possible).	
	b. Transfer to preservative (use steps of 20% concentration).	
	c. Hydration or dehydration of the specimen.	
3. Storage Environment	a. Temperature and fluctuations (18–21°C is preferred, minimize fluctuations).	
	b. Light (intensity and cumulative exposure)—reduce as much as possible.	
	c. Ultraviolet radiation (intensity and cumulative exposure)—reduce as much as possible.	
	d. Relative humidity and fluctuations (<50% is preferred, minimize fluctuations).	
	e. Vibrations (minimize).	
	f. Integrity of the container seal.	i. Preservative evaporation ii. Contamination of the preservative
	g. Handling and use of the specimen.	i. Physical damage ii. Dehydration iii. Rehydration

Table 8. Timeline of the Known Introduction of Chemicals in Fluid Preservation

Chemical Name	*Date First Recorded*	*CAS Number*
Absolute alcohol (100% ethyl alcohol)	1899	64-17-5
Acetic acid	1899	64-19-7
Acetone	1976	67-64-1
Acrolein (C_3H_5O)	1984	107-02-8
Alcohol (ethyl alcohol)	1662	64-17-5
Alum (aluminum sulfate)	1817	10043-01-3
Aluminum sulfate	1828	10043-01-3
Ammonia	1921	8013-59-0
Ammonium sulfate (diazanium sulfate)	1968	7783-20-2
Amyl chlorohydrin (amylocaine; stovaine)	1979	644-6-8
Arsenious acid (arsorous acid)	1892	13464-58-9
Baking soda (sodium bicarbonate; bicarbonate of soda)	1968	144-55-8
Benzamine hydrochloride (anilinium chloride)	1979	142-04-1
Beverage alcohol (aguardiente, arrack, rum, brandy, whiskey). Aguardiente and rum are made from sugar cane; arrack (arak, or rack) is made from coconut flowers or sugar cane; brandy is distilled from wine; whiskey is made from fermented grain mash.	Prehistory	None; see CAS number for ethyl alcohol
Bicarbonate of soda (baking soda)	1911	144-55-8
Bichromate of potash	1884	7778-50-9
Bichromate of potassium	1899	7778-50-9
Bichloride of mercury	1884	7487-94-7
Borax (sodium tetraborate decahydrate)	1944	1303-96-4
Brine = a salt solution (usually sodium chloride) of about 3.5% to 26%	Prehistory	None; see CAS number for sodium chloride
Butyl hydroxytoluene (BHT; Ionol-40)	1969	128-37-0
Cacodylate	1985	75-60-5
Calcium chloride	1884	10043-52-4
Camphor	1942	76-22-2

Camphorated fluid (water, wine, creosote, chalk)	1884	8021-39-4
Cane sugar	1919	57-50-1
Carbolic acid (phenol, benzenol, phenylic acid)	1892	108-95-2
Chloral hydrate (2,2,2-trichloroethane-1,1-diol)	1899	302-17-0
Chloretone	1966	6001-64-5
Chloride of lime	1831	8031-48-9
Chloride of zinc	1884	7685-85-7
Chloroform	1944	865-49-6
Chromic acid	1899	7738-94-5
Clove oil	1966	8000-34-8
Cobalt nitrate	1962	10141-05-6
Coconut oil	1892	8001-31-8
Cocaine hydrochloride	1944	53-21-4
Cocaine powder	1899	53-36-2
Copper acetate (cupric acetate)	1968	142-71-2
Copper sulfate	1899	7758-98-7
Corrosive sublimate (mercuric chloride; mercury dichloride; sublimate)	1852	7487-94-7
Cupric chloride	1985	7447-39-4
Cyanide	1955	74-90-8
Dibasic sodium phosphate	1951	7558-79-4
Dioxane	1964	123-91-1
Dowicil [1-(3-chlorallyl)5,7-triaza-1-azoniaadamantane chloride]	1979	none
Epsom salts (magnesium sulfate)	1895	7487-88-9
Ether	1955	7578-38-4
Ethyl alcohol	1662	64-17-5
Ethyl bromide (bromoethane)	1955	74-96-4
Ethylene glycol	1965	107-21-1
Ethylene glycol monopheryl ether	1956	122-99-6
Eucaine (ß-eucaine hydrochloride)	1979	500-34-5

(continued)

Table 8. ***(Continued)***

Chemical Name	*Date First Recorded*	*CAS Number*
Formaldehyde	1893	50-00-0
Glacial acetic acid (= undiluted acetic acid)	1899	64-19-7
Gluteraldehyde	1976	90045-36-6
Glycerine (glycerin, glycerol)	1883	56-81-5
Hexamethyltetramine (hexamine)	1953	100-97-0
Hexamine	1953	100-97-0
Hydrochloric acid	1884	7647-01-0
Iodine	1932	7553-56-2
Ionol (2, 6-ditert-butyl 4-methylphenol)	1976	9713-41-6
Isopropyl alcohol	1928	8013-70-5
Kalium	1935	7440-09-7
Kerosene	1964	8008-20-6
Lactic acid	1985	50-21-5
Lead acetate	1968	301-04-2
Magnesium chloride	1939	7786-39-3
Magnesium sulfate (Epsom salts)	1895	7487-88-9
Menthol	1955	89-78-1
Mercuric chloride (corrosive sublimate)	1852	7487-94-7
Mercury acetate	1968	1600-27-7
Methyl alcohol (methanol, wood alcohol)	1884	67-56-1
Mineral oil	1924	8042-47-5
Molasses (a byproduct of sugar cane	1829	none
Monobasic sodium phosphate	1951	89140-32-9
MS222 (ethyl-m-aminobenzoate)	1976	886-86-2
Natrium bisulphate	1935	7681-38-1
Nembutol (pentobarbital)	1966	76-74-4
Nitre (potassium nitrate; saltpetre; saltpeter)	1831	7757-79-1

Nitric acid	1899	7697-37-1
Osmic acid (osmium tetroxide)	1899	20816-12-0
Osmium tetroxide (osmic acid)	1976	20816-12-0
Oxyquinoline sulfate (=hydroxyquinoline sulfate)	1930	3819-18-9
Para-hydroxybenzoic acid	1956	99-96-7
Paraffin oil	1968	86742-46-7
Paraformaldehyde	1976	69476-52-8
Phenol (carbolic acid; benzenol, phenylic acid)	1942	108-95-2
Picric acid	1899	88-89-1
Picro-sulfuric acid	1895	unknown
Platinum chloride	1955	10025-65-7
Potassium bichromate	1899	7778-50-9
Potassium chloride	1933	7447-40-7
Potassium dichromate	1979	7778-50-9
Potassium iodide	1932	7681-11-0
Potassium nitrate (nitre (saltpetre; saltpeter)	1939	7757-79-1
Propylene phenoxetol	1968	770-35-4
p-toluenesulfonic acid	1976	104-15-4
Pyroligeneous acid	1899	8030-97-5
Propionic acid (propanoic acid)	1976	79-09-4
Rock salt (see sodium chloride)	1852	7647-14-5
Sal armoniak (sal ammoniac, ammonium chloride)	1665	12125-02-9
Salicyclic acid	1939	69-72-7
Saltpeter (nitre; potassium nitrate)	1884	7757-79-1
Soda sulfate (sulfate of soda; sodium sulfate)	1884	7757-82-6
Sodium acetate	1939	127-09-3
Sodium arsenate	1892	10103-60-3
Sodium bicarbonate (baking soda)	1968	144-55-8
Sodium chloride	1852	7647-14-5
Sodium citrate	1968	8055-55-8

(continued)

Table 8. ***(Continued)***

Chemical Name	*Date First Recorded*	*CAS Number*
Sodium fluoride	1935	7681-49-4
sodium glycerophosphate	1995	154804-51-0
Sodium hydroxide	1979	1310-73-2
Sodium hydrosulfite (sodium dithionite)	1942	7775-14-6
Sodium silicate	1985	6834-92-0
Sodium sulfite	1928	7757-83-7
Sodium tetraborate (borax)	1976	1303-96-4
Spirit of wine (ethyl alcohol)	1662	64-17-5
Spirit or spirits (ethyl alcohol)	1662	64-17-5
Stovaine (amyl chlorohydrin; amylocaine)	1979	644-6-8
Strontium	1976	7440-24-6
Sublimate (corrosive sublimate; mercuric chloride; mercury d chloride)	1852	7487-94-7
Sucrose	1984	57-50-1
Sulfate of alumina (aluminum sulfate)	1828	10043-01-3
Sulfur dioxide	1928	7446-09-5
Sulfuric acid	1829	7664-93-9
Tertiary butyl alcohol (tert-butyl alcohol)	1968	75-65-0
Thiourea	1968	62-56-6
Tincture of iodine	1899	None; see iodine
Tobacco (an nicotine water)	1899	none
Trichloroacetic acid	1976	76-03-9
Triethanolamine	1979	102-71-6
Trisodium phosphate	1951	7601-54-9
Turpentine	1955	9004-90-7
Urethane	1966	51-79-6
Vinegar (active ingredient acetic acid)	Pre-history	None; see acetic acid
Wood alcohol (methanol; methyl alcohol)	1919	67-56-1
Wood tar	1930	98-91-1
Xylene	1924	1330-20-7
Zinc chloride	1985	7646-85-7
Zinc sulfate	1985	7733-02-0

Table 9. Anatomical Fixation and Preservation Techniques

Reference	*Organ or Tissue*	*Preservation Techniques*
Abbott 1918	Postmortem tissues with war wounds for shipment to museum	Prepare preservative solution (soda sulfate 20 pts, soda bicarbonate 20 pts, soda chloride 20 pts, potassium nitrate 38 pts, potassium sulfate 2 pts, chloral hydrate 100 pts, formalin 100 pts, and water 3,200 pts) and divide into three containers. Wash specimens, then preserve in container #1 for 2 hr; transfer to container #2 for 12 hr; transfer to container #3 for 24–48 hr. Remove from solution, wrap in cotton wool or lint soaked in preservative, seal in impermeable cover, and ship.
Craig 1916	Anatomical preparations in general	Fix in Solution I (200 cc formaldehyde, 15 cc potassium nitrate, 30 cc potassium acetate, and 1000 cc water); transfer to Solution II (NaCl, saturated solution); then to Solution III (60 g potassium acetate, 30 g glycerin, and 1000 cc water); and finally to Solution IV (100 g potassium acetate, 250 g glycerin, 1000 cc water, and 5–10 cc formaldehyde).
Delafield and Prudden 1904	Anatomical preparations in general	Fix and harden tissue pieces of no larger than 2 cc in 80–95% ETOH for 24 hr; change alcohol; after 3 d; transfer to 97%-absolute alcohol to complete hardening.
		Submerge in solution of 5 pts formaldehyde and100 pts water for 24 hr; replace with fresh solution and leave for another 24 hr.
		Fix and harden in Müller's fluid (2 pts potassium bichromate, 1 pt sulfate, and 100 pts water).
		Fix and harden in Orth's Fluid (100 pts Müller's fluid and 10 pts commercial formaldehyde).
		Fix and harden in 1% aqueous osmic acid for 24 hr; wash in water and store in 80% ETOH.
		Fix and harden in freshly mixed Flemming's Osmic Acid Mixture (15 pts 1% chromic acid, 4 pts 2% osmic acid, and 1 pt glacial acetic acid) for 24 hrs, wash thoroughly, transfer to 70% ETOH.

(continued)

Table 9. *(Continued)*

Reference	*Organ or Tissue*	*Preservation Techniques*
		Fix and harden in Lang's Solution (5 g mercuric chloride, 6 g sodium chloride, 5 cc glacial hydric acetate, and 100 cc water) for 1–3 hrs.
		Fix in Zenker's fixative (2 pts potassium bichromate, 1 pt sodium sulfate, 5 pts mercuric chloride, 5 pts glacial hydric water, and 90 cc water) for 24–48 hrs; wash thoroughly and preserve in 80% ETOH.
		Fix in Kaiserling's solution (200 cc formaldehyde, 1000 cc water, 15 g nitrate of potassium, and 80 g acetate of potassium); drain and place in 80% ETOH for 1-6 hr, then in 95% ETOH for 1–2 hrs to restore color; preserve in mixture of 200 g acetate of potassium, 400 cc glycerin, and 2000 cc water.
Edwards and Edwards 1959	For anatomical preparations in general	Fix in 5% formalin solution.
		Kaiserling's Method—Fix in Solution I (800 mL formalin, 85 g potassium acetate, 45 g potassium nitrate, 4,000 mL distilled water). Revive color with Solution II (80–95% alcohol). Preserve in Solution III (3,000 g glycerin, 2,000 g potassium acetate, 9,000 mg distilled water).
		Pick's Method—Fix in Solution I (100 pts formalin 5%, 5 pts artificial Karlsbad salts by weight [20 g sodium sulfate, 1 g potassium sulfate, 9 g sodium chloride, 18 g sodium bicarbonate]). Drain specimen and place in 80–85% alcohol to revive color. Preserve in Solution III (2700 g sodium acetate, 5400 mL glycerin, 9000 mL distilled water). This method first published by Pick in 1900.
		Judah's Method—Fix in Solution I (170 g potassium acetate, 90 g potassium nitrate, 1,600 mL formalin, 8,000 mL distilled water). Revive color in 95% alcohol. Preserve in Solution III (4 lbs potassium acetate, 2,000–4,000 mL glycerin, 10,000 mL distilled water, 20 mL carbolic acid). Described as "definitely an improvement" to the methods of Kaiserling and Pick.
		Modified Judah's Method—Fix in Solution I (170 g potassium acetate, 90 g potassium nitrate, 1,600 mL formalin, 8,000 mL distilled water). Revive colors in methyl alcohol for 2–3 hr. Preserve in Solution III (45 mL formalin, 1,800 g potassium acetate, 3,600 mL glycerin, 9,000 mL distilled water).

		Hydrosulfite Method for Color Preservation—Fix three months in Solution I (80–100 mL formalin, 40 g sodium acetate, 1000 mL water); preserve in Solution II (20 mL formalin, 40 g acetate, 1,000 g water (reportedly developed by Wentworth during World War II to avoid the use of glycerin).
		Wentworth's 1957 Revised Hydrosulfate Method—Fix one month in Solution I (100 mL formalin, 40 g sodium acetate, 1000 mL water); if more acidic than pH 6.5, put specimen in Solution II (10 mL formalin, 40 g sodium acetate, 1 g sodium phosphate tribasic, 1000 mL water). Preserve in Solution III (10 mL formalin, 100 g sodium acetate, 1 g sodium phosphate, 200 mL glycerin, to 1000 mL water).
	Amyloid degeneration	After fixing, wash for 8 hr and place in a solution of 1 g iodine, 2 g potassium iodide, and 100 mL distilled water; mount in liquid paraffin.
	Fatty foci	Fix, wash 2 hr, place in saturated solution of Sudan III or Scarlet R for 1 hr, wash in alcohol and mount in 10% formalin.
	Fetuses	Fix in 5% formalin; for a fetus greater than 6 months inject via umbilicus and mount in 10% formalin with 3 g sodium hydrosulfite per kilo of specimen.
	Free Iron	Fix, wash, transfer to saturated potassium ferrocyanide in distilled water for 4-6 hr; transfer to 1% hydrochloric acid to develop blue color; mount in 10% formalin. When color fades, place specimen in 20% hydrogen peroxide until color returns, and remount.
	Gout specimen	Fix and dehydrate in absolute alcohol for 14 days, mount in undiluted glycerin.
	Melanotic sarcoma specimen	Fix, slice, and transfer to methylated spirit for 12 hr; transfer to acetone for 12-18 hr; transfer to xylol for 2 hr; mount in liquid paraffin.
Fish 1893	Brains	Harden in a mixture of 6 g potassium bichromate, 4 g potassium nitrate, and 180 cc water; decolorize with absolute alcohol or hydrogen peroxide.
		Harden in a solution of 6 oz bichloride of mercury and 0.5 gal water.
		Harden in a solution of 2 g mercuric chloride, 90 cc alcohol and 2 cc hydrochloric acid. Camphor or iodine may be used to increase the solubility of the mercuric chloride.
		Harden in solution of 200 g zinc chloride, 3000 cc of 70% ETOH, and 1000 cc glycerin.
		Harden in 10% solution of zinc chloride.

(continued)

Table 9. ***(Continued)***

Reference	*Organ or Tissue*	*Preservation Techniques*
		Harden in equal pts saturated aqueous potassium bichromate and 10% aqueous zinc chloride.
		Harden in saturated aqueous mercuric chloride.
		Harden in equal pts saturated potassium bichromate and mercuric chloride.
		Harden in saturated aqueous solution of potassium bichromate.
		Harden and store in solution of 400 cc water, 400 cc 95% alcohol, 250 cc glycerin, 20 g zinc chloride, and 20 g sodium chloride.
		Harden in 0.2% picric acid in 50% alcohol, then stage up gradually to 95% alcohol (Gage method).
Gannal and Harlan 1840 [Harlan's translation of Gannal 1838]	Anatomical preparations in general	Preserve in spirits of wine, 14-18° Baume [95.6% ethyl alcohol].
		Preserve in solution of 2 lb very pure water, 1 lb alcohol, and 6 oz sulphate of alumina (recommended by Nicholas).
		Preserve in 8 oz alum, 1 lb common water, and ½ lb alcohol. To prepare preservative, pulverize alum, add hot water, filter when cool, and add alcohol (recommended by George Graves).
		Preserve in solution of 1 lb common water, 1 lb alcohol, and 12 oz alum (recommended by George Graves for mixing in cold water).
	Specimens of normal anatomy, of pathological anatomy, and of natural history	Preserve in 1 lb alum, 1 lb nitre, 1 lb sea salt, 4 lb common water, and 1 lb alcohol (recommended by Abbe Manesse).
Hangay and Dingley 1985a	To preserve color in anatomical specimens	Fix in Solution A (3 g potassium acetate, 1 g potassium nitrate, 75 ml formalin, 100 mL water). Revive color in Solution B (95% alcohol). Preserve in Solution C (30 g potassium acetate, 60 ml glycerin, 100 mL water). Said to be the "July 1896 formula of Kaiserling."
		Fix in Solution A (30 g potassium acetate, 30 g potassium nitrate, 400 mL formalin, 2 L water). Revive color in Solution B (80% alcohol). Preserve in Solution C (2.72 kg potassium acetate, 5 L glycerin, 2 L water, 5 g phenol crystals).

		Fix in Solution A (85 g potassium acetate, 45 g potassium nitrate, 800 mL formalin, 4 L water). Revive color in Solution B (95% alcohol). Preserve in Solution C (30 g potassium acetate, 60 mL glycerin, and 100 mL water). Said to be the 1898 Kaiserling formula.
		Color preservative—140 g sodium chloride, 80 g sodium bicarbonate, 625 g chloral hydrate, 512 mL formaldehyde, 20 L distilled water.
		Fixative—600g potassium acetate, 300 g potassium nitrate, 4 L formaldehyde, 20 L water.
		Preservative solution—2 kg potassium acetate, 4 L glycerin, 5 g thymol, 20 L water (Klotz-Kaiserling 1946 formula).
		Fix in Solution 1 (100 mL formaldehyde, 40 g sodium acetate, 1 L water). Preserve in Solution 2 (20 mL formaldehyde, 40 g sodium acetate, 1 g trisodium phosphate, 200 mL glycerin, 3 g sodium hydrosulfite, 1 L distilled water). Described as Wentworth's 1942 formula.
		Fix in Solution 1 (80–100 mL formaldehyde, 40 g sodium acetate, 1 L water). Preserve in Solution 2 (20 mL formaldehyde, 40 g sodium acetate, 1 L water, with 3g/kg specimen sodium hydrosulfite). Wentworth's 1947 formula.
Judah 1918	Post-mortem specimens of war injuries	Preserve in solution of 125 g modified Carlsbad salts, 125 g chloral hydrate, 125 g formalin, and 4000 cc water.
Judah 1922	Color preservation in anatomical specimens	Fix in Solution I (potassium acetate 170 g, potassium nitrate 90 g, 40% formalin 1,600 cc, and 8,000 cc water); restore color in Solution II (95% alcohol; do not let solution go below 80%); preserve in Solution III (potassium acetate 4 lbs, glycerin 2,000–4,000 cc, and water 10,000 cc, "to which 20 cc carbolic acid may be added as a preservative").
Kaiserling 1896	Color preservation in anatomical specimens	Solution I (200 mL formalin, 30 g potassium acetate, 15 g potassium nitrate, 1000 mL distilled water). Revive color with Solution II (80% alcohol). Preserve in Solution III (400 mL glycerin, 200g potassium acetate, 2000 mL distilled water).

(continued)

Table 9. ***(Continued)***

Reference	*Organ or Tissue*	*Preservation Techniques*
Kaiserling 1896	Color preservation in anatomical specimens	Solution I (750 mL formalin, 30 g potassium acetate, 10 g potassium nitrate, 1000 mL distilled water). Revive color with Solution II (80-95% alcohol). Preserve in Solution III (1 pt glycerin, 30% potassium acetate, 1 pt distilled water).
Kaiserling 1900	Color preservation in anatomical specimens	Solution I (800 mL formalin, 85 g potassium acetate, 45 g potassium nitrate, 4000 mL distilled water). Revive color with Solution II (absolute alcohol, or 80-95% alcohol). Preserve in Solution III (3000 mL glycerin, 2000 g potassium acetate, 9000 mL distilled water).
Parsons 1831	Lungs	Immerse in alcohol, inject through the trachea and tie off; after a few days, slice longitudinally and suspend in a glass container in alcohol.
		Fill small lung specimens with quicksilver (mercury), inject through trachaea, allow to dry for several days, then submerge in a container in oil of turpentine.
	Intestines	Inject with alcohol.
	Anatomical and pathological preparations	Preserve in alcohol.
		Preserve in alcohol with metallic, alkaline, or earthy salt (muriate of mercury, nitrate of potash, muriate of soda, muriate of ammonia, or muriate of alum).
		Preserve in aqueous salt mixture (muriate of mercury, nitrate of potash, muriate of soda, muriate of ammonia, or muriate of alum).
	Cartilage, fibrocartilage, tendons, fibrous membranes, heads	Preserve in oil of turpentine.
	Anatomical and pathological preparations	Preserve in acid (sulfuric, nitric, muriatic, acetic, pyroligneous) diluted with distilled water.
		Preserve in acid (sulfuric, nitric, muriatic, acetic, pyroligneous) and alcohol mixture.
		Preserve in vinegar with arsenic or corrosive sublimate (mercuric chloride).
	Brains and nerves	Preserve in alcohol and sulfuric, nitric, or muriatic acid.
	Brains and muscles, lungs, liver, ligaments	Preserve in alcohol and volatile oil mixtures.

	Anatomical and pathological preparations	Preserve in ammonia.
	Morbid anatomical specimens	Remove diseased part; rinse with muriate of ammonia, wrap in old linen and immerse in 1 pt muriate of ammonia and 2 pts spirit of wine; after 1-2 d, remove linen and return specimen to alcohol.
	Brain, spinal marrow	Preserve in 2 pts oxymuriate of mercury and 16 pts alcohol.
Pole 1813	Various anatomical specimens	Macerate specimen in water with daily changes until specimen loses bloody color; remove before it begins to putrefy. Dissect parts intended for preservation from unwanted tissue and suspend specimen in position for hardening in alcohol, filling any cavities with curled hair, wool, or cotton. Once preserved, mount in jar of filtered spirits.
Spence 1967	Organs from embalmed cadavers	Preserve in solution of 10 mL formaldehyde, 40 mL alcohol, and 50 mL tap water.
	Unfixed viscera	Rinse in running water for several hours; place in solution of 10% formalin in tap water until fixation is complete (larger specimens may need to be injected).
	Brain	Submerge in solution of 10% formalin, suspend by a string; after fixation may be transferred to solution of 75% alcohol and 25% tap water.
	Small animals	Preserve in solution of 100 mL propylene phenoxetol, 1 L glycerin, and 9 L tap water. Use preservative volume 3x that of specimen.

Table 10. Clearing and Staining Techniques

Source	*Chemicals and Technique*
Davis and Gore 1936	Stain with a mixture of 10 cc of stock solution (saturated solution of alizarin in 5 cc glacial acetic acid, 10 cc glycerin, and 60 cc 1% chloral hydrate) in 1 L of 4% potassium hydroxide. Clear alcohol-preserved specimens in a solution of 3–4% potassium hydroxide; use 1% potassium hydroxide for fresh specimens. Rinse in hydrogen peroxide (if necessary) and expose to sunlight to bleach. Stage through solutions of 20 pts glycerin, 3 pts 4% potassium hydroxide, and 77 pts hydrogen hydroxide; then 50 pts glycerin, 3 pts 4% potassium hydroxide, and 47 pts hydrogen hydroxide; then 75 pts glycerin, and 24 pts hydrogen hydroxide; then 1 pt glycerin, 3 pts 4% potassium hydroxide, and 77 pts hydrogen hydroxide; store in 100% glycerin.
Dawson 1926	Fix in 95% ethyl alcohol; clear in 1% potassium hydroxide, then transfer to solution of 1 pt alizarin and 10,000 pts 1% potassium hydroxide. Clear again in solution of 79 pts water, 20 pts glycerin, and 1 pt potassium hydroxide or step through increasing concentrations of glycerin to clear. Store in 100% glycerin.
Dingerkus and Uhler 1977	Fix in 10% formalin; wash in distilled water; stain in solution of 10 mg alcian blue, 80 mL 95% ethyl alcohol, and 20 mL glacial acetic acid. Wash in 95% ethyl alcohol, stage through 75%, 40%, and 15% ethyl alcohol to distilled water. Macerate in solution of 30 mL saturated sodium borate, 70 mL distilled water, and 1 g trypsin. Stain in solution of 0.5% potassium hydroxide with alizarin red. Transfer through solutions of 75% 0.5% potassium hydroxide and 25% glycerin; 50% 0.5% potassium hydroxide and 50% glycerin; and 25% 0.5% potassium hydroxide and 75% glycerin to 100% glycerin. Use a few drops of hydrogen peroxide to bleach if necessary. Store in 100% glycerin with a few crystals of thymol.
Evans 1948	Fix in 10% formalin; place in 2% potassium hydroxide or 2% sodium hydroxide, bleach in 3 drops of 3% hydrogen peroxide per 30 cc fluid volume if necessary. Stain bone in solution of alizarin in 2% potassium hydroxide; stain cartilage in a solution of 100 cc of 70% alcohol, 2 cc of 0.5% hydrogen chloride, and 0.25 g toluidine blue. Stage into solution of 1 pt glycerin and 1 pt 1% hydroxide to clear; store in 100% glycerin.
Hollister 1934	Use fresh specimens fixed in 70% ethyl alcohol. Stain in solution of 1 cc alizarin red and 1000 cc potassium hydroxide; clear in a pure solution of 1–4% potassium hydroxide and exposure to ultraviolet radiation, then transfer to solution of potassium hydroxide and 40 pts glycerin; preserve in pure glycerin.
Jensh and Brent 1966	Fix in 10% formalin; dehydrate in acetone; stain in solution of 1% potassium hydroxide with 6 mg alizarin red S per liter of solution; wash in cold water; transfer to solution of 1 pt benzyl alcohol, 2 pts ethanol, and 2 pts glycerin; store in 100% glycerin.
Lundvall 1905	Introduced alizarin for staining bone (reported in Davis and Gore 1936).

Maisano 2008	Fix in formalin, dehydrate in 95–100% alcohol. Stain for cartilage in solution of alcian blue powder in 30% glacial acetic acid and 70% ETOH; neutralize in saturated sodium borate solution; digest in saturated pancreatin in 65% distilled water and 35% sodium borate. Stain for bone in solution of alizarin red in 0.75% potassium hydroxide, then place in pancreatin solution again. Complete final clearing in 30% glycerin and 70% potassium hydroxide; transfer to solution of 60% glycerin and 40% potassium hydroxide; store in 99% glycerin with a few thymol crystals.
Moore 2006	Specimens are generally fixed in 70% IMS (industrial methylated spirits), macerated in 2–5% aqueous potassium hydrate, stained in alizarin red in alcohol, and stored in 5–10% glycerin for several weeks before transfer to 70–100% glycerin with a crystal of thymol (*Note*: this paper does not describe clearing and staining technique in detail, but rather the conservation of cleared and stained specimens).
Noble 1917	Dip the tips of the digits in a solution of glycerin and hydrogen peroxide, dehydrate, and clear in cedar oil to see bone. Wash in 70% ETOH to restore specimens "to nearly normal condition" (Noble 1917, 795).
Potthoff 1984	Fix specimens in 10–15% formalin buffered with marble chips; dehydrate in solution of 50 pts 95% ETOH and 50 pts distilled water; transfer to solution of absolute (100%) alcohol. Stain for cartilage in 70 pts absolute alcohol, 30 pts glacial acetic acid with 20 mg of alcian blue powder per 100 mL (some adjustments in ingredients necessary for large specimens). Neutralize in saturated solution of sodium borate. Bleach (if necessary) in 15 pts 3% hydrogen peroxide and 85 pts 1% potassium hydroxide. Digest tissue with trypsin and stain for bone as described by Taylor (1967).
Schultze 1897	Recommends potash and glycerin to clear tissues, later use of potassium hydroxide (system developed for use with human embryos).
Springer and Johnson 2000	Substitute a solution of 75% ethyl alcohol with dissolved alizarin red S crystals for the potassium hydroxide and alizarin red S solution to avoid damage to muscle and cartilage. Specimens not previously subjected to potassium hydroxide may also be stained with a solution of 75% ethyl alcohol with dissolved alcian blue powder.
Taylor 1967a	Fix in solution of 1 pt formaldehyde and 9 pts water buffered with 0.5 teaspoon borax per quart solution, then wash in water (specimens may be stored in 70 ethyl alcohol if desired). Bleach in solution of 10–20% hydrogen peroxide and 80–90% stock solution of 0.5–1.0% potassium hydroxide in distilled water, exposed to light. Buffer in saturated solution of sodium borate and distilled water, then add about 0.45 g trypsin per 400 mL of solution. Stain in solution of a small amount of alizarin red S powder in stock solution of 0.5–1.0% potassium hydroxide in distilled water. Rinse in distilled water or borax solution. Transfer through graded ethanol concentrations if degreasing is necessary; remove guanine with 2–4% potassium hydroxide if necessary. Stage through 40% and 70% glycerin and distilled water or stock solution of 0.5–1.0% potassium hydroxide in distilled water. Store in 100% glycerin with a few crystals of thymol.

(continued)

Table 10. *(Continued)*

Source	*Chemicals and Technique*
Taylor 1967b	Fix in 10% formalin with 0.5 teaspoon of sodium borate per quart of solution. Soak formaldehyde from specimen and store in 70% ETOH. Bleach in 10 pts 3% hydrogen peroxide and 90 pts 1% potassium hydroxide. Buffer in 30 pts saturated sodium borate and 70 pts distilled water, then add 0.25 teaspoon pancreatin (purified trypsin); change solution as necessary. Stain in alizarin red S in 0.5–1.0% potassium hydroxide. Stage specimen to 100% glycerin for storage.
Taylor and van Dyke 1985	Fix in solution of 1 pt formaldehyde and 9 pts water; formalin may be buffered with 4 g sodium phosphate monobasic and 6.5 g anhydrous sodium phosphate per liter, or with powdered limestone. Transfer to 0.5–1.0% potassium hydroxide (or potassium hydroxide with alizarin red S). Transfer specimens for double-staining to 50–70% ETOH, then to 90% ETOH, then to absolute alcohol; pass through xylene if degreasing is desired; transfer to solution of 9–30 mg alcian blue powder in 40 pts glacial acetic acid and 60 pts 100% ethyl alcohol. Neutralize in saturated sodium borate solution or 0.5–1.0% potassium hydroxide in distilled water. Bleach in solution of 10–20 pts hydrogen peroxide with 80–90 pts 0.5–1.0% potassium hydroxide, while exposing specimens to light source. Clear in 0.5–1.0% potassium hydroxide with 0.45 g trypsin powder per 400 mL liquid. Stain with alizarin red S in 0.5–1.0% potassium hydroxide. For final clearing rinse in distilled water or soak in mixture of 3 pts saturated borate solution and 7 pts distilled water until clear. Transfer to solution of 40 pts glycerin and 60 pts potassium hydroxide solution; then 70 pts glycerin and 30 pts potassium hydroxide solution, and store in 100% glycerin.
Wassersug 1976	Fix in 10% formalin buffered with 0.1 g magnesium carbonate per 200 mL solution. Wash in running tap water, rinse in deionized water. Stain for cartilage in solution of 9 mg alcian blue, 60 mL 100% alcohol, and 40 mL glacial acetic acid. Dehydrate in 100% alcohol. Macerate in 0.5% potassium hydroxide with 3–4 drops alizarin red S per 100 mL of solution. Pass through graded series of glycerin concentrations up to 100% glycerin (this technique is a modification of Simons and Van Horn 1971).

Note: this table provides a summary of steps for comparative purposes only; for specific instructions on specimen preparation, mixing of solutions, time specimens should spend in solution, and other information, refer to the original publication.

Table 11. Fluid Density Scales

Name	*Scale*	*Notes*
API	> 10°API floats on water < 10°API sinks in water Light oil = > 31.1°API Medium oil = 22.3-31.1°API Heavy oil = < 22.3°API Extra heavy oil = <10.0°API	Developed in 1921 to describe density standards for petroleum products.
Baumé	0° Bé = water with 10% NaCl 10° Bé = pure water 31° Bé = 70% ETOH 41 ° Bé = 95% ETOH	Developed by Antoine Baumé in 1768.
Salometer	0° = pure water 100° = saturated brine (26.395% salt) Salt solutions are classed as: Brackish = 0.05-3.00% Saline = 3.00-5.00% Brine > 5.00%	Developed to express salt concentration, based on a percentage of salt in a liquid
Specific gravity	The ratio of the density of a liquid compared to the density of a standard (usually pure water, which is 1.00 g/cm^3)	Density measurements are temperature dependent; readings from hygrometers and density meters must be adjusted accordingly.
Tralle	Percent ethyl alcohol by weight in water, from 0% to 100%	Tralle Proof Scale developed to indicate proof of alcohol from 0 to 200 at 60°F.
Twaddle	Specific gravity range of 1.000 to 2.000, divided into 200 equal parts, so that 1°Twaddle = 0.005 specific gravity	Used for liquids heavier than water.

Table 12. Criteria for Identifying Alternative Preservative Fluids (Based in Part on van Dam 2003)

Factor	*Desired Characteristics*
Fixation	Does not necessarily have to have fixative properties (e.g., may not form cross-links)
Disinfectant properties	Effective as a biocide
Specimen color	Preserves colors (color preservation is related to shrinkage, swelling, and pigment solubility)
Water balance	Causes minimal dehydration or hydration (shrinkage and swelling of tissues)
Solubility	Soluble in water but not soluble in fat (to reduce migration of lipids)
DNA	Non-cross-linking; preserves DNA
pH	Neutral to slightly acidic for best preservation of proteins; neutral to slightly alkaline for best preservation of color
Tissue stability	Minimizes demineralization, extraction of lipids, and extraction of proteins from tissues
Surface tension	Low evaporative potential
Density	Less than 1.00 (less than the density of water)
Stability	Will remain stable for decades to centuries
Color	Colorless
Turbidity	Clear
Toxicity	Low toxicity to humans
Cost and availability	Low unit cost, widely availability

Table 13. Summary of Fixative and Preservative-Induced Changes in Invertebrates. (See Table 14 for Organization by Taxonomic Group)

Reference	*Organisms*	*Fixative or Preservative*	*Comments*
Ahlstrom and Thrailkill 1963	Plankton	3% buffered formaldehyde	Decreased live volume by 15–87%
Båmstedt 1998	Tentaculate ctenophores (*Pleurobrachia pileus*)	4% formaldehyde and seawater	Average loss of 31% in length
Beers 1976a	Actinopods	Formaldehyde and seawater	Skeletons of actinopod sarcodinians in marine plankton samples are usually dissolved to some degree; to minimize this loss, store preserved specimens at low temperature, add strontium ($SrCl_2$) to stabilize the fixative, or preserve the specimens directly in ethanol.
Borutzky 1934	Benthic animals	2–10% formaldehyde	Caused weight loss of 3–10%
Britt 1953	Aquatic nymphs (mayflies)	Fixed in KAAD (kerosene, alcohol, glacial acetic acid, and dioxane) for 24 hr; preserved in 95% alcohol	KAAD increased the length in uninjured specimens by 4.18% and decreased length in injured specimens by 12.33%.
Champalbert and Kerambrun 1979	Copepods (*Pontella mediterranea*)	5% formaldehyde with borax buffer	20–25% weight loss with fixation
de Lafontaine and Leggett 1989	Planktonic hydromedusae (*Staurophora mertensi, Catablema vesicarium*)	10% v/v formaldehyde in seawater, buffered with sodium tetraborate	Mean diameter of umbrella shrank to 67.8%–84.9% of live size; weight reduced to 30.5–37.9% of fresh weight
DiStefano et al. 1994	Virile crayfish (*Orconecte virilis*)	Isopropanol 100%	After 180 days, carapace length change of –0.10%; chela width change of –2.20%; wet weight change of –13.53%
		Isopropanol 75%	After 180 days, carapace length change of –0.81%; chela width change of –1.55%; wet weight change of –10.40%

(continued)

Table 13. *(Continued)*

Reference	*Organisms*	*Fixative or Preservative*	*Comments*
		Isopropanol 50%	After 180 days, carapace length change of –0.37%; chela width change of –1.74%; wet weight change of –6.23%
		Isopropanol 25%	After 180 days, carapace length change of –0.16%; chela width change of –2.36%; wet weight change of –4.17%
		Ethanol 100%	After 180 days, carapace length change of –0.19%; chela width change of –1.96%; wet weight change of –19.90%
		Ethanol 75%	After 180 days, carapace length change of –0.12%; chela width change of –1.26%; wet weight change of –10.95%
		Formalin 10%	After 180 days, carapace length change of –0.23%; chela width change of –2.63%; wet weight change of –0.31%
		Freezing (to –15°C)	After 180 days, carapace length change of –1.00%; chela width change of –1.67%; wet weight change of –6.82%
Donald and Paterson 1977	Chironomid larvae	10% formaldehyde and 70% ethanol	Weight gain of up to 32% in formaldehyde, followed by weight loss in many species
Fancett 1986	Sycphomedusa (*Aurelia aurita*)	NA	Shrinkage of 7–10% in umbrella diameter
Fudge 1968	Zooplankton	Formaldehyde, trichloracetic acid, and ethanol	Compared biomass of frozen and freeze-dried zooplankton with specimens preserved in various fluids; freeze-drying produced the best results. Formalin-preserved specimens lost 68–59% of dry weight.

Giguère et al. 1989	Zooplankton	75% ethanol, 5% and 10% formaldehyde and seawater buffered with borax	Chemical preservation reduced dry weight by 37–43%; no significant differences between preservative or length of time in preservative
Grandperrin and Caboche 1968	Salps, doliolids, siphonophores	10% v/v formaldehyde	Wet weight losses of 39% to 52%
Hendrickx et al. 2003	Isopoda	5% formaldehyde	Storing isopods in 4% formaldehyde significantly reduced their metal concentration (particularly copper), and caused significant weight loss in the specimens
Hopkins 1968	Euphausiids (*Nematoscelis difficilis*)	10% hexamine buffered freshwater and seawater formaldehyde; 70% ethanol	Carbon loss of 17% in seawater formaldehyde, 20% in freshwater formaldehyde, and 23% in ethanol; loss of 12.2% of total organic matter in seawater formaldehyde; nitrogen loss of 19.2–21% in formaldehyde and ethanol; nitrogen loss of 29.7% in freshwater formaldehyde
Howmiller 1972	Tubificid worms (*Limnodrilus* spp.)	10% formaldehyde; 70% ethanol with 5% glycerin; and 70% isopropyl with 5% glycerin	After initial preservation, decrease to 97% (formaldehyde), 86% (ethanol), and 81% (isopropyl) of live weight. After 24 hr, decrease to 90% (formaldehyde), 73% (ethanol), and 62% (isopropyl) of initial wet weight. After 44 days in preservative, specimens showed decrease of 76% (formaldehyde), 44% (ethanol), and 26% (isopropyl) of initial wet weight.
	Chironomid larvae (*Anatopynia*)	10% formaldehyde; 70% ethanol with 5% glycerin; and 70% isopropyl with 5% glycerin	After 57 days in preservative, decrease to 58% (formaldehyde), 51% (ethanol), and 42% (isopropyl) of initial wet weight

(continued)

Table 13. *(Continued)*

Reference	*Organisms*	*Fixative or Preservative*	*Comments*
Lappalainen and Kangas 1975	Littoral benthos	4% formaldehyde in seawater buffered with hexamine	Loss in weight after >3 months in preservative of 1–18% or gain in weight of 1–7% (depending on taxon)
Lebedeva and Kozlova 1969	Crustaceans (*Cladocera*)	Formaldehyde	Depending on the species, postfixation weights increased by 30–60%; decreased 20–40%; or remained the same.
Leuven et al. 1985	Aquatic macro-invertebrates	70% ethanol, 4% and 10% buffered formaldehyde	Loss of 7.2–21.9% dry weight following preservation in 70% ethanol; no significant loss in dry weight with preservation in formaldehyde
Madin et al. 1981	Salps	5% v/v formaldehyde	Volume loss of 52.2% after 8 months
Mills et al. 1982	Marine annelids, nematodes, bivalves, amphipods, and isopods	70% ETOH	Most organisms increased weight in formaldehyde; all test organisms lost weight in 70% ethyl alcohol
Modigh and Castaldo 2005	Ciliates	2% formaldehyde and 1% Lugol's solution	Poor preservation in formaldehyde compared to Lugol's solution; poor preservation in commercial Lugol's solution compared to laboratory mixed Lugol's solution
Möller 1980	Sycphomedusa (*Aurelia aurita*)	Formaldehyde	Shrinkage of 15–30% in umbrella diameter
Mutlu 1995	Medusae (*Aurelia aurita*) and cydipid ctenophore (*Pleurobrachia pileus*)	2% formaldehyde buffered with borax	After 2 months in preservative, loss of 26% in bell diameter and 70% in weight
Mutlu and Bingel 1999	Tentaculate ctenophores (*Pleurobrachia pileus*)	5% formaldehyde buffered with borax	"Some loss" in length and weight

Nishikawa and Terazaki 1996	*Thalia democratica* and *Dolioletta gegenbauri*	5% formaldehyde and seawater buffered with sodium tetraborate	*Thalia democratica* shrank to 93% of live length; *Dolioletta gegenbauri* shrank to 86% of live length
Omori 1978	Copepods	4% formaldehyde and seawater buffered with borax or hexamine	Specimens fixed in borax buffered formaldehyde lost 21% of dry weight, 34% of carbon, and 49% of nitrogen. Specimens fixed in hexamine buffered formaldehyde lost more weight than those in borax buffers.
Oorthuyzen and Sadée 1982	Tentaculate ctenophores (*Pleurobrachia pileus*)	Formaldehyde	20% shrinkage
Reese et al. 1996	Mites (*Psoroptes cuniculi*)	70% ETOH; 70% ETOH with 5% glycerol; 95% ETOH; 95% ETOH with 5% glycerol; 2%, 5%, and 10% formaldehyde	13 morphological characters were evaluated; chitinous characters were not affected. Alcohol and formaldehyde fixation significantly reduced body measurements. Fixed mites mounted in Hoyer's medium increased in body length and width; unfixed mounts in Hoyer's significantly increased body length.
Ross 1953	Snail (*Helix aspersa*) spermatocyte cells	Picric acid, ethyl alcohol, potassium dichromate	Shrinkage in diameter of 40 ± 6%; shrinkage in volume 78 ± 6%
		Chromic acid, mercuric chloride, acetic acid, trichloracetic acid	Shrinkage in diameter of 34 ± 4%; shrinkage in volume 71 ± 5%
		Neutral formalin Aoyama	Shrinkage in diameter of 30 ± 4%; shrinkage in volume 66 ± 6%
		Osmium tetroxide, Zenker, Flemming, Bouin, Helly, Mercuric-acetic	Shrinkage in diameter of 27 ± 9%; shrinkage in volume 61 ± 14%
		Mercuric-formol	Shrinkage in diameter of 22 ± 4%; shrinkage in volume 53 ± 7%

(continued)

Table 13. ***(Continued)***

Reference	*Organisms*	*Fixative or Preservative*	*Comments*
		Sanfelice	Shrinkage in diameter of 13 ± 5%; shrinkage in volume 34 ± 11%
Schram et al. 1981	Crustaceans (Ceriodaphnia lacustris)	3% formaldehyde	Loss of 25% of dry weight after 15 days; loss of 47% after 45 days
Stanford 1973	Common benthic insect larvae	70% ETOH	*Hydropsyche occidentalis* gained weight for 7 days; *Brachcentrus* and *Pternoarchys californica* lost weight, *Altherix variegata* showed drastic dehydration
Stoecker et al. 1994	Microzooplankton	2% hexamine buffered formaldehyde; 10% acid Lugol's solution; 5% Bouin's solution	Plankton ciliate counts were 56% higher in acid Lugol's solution than in formaldehyde; fixation in Lugol's solution resulted in higher cell counts than fixation with formaldehyde.
Thibault-Botha and Bowen 2004	Ctenophora (*Pleurobrachia bachei*)	4% formaldehyde in seawater	Reported shrinkage of 5–80% from live weight, depending on the size of the specimen (small specimens shrank more than large specimens)
Voight 2001	*Graneledone* (deep-sea octopus)	Fixation in 8–10% formalin followed by storage in 70–75% ETOH	Preservation can alter morphology, particularly shape of mantle and head, impairing reliability of standard measurements.

Van der Veer and Oorthuysen 1985	Sycphomedusa (*Aurelia aurita*)		Shrinkage of 7–10% of umbrella diameter
von Schiller and Solimini 2005	Mayflies (*Baetis* and *Ephemerella*)	Ethanol (70% and 95%); formaldehyde (5% and 10%)	*Baetis* showed an average shrinkage of 17% in 95% ETOH, 8% in 70% ETOH, and 6% in 10% formaldehyde, no significant changes in 5% formaldehyde; *Ephemerella* showed no significant changes in ETOH, but 5% shrinkage in 10% formaldehyde
Wiederholm and Eriksson 1977	Benthic invertebrates	70% ethanol	Wet weight after 2 months in preservative 80–111% of fresh weights, depending on taxon
Williams and Robbins 1982	Copepods (*Calanus helgolandicus*)	Fresh, frozen, and formaldehyde fixed	Frozen samples lost 57% of dry weight; fixed samples lost 38%. Both lost nitrogen-rich material.
Yip 1982	Tentaculate ctenophores (*Pleurobrachia pileus*)	4% formaldehyde in seawater	8–13% shrinkage in length immediately after preservation; 15–22% shrinkage after 10 months in preservative; 17–31% shrinkage after 12 months in preservative

Table 14. Summary of Fixative and Preservative-Induced Changes in Invertebrates by Taxonomic Group (see Table 13 for Details)

Group	*Organisms*	*Change*	*Reference*
Benthic invertebrates	Various littoral benthic invertebrates	Weight increase and decrease	Lappalainen and Kangas 1975
	Various benthic invertebrates	Wet increase	Wiederholm and Eriksson 1977
	Benthic animals	Weight loss	Borutzky 1934
	Benthic larvae	Variable gain and loss of weight	Stanford 1973
Annelida	Annelids	Weight increase in formaldehyde and loss in alcohol	Mills et al. 1982
	Hirudinea	Weight loss	Leuven et al. 1985
	Tubificid worms	Significant weight loss	Howmiller 1972
Arthropoda	Chironomid larvae	Weight gain followed by weight loss	Donald and Paterson 1977
		Significant weight loss	Howmiller 1972
	Glyptotendipes sp.	Shrinkage in ethanol, no significant loss in formalin	Leuven et al. 1985
	Crayfish	Length and weight decrease; embrittlement	DiStefano et al. 1994
	Crustaceans	Variable weight loss or gain	Lebedeva and Kozlova 1969
		Decrease in weight	Schram et al. 1981
		Loss of organic matter	Hopkins 1968
	Copepods	Decrease in weight	Champalbert and Kerambrun 1979
		Decrease in weight	Omori 1978
		Decrease in weight with preservation and freezing	Williams and Robbins 1982
	Isopods	Significant weight loss and mineral loss	Hendrickz et al. 2003
		Weight increase in formaldehyde and decrease in alcohol	Mills et al. 1982

		Weight decrease	Leuven et al. 1985
	Amphipods	Weight increase in formaldehyde; loss in alcohol	Mills et al. 1982
	Mayflies	Length increase in uninjured, decrease in injured specimens	Britt 1953
		Variable shrinkage	Von Schiller and Solimini 2005
	Mites	Significant reduction in body measurements	Reese et al. 1996
Cnidaria	Hydromedusae	Dimensional and weight decrease	de Lafontaine and Leggett 1989
	Medusae	Decrease in bell diameter and weight	Mutulu 1995
Coelenterata	Scyphomedusa	Shrinkage of umbrella	Fancett 1986
		Shrinkage of umbrella	Möller 1980
		Shrinkage in umbrella diameter	Van der Veer and Oorthuysen 1985
Ctenophora	Ctenophores	Decrease in length	Båmstedt 1998
		Decrease in weight	Thibault-Botha and Bowen 2004
	Cydipid ctenophores	Decrease in bell diameterand weight	Mutlu 1995
	Tentaculate ctenophores	Some loss in length and weight	Mutul and Bingel 1999
		Shrinkage	Oorthuyzen and Sadée 1982
		Shrinkage in lenth	Yip 1982
Mollusca	Bivalves	Weight increase in formaldehyde; loss in alcohol	Mills et al. 1982
	Gastropoda	Weight loss	Leuven et al. 1985
		Shrinkage in diameter	Ross 1953
	Octopus	Altered morphology	Voight 2001

(continued)

Table 14. ***(Continued)***

Group	*Organisms*	*Change*	*Reference*
Nematoda	Nematodes	Weight increase in formaldehyde; loss in alcohol	Mills et al. 1982
Protists	Plankton	Decrease in volume	Alhlstrom and Thrailkill 1963
	Zooplankton	Decrease in weight	Giguère et al. 1989
		Decrease in weight	Fudge 1968
	Ciliates	Variable preservation	Modigh and Castaldo 2005
	Microplankton	Reduction in ciliate counts	Stoecker et al. 1994
Protozoa	Actinopods	Loss of skeleton	Beers 1976a
Tunicata	*Dolioletta gegenbauri*	Decrease in length	Nishikawa and Terazaki 1996
	Thalia democratica	Decrease in length	Nishikawa and Terazaki 1996
	Doliolids	Decrease in weight	Grandperrrin and Caboche 1968
	Salps	Decrease in weight	Grandperrrin and Caboche 1968
		Decrease in volume	Madin et al. 1981
	Siphonophores	Decrease in weight	Grandperrrin and Caboche 1968

Table 15. Summary of Fixative and Preservative-Induced Changes in Vertebrates (See Table 16 for Organization by Taxonomic Group)

Reference	*Organisms*	*Fixative or Preservative*	*Comments*
Al-Hassan et al. 2000	Fishes Perciformes: Mullidae:	5% formalin mixed with tap water	TL no change; SL increased 1.6339%; HL decreased 6.0869%
TL = total length SL = standard length HL = head length	*Mullus barbatus*	5% formalin mixed with distilled water	TL increased 0.2747%; SL no change; HL decreased 6.9767%
		10% formalin mixed with tap water	TL increased 0.5911%; SL increased 1.2346%; HL increased 5.3191%
		70% alcohol mixed with tap water	TL decreased 2.0000%; HL decreased 5.0955%; HL decreased 6.7392%
		70% alcohol with distilled water	TL decreased 2.3256%; SL decreased 2.9940%; HL decreased 10.4167%
	Fishes Perciformes: *Mullus*	Freezing	TL decreased 2.3585%; SL decreased 1.2789%; HL decreased 18.3673%
	surmuletus	5% formalin mixed with tap water	TL increased 1.4103%; SL increased 0.8130%; HL decreased 6.7568%
		5% formalin mixed with distilled water	TL increased 0.9202%; SL decreased 0.1504%; HL decreased 5.623%
		10% formalin mixed with tap water	TL increased 0.5952; SL decreased 0.7519%; HL decreased 2.8947%
		70% alcohol mixed with tap water	TL decreased 3.0726%; SL decreased 3.1739%; HL decreased 7.6923%
		70% alcohol with distilled water	TL decreased 4.444%; SL decreased 4.2857%; HL decreased 10.0000%
		Freezing	TL decreased 4.0119%; SL decreased 1.6000%; HL decreased 2.8571%
Amosov 1960	Pond carp and Amur wild carp Cypriniformes: Cyprinidae	4% formaldehyde for 1 month followed by soaking in water	Increase of 3.6–8.5% in weight
Armstrong and Stewart 1997	Salmon Salmoniformes: Salmonidae	Freezing	Shrinkage of 6–11% in length after freezing (–17°C for 7 days) and thawing

(continued)

Table 15. ***(Continued)***

Reference	*Organisms*	*Fixative or Preservative*	*Comments*
Bailey 1952	Sculpin Scorpaeniformes: Cottidae: *Cottus bairdi punctulatus*	10% formaldehyde	Shrinkage of approximately 5% after 3 months in preservative
Bayer and Counihan 2001	White sturgeon larvae Acipenseriformes: Acipenseridae	Fixed in 10% unbuffered formaldehyde	Notochord shrinkage of 0.4% to 3.4% for larvae of 1–14 days old; total length shrinkage of 0.4% to 2.7% after 95 days
		Preserved in 95% ethyl alcohol (no fixation)	Notochord shrinkage of 2.9% to 2.7% for larvae of 1–14 days old; total length shrinkage of 3.2% to 2.4% after 95 days
Bernard and Wynn 1964	< 0.5 mm slices of rat liver, kidney, and muscle tissue Rodentia: Muridae: *Rattus* sp.	10% formalin	Change as percent of original weight: liver 121%, kidney 125%, muscle 128%
		20% formalin	Change as percent of original weight: liver 107%, kidney 104%, muscle 105%
		30% formalin	Change as percent of original weight: liver 103%, kidney 98%, muscle 98%
		10% neutral buffered formaldehyde	Change as percent of original weight: liver 117%, kidney 102%, muscle 119%
		10% formaldehyde 0.12 M saline	Change as percent of original weight: liver 102%.
		10% formaldehyde 0.15 M saline	Change as percent of original weight: liver 99%, kidney 93%, muscle 101%
		10% formaldehyde 0.18 M saline	Change as percent of original weight: liver 95%
		10% formaldehyde 0.25 M saline	Change as percent of original weight: liver 90%

Billy 1982	Fish Perciformes: Chiclidae: *Sarotherodon mossambicus*)	10% formaldehyde followed by 37.5% isopropyl alcohol	All specimens increased in slightly in length and greatly in weight after 5 days in formaldehyde; length decreased to near live length after 65 days in isopropanol, but weight remained higher
Bininda-Edmonds and Russell 1992, 1993	Bats Chiroptera: Vespertilionidae: *Myotis lucifugus*	Formaldehyde fixation; preservation in 70% ETOH and 45% isopropyl	No significant difference in area or wing mass between specimens preserved in ethyl alcohol and isopropyl alcohol
Blaxter 1971	Clyde herring larvae	4% formalin in seawater, 15% salinity	Shrinkage of 14% after 3 months
	Clupeiformes: Clupedae	4% formalin in seawater, 34% salinity	Shrinkage of 13% after 3 months
		2% formalin in seawater, 15% salinity	Shrinkage of 11% after 3 months
		2% formalin in seawater, 34% salinity	Shrinkage of 14% after 3 months
		10% formalin in seawater, 15% salinity	Shrinkage of 7% after 3 months
Bliss 1872	Cyprionid fish	Alcohol	Vermillion spot appeared on abdomen in specimens preserved in alcohol
Burgner 1962	Red salmon smolts Salmoniformes: Salmonidae	10% formaldehyde	5% shrinkage in first 24 hr in formaldehyde; 6% shrinkage after 5 months
Clutter and Whitesel 1956	Sockeye smolts Salmoniformes: Salmonidae	Formaldehyde	Decrease in length of 5%, 2% weight loss after 25–180 days
Cunninghan et al. 2000	Silverside larvae	80% ethyl alcohol	After 21 d, shrinkage of 3.9%
	Atheriniformes:	100% ethyl alcohol	After 21 d, shrinkage of 4.1%
	Atherinopsidae:	5% buffered formalin	After 21 d, shrinkage of 2.2%
	Menidia beryllina	10% buffered formalin	After 21 d, shrinkage of 3.2%

(continued)

Table 15. *(Continued)*

Reference	*Organisms*	*Fixative or Preservative*	*Comments*
Dabrowski and Bardega 1982	Vendace larvae Salmoniformes: Salmonidae: *Coregonus albula*	Formaldehyde, glycerin, and ethanol	After 285 days, slight increase (1.8%) in total length.
		10% neutral buffered formaldehyde	After 285 days, length decreased 13.2%
		Glutaraldehyde and sodium cacodylat	After 285 days, length decreased by 9.2%
Davis 1978	Atlantic herring Clupeiformes: Clupedae	5% formaldehyde in seawater	Shrinkage of 5–14% in length following preservation
Deichmann et al. 2009	Frogs (14 species) Anura: Bufonidae and Hylidae	Fixed in 10% formalin; preserved in either 70% ethanol or 55% isopropanol	Shrinkage in snout-urostyle length varied between 0.31 to 5.62%, depending on the species (the study did not evaluate the differences between ethanol and isopropyl preservation)
DiStefano et al. 1994	Central stonerollers Cypriniformes: Cyprinidae: *Campostoma anomalum*	Isopropanol 100%	After 180 days, length change of –2.48%; weight change of –37.18%
		Iscpropanol 75%	After 180 days, length change of –2.40%; weight change of –32.83%
		Isopropanol 50%	After 180 days, length change of –2.62%; weight change of –19.57%
		Isopropanol 25%	After 180 days, length change of –1.29%; weight change of –14.41%
		Ethanol 100%	After 180 days, length change of –1.18%; weight change of –28.43%
		Ethanol 75%	After 90 days, length change of –1.20%; weight change of –16.86%
		Formalin 10%	After 180 days, length change of –2.56%; weight change of –12.21%
		Freezing (to –15°C)	After 180 days, length change of –2.88%; weight change of –9.65%

Engel 1974	Perch Perciformes: Percidae: *Perca flavescens* and cisco Salmoniformes: Salmonidae: *Coregonus artedii*	10% formaldehyde; freezing	Formaldehyde and freezing reduced length by 0.7–2.1%; formaldehyde increased weight by 1.8–5.0%.
Farris 1963	Sardine larvae Clupeiformes: Clupeidae: *Sardina caerulea*	3% formaldehyde buffered with borax	Greatest shrinkage occurred in first 30 min in formaldehyde; fish specimens shrank to 88.7 to 93.2% of live length.
Fowler and Smith 1983	Silver hake	4% neutral formalin-seawater (31–32%)	After 337 days, shrinkage of 4.3%
	Perciformes: Mullidae:	95% ethanol	After 209 days, shrinkage of 7.0%
	Merluccis bilinearis	Freezing	After 338 days, shrinkage of 1.4%
Fry 1985	Bee eaters Coraciiformes: Meropidae: *Merops*	Immersion in alcohol	Color changes in feathers of green to blue green; scarlet to pale bluff; bright pink to buffy white; carmine to gingery brown; suppression of yellows and reds; no effect on blue, orange-buff, or black
Geldmacher and Wieland 1999	Baltic cod eggs Gadiformes: Gadidae: *Gadus morhua callarias*	4% buffered formaldehyde in seawater	Exposure to formaldehyde caused formation of dark enclosures inside the egg, and shrinkage and darkening of the blastodisc/embryo and the yolk sac
Gibbs et al. 1974	Myctophid fish Actinoplterygii: Myctophiformes	Fixed in 10% formalin, rinsed, and preserved in 40% isopropyl or 70% ETOH	Resulted in variable higher concentrations of cadmium, copper, zinc, and in some species, lead; lower concentrations of mercury and lead for some species
Guillette et al. 1988	Lizard eggs	10% neutral buffered formaldehyde	Net change in weight 34.8%
	Squamata:	Bouin's picric acid	Net change in weight 39.3%
	Phrynosomatidae:	Formaldehyde to 70% ETOH	Net change in weight 34.1%
	Sceloporus undulatus	Bouin's to 70% ETOH	Net change in weight 27.3%

(continued)

Table 15. ***(Continued)***

Reference	*Organisms*	*Fixative or Preservative*	*Comments*
Hay 1981	Pacific herring larvae Clupeformes: Clupeidae: *Clupea harengus pallasi*	4% formaldehyde in seawater	Larvae fixed immediately after towing shrank an average of 12–18% in body length, depending on the time interval of the tow; shrinkage of 30–43% was measured in specimens with delayed fixation; shrinkage for specimens that were not towed averaged 5%
Hay 1982	Herring larvae Clupeformes: Clupeidae: *Clupea harengus pallasi*	2–5% formaldehyde at varying salinities	2–5% formaldehyde resulted in 2–10% shrinkage; 20–30% formaldehyde caused 3–5% shrinkage
Hay 1984	Herring eggs and larvae Clupeformes: Clupeidae: *Clupea harengus pallasi*	4–20% formaldehyde of 0–28% salinity	Weight loss varied inversely with salinity and formaldehyde concentration; larval weight loss 19–36%
Heming and Preston 1981	Chinook salmon yolk and tissue Salmoniformes: Salmonidae: *Oncorhynchus tshawytscha*	5% neutral formaldehyde	Yolk and tissue weight increased significantly for 2 d, then gradually decreased to a constant level after 10 d; After 50 d in formalin yolk weight increased 9.5%, tissue weight increased 1.3%
Hile 1936	Cisco Salmoniformes: Salmonidae: *Leucichthys artedi*	10% formaldehyde followed by preservation in 70% alcohol	Shrinkage of 1.9%
Hile 1941	Rock bass Actinopterygii: Centrarchidae: *Amblophites rupestris*	Fixed in 10% formalin; preserved in 70% ETOH	Ratio of preserved length to live length from first site was 1.000:1.019. Ratio for second site was 1.000:1.005 (fish under 350 mm); 1.000:1.010 (fish 250–349 mm); and 1.000:1.028 (fish 150–249 mm)

Hoar 1939	Atlantic salmon Actinopterygii: Salmonidae: *Salmo salar*	Formaldehyde	Decrease in length, increase in weight
Jawad 2003	Perciformes: Carangidae: *Alepes djeddaba*	5% formaldehyde mixed with tap water	6.27% shrinkage
		5% formaldehyde mixed with distilled water	1.28% shrinkage
		10% formaldehyde mixed with tap water	5.41% shrinkage in tap water
		10% formaldehyde mixed with distilled water	3.13% shrinkage
		70% ethanol mixed with tap water	0.88% shrinkage
		70% ethanol mixed with distilled water	2.20% shrinkage
Jennings 1991	Bass Perciformes: Moronidae: *Dicentrarchus labrax*	Fixed in 4% buffered formalin; preserved in 70% ethyl alcohol	Larvae shrank significantly due to postmortem retention in nets before preservation; time in fixative or preservative had no significant effect on preserved length
Jones and Geen 1977	Spiny dogfish Squaliformes: Squalidae: *Squalus acanthias*	5% formaldehyde isotonic to elasmobranch body (salinity 16%)	Eggs—volume increased 3.6%; weight increased 8.5% Embryos—weight increased 8.1% then decreased; length decreased to 97.7% of live length; Adult—body length decreased by 5.1%
		5% formaldehyde mixed with freshwater	Body length decreased by 4.2%
		5% formaldehyde mixed with seawater	Body length decreased by 5.6%
		Frozen	Body length decreased by 1.4%
Klauber 1943	Snakes Squamata: Colubridae	Unknown	Notes "some differential shrinkage between body and tail, resulting from preservation" (Klauber 1943, 20); 2.09–3.15% shrinkage in body length in alcohol

(continued)

Table 15. ***(Continued)***

Reference	*Organisms*	*Fixative or Preservative*	*Comments*
Kristoffersen and Salvanes 1998	Fish Stomiiformes: Sternoptychidae: *Maurolicus muelleri* and Myctophiformes: Myctophidae: *Benthosema glaciale*	4% formaldehyde with seawater	Loss of 12.9–15.9% of body weight and 0.8–1.1% of body length
		80% ethanol	Loss of 36.9–39.4% of body weight and 2.3–3% of body length
Kruse and Dalley 1990	Capelin Osmeriformes: Osmeridae: *Mallotus villosus*	5% buffered formaldehyde	After 24 wks, shrinkage of 5.2% total length; 5.3% standard length
		100% anhydrous alcohol	After 24 wks, shrinkage of 14.4% total length; 13.6% standard length
Lai 1963	Fish Cypriniformes: Cyprinidae: *Pimephales promelas*	7.5%, 10%, and 15% formaldehyde	No significant differences in proportional measurements
		Transfer of fixed specimens from 40% isopropanol to 70% ethanol	No significant differences in proportional measurements
		Transfer of fixed specimens from 70% ethanol to 40% isopropanol	Significant change in proportional measurements
Lazell 1972	Lizards Sqamata: Polychrotidae: *Anolis*	Preserved in 85–95% ETOH	Shrinkage of 3.3–15.5 % noted after storage in alcohol
Lee 1982	Toads Anura: Bufonidae: *Bufo marinus* (*Rhinella marinus*)	10% formaldehyde; 24 hr in water; 70% ETOH; measured after 6 months and 14 months in ETOH	Means of six morphometric characters were significantly larger and eight were significantly smaller after preservation; preservation caused shrinkage in SVL of 6.19%
Leslie 1983	Freshwater young fish	5% formaldehyde buffered with marble chips; 4% neutral formaldehyde, 60% ETOH, 10% Borealene, or Davidson's B (1 part formaldehyde, 1 part glycerol, 5 parts ETOH, 3 parts water)	Found that fixation and preservation did not induce immediate changes in length, but that over time specimens both shrank and lengthened in the various solutions.

Leslie and Moore 1986	Freshwater fish	4% formaldehyde, 60% ethanol, Davidson's B (30 cc 95% ethyl alcohol, 20 cc neutralized formaldehyde, 10 cc glycerin, 30 cc distilled water), and 10% Borealene	No significant change in body length for first 45 d; but some species showed shrinkage of 2.5–4% after 45 d
Lockwood 1973	Plaice Pleuronectiformes: Pleuronectidae: *Pleuronectes platessa*	4% neutral formaldehyde in seawater	Reduction in length of 2–6%; most shrinkage occured in the first two weeks of exposure to the preservative
Lockwood and Daly 1975	Flatfish Pleuronectiformes: Pleuronectidae: *Pleuronectes platessa*	4% neutral formaldehyde in seawater	Rapid loss of 1–4% length in the first few days of exposure to preservative, over time lost 14–29% of initial weight; smaller specimens underwent greater biomass loss than larger specimens
Mabee et al. 1998	Fish Perciformes: Cichlidea: *Tilapia mossambica*	Fixed in 10% formalin buffered with phosphate; transferred directly to 70% ethyl alcohol; cleared and stained following Dingerkus and Uhler (1977)	Shrinkage of 2.30% after 2 d in alcohol or 3.04% after 30 d in alcohol; shrinkage after clearing and staining of 1.42–1.45%, resulting in overall shrinkage of 3–6% from live length to length after clearing and staining
Martin 1978	Lizards Squamata: Phrynosomatidae: *Urosaurus ornatus*	Fixed in 10% formaldehyde; preserved in 60% isopropanol	Clutch weight to total body weight ratios significantly higher following formalin fixation; after 5 months in isopropyl, ratios significantly higher than in formalin
McMahon and Tash 1979	Otoliths from green sunfish Perciformes: Centrarchidae	10% formaldehyde (unbuffered or with $NaHCO_3$); and hydrochloric acid	Otoliths in both formaldehyde solutions became opaque and lost surface characters after 12 hr
Mills 1975	Fish	Review of measurements of changes caused by freezing fish	Although many changes are known to occur when fish are frozen, no single test is adequate to assess freezing.

(continued)

Table 15. *(Continued)*

Reference	*Organisms*	*Fixative or Preservative*	*Comments*
Moku et al. 2004	Myctophid fish larvae Myctophiformes: Myctophidae: *Diaphus* spp.	5% formaldehyde	After 140 d, shrinkage to 98.4% of initial length
		10% formaldehyde	After 140 d, shrinkage to 98.9% of initial length
		70% ethyl alcohol	After 140 d, shrinkage to 97.6% of initial length
		90% ethyl alcohol	After 140 d, shrinkage to 94.9% of initial length
		70% isopropyl alcohol	After 140 d, shrinkage to 91.7% of initial length
Morkert and Bergstedt 1990	Sea lamprey larvae Petromyzontiformes: Petromyzontidae: *Petromyzon marinus*	5% formaldehyde	Average shrinkage of 3.8%
		10% formaldehyde	Average shrinkage of 4.3%
Nelson 1971	Frogs Anura: Microhylidae: *Gastrophryne* and *Hypopachus*	Fixed in 10% or 20% formaldehyde; preserved in 50% isopropanol	20% formaldehyde fixation resulted in more pustular skin than 10% formaldehyde fixation
Nowak and Popek 2012	Fish Cypriniformes: Cyprinidae: *Leucaspius delineates*	Formalin fixative to 70% ethanol	Shrinkage in total length of –1.98%; head length –4.36%; body depth –1.92%, and body width –5.37%
		Formalin fixative to 50% and then 70% ethanol	Shrinkage in total length of –1.85%; head length –2.08%; body depth –2.39%, and body width –3.84%
		Formalin fixative to 20%, 40%, 60%,and 70% ethanol	Shrinkage in total length of –1.68%; head length –3.17%; body depth –2.30%, and body width –2.62%

Parker 1963	Fishes Salmoniformes: Salmonidae: *Oncorhynchus*	Fixation in 3.8% formaldehyde in fresh or seawater	Shrinkage to 97% of live length occurred within 12 hr; to 96% of live length after 30–40 days, regardless of ionic strength of solution. In freshwater formaldehyde, weight increased then decreased to 105–112% of live weight. In 23% salinity (seawater) formaldehyde, 97–135% of live weight
Patten and Philpott 1921	Pig embryos Artiodactyla: Suidae: *Sus*	Zenker's fluid, Orth's fluid, and Tellyesnicky's fluid	Shrinkage of 20–30%
		10% formaldehyde	Shrinkage of 5%
		Formol-alcohol	Initial swelling, followed by significant shrinkage
		Bouin's fluid	Shrinkage of 5%
Paulet and Kaiser 2004	Goldfish Cypriniformes: Cyprinidae: *Carassius auratus*	10% buffered formaldehyde	Average loss of 17% in length and 14% of body depth after 36 h in preservative
Pearcy 1962	Flounder Pleuronectiformes: Pleuronectidae: *Pseudopleuronectes americanus*	Neutral formaldehyde	3.7% reduction in length of juveniles after 12 months in preservative
Prum et al. 1994	Velvet asity Passeriformes: Philepittidae: *Philepitta castanea*	Formaldehyde and gluteraldehyde	Blue color on face in life better preserved by 2.5% gluteraldehyde fixative; 10% formalin fixative followed by 70% ethanol preservation changed green to violet blue or blue
Quester and Schröder 1997	Human brain stem	4% formaldehyde, unbuffered	Shrinkage of 1–8% in length
Radtke 1989	Atlantic cod Gadiformes: Gadidae: *Gadus morhua*	96% ethanol	No significant shrinkage of larvae placed alive in 96% ethanol, but found considerable shrinkage occurred at death without preservation.

(continued)

Table 15. *(Continued)*

Reference	*Organisms*	*Fixative or Preservative*	*Comments*
Radtke and Waiwood 1980	Larval cod Atlantic cod Gadiformes: Gadidae: *Gadus morhua*	60% ETOH for 60 days; transferred to 95% ETOH	Larval body shrinkage of 9–15% within 15 min of death, and 11–20% shrinkage when preserved in ethanol
Reed 2001	Snakes	Fixation followed by alcohol preservation	Significant shrinkage of 6–7% in snout-vent length and total length, but no loss of body mass
Rosenthal and von Westernhagen 1976	Siganid fish Perciformes: Siganidae: *Siganus canaliculatus*, *Siganus guttatus*, and *Siganus vermiculatus*	4% formaldehyde in seawater	No significant changes in weight or length after 20 days in preservative
Rosenthal et al. 1978	Red Sea Bream Perciformes: Sparidae: *Chrysophrys major*	2% formaldehyde in seawater	Shrinkage of 16.78% in total length and 17.10% in standard length after 20 days (most shrinkage took place in the first 10 days in preservative)
		4% formaldehyde in seawater	Shrinkage of 16.63% in total length and 16.77% in standard length after 20 days (most shrinkage took place in the first 10 days in preservative)
Sagnes 1997	Grayling Salmoniformes: Salmonidae	4% formaldehyde	Variable decreases in body length, but increases in body width dimensions
Sameoto 1971	Herring larvae Clupeiformes: Clupedae: *Culpea harengus*	5% buffered formaldehyde in seawater	Weight loss of an average of 3.8% after 1 month in preservative

Schnack and Rosenthal 1978	Pacific herring larvae Clupeiformes: Clupedae	4% formaldehyde in seawater	Standard lengths shrank an average of 5% during first 2 months in preservative; some groups shrank 6–7%, newly hatched larvae shrank 10% after 1.5 years
Scott and Aquino-Shuster 1989	Frogs Anura: Ranidae: *Rana pipiens* (*Lithobates pipiens*)	Frozen (–15°C); thawed; fixed in 10% unbuffered formaldehyde; preserved in 40% isopropyl alcohol	Frogs that had been frozen were "softer and grayer" than other specimens and showed loss of epidermis. Thawed frozen specimens were more flaccid than nonfrozen specimens.
	Snakes (*Thamnophis elegans*)	Frozen (–15°C); thawed; fixed in 10% unbuffered formaldehyde; preserved in 50% isopropyl alcohol	Snakes that had been frozen "were considerably darker" than others; several had burst gallbladders. Frozen snakes were 1–3% longer than pretreated snakes, fixed and preserved snakes 1–3% shorter. Thawed frozen specimens were more flaccid than nonfrozen specimens.
Serns 1982	Walleye eggs Perciformes: Percidae: *Stizostedion vitreum vitreum*	10% unbuffered formaldehyde	Egg size did not change significantly during a 20-month exposure to 10% unbuffered formaldehyde.
Shetter 1936	Trout Salmoniformes: Salmonidae: *Salvelinus fontinalis* and *Salmo trutta*	10% formaldehyde fixation, preservation in 70% ETOH	A correction factor of 1.027 is suggested to compensate for shrinkage due to rigor mortis; a correction factor of 1.054 is suggested for comparison of preserved fish to live measurements.
Sigler 1949	White bass Perciformes: Moronidae: *Lepibema chrysops*	10% formaldehyde followed by 70% alcohol	Calculated a weight correction factor of 0.947 for formaldehyde preserved specimens and 1.147 for alcohol preserved specimens

(continued)

Table 15. ***(Continued)***

Reference	*Organisms*	*Fixative or Preservative*	*Comments*
Smith 1955	Snakes Squamata: Colubridae: *Thamnophis rufipunctatus*	Fixation presumably 10% formaldehyde; preservation presumably 70% ETOH	Five years in storage resulted in loss of yellows and reds, which changed the color pattern of the specimens significantly.
Sterns 1982	Walleye eggs Perciformes: Percidae	10% formaldehyde	Recorded no significant change in size of preserved walleye eggs compared to fresh eggs.
Stevenson 1962	Pacific herring Clupeiformes: Clupedae	"About 5%" formaldehyde	Shrinkage measured of "about 2%"
Stickland 1975	Beef muscle Artiodactyla: Bovidae: *Bos*	Carroy's solution (without chloroform)	Muscle fiber diameter 73.9% of fresh measurement
		Susa solution	Muscle fiber diameter 72.8% of fresh measurement
		Baker's formalin	Muscle fiber diameter 82.7% of fresh measurement
		5% formalin	Muscle fiber diameter 69.3% of fresh measurement
		10% formalin	Muscle fiber diameter 69.0% of fresh measurement
		10% formalin saline	Muscle fiber diameter 69.8% of fresh measurement
		Bouin's solution	Muscle fiber diameter 76.9% of fresh measurement
		2.5% glutaraldehyde	Muscle fiber diameter 70.7% of fresh measurement
		Zenker solution	Muscle fiber diameter 62.0% of fresh measurement
Stobo 1972	Yellow perch Actinopterygii: Percidae	10% formaldehyde	Most shrinkage occurred in first 5 days; 0.05% (larger fish) to 1.35% (smaller fish); weights initially increased, then decreased

Stuart 1995	Lizards Squamata: Teiidae: *Cnemidophorus uniparens*	10% or 20% formaldehyde buffered with monobasic and dibasic sodium phosphates	Greater length of time in 10% formaldehyde (over one week) caused increasing darkening; 20% formaldehyde caused darkening after less than 1 week
Takizawa et al. 1994	Larval fish Perciformes: Sparidae: *Pagrus major*, Pleuronectiformes: Paralichthyidae: *Paralichtys olivaceus*, Osmeriformes: Plecoglossidae: *Plecoglossus altivellis*	10% formaldehyde in distilled water	Shrinkage in total length of 1–16%; loss of 2–31% of live weight in preservative
Tarkhan 1931	Cells from cat liver Carnivora: Felidae: *Felis*	5% and 10% formaldehyde, mercuric chloride, picric acid, acetic acid, and trichloracetic acid	No shrinkage or swelling in formaldehyde for 24–48 hr; mercuric chloride, picric acid, and absolute alcohol caused significant shrinkage; acetic acid and trichloracetic acid caused swelling
Theilacker 1980	Anchovy Clupeiformes: Engraulidae: *Engraulis mordax*	Bouin's fixative, 5% buffered formaldehyde, and 80% ethanol	Formaldehyde and Bouin's produced 8% shrinkage in length, alcohol produced no shrinkage; similar shrinkages reported for other body measurements
Treasurer 1992	Perciformes: Percidae: *Perca fluviatilis*	4% and 10% neutralized formaldehyde	After 24 hr in 4% formalin, specimens shrank to 98–98.4% of live values; after 8 weeks, specimens shrank to 96.9–97.9% of live values. After 24 wks in 10% formalin, specimens shrank to 96.5–98.3% live values

(continued)

Table 15. ***(Continued)***

Reference	*Organisms*	*Fixative or Preservative*	*Comments*
Tucker and Chester 1984	Southern flounder Pleuronectiformes: Pleuronectidae: *Paralichthys lethostigma*	Formaldehyde (4, 7%, 10%) of 0% or 35% salinity	Formaldehyde with salt water caused significantly more shrinkage than formaldehyde with freshwater; unbuffered 4% formaldehyde produced the least change from live weight; buffered and unbuffered 10% formaldehyde in salt water caused least postfixation changes (92.7–94.8% of live weight)
Van Oosten 1929	Cisco Salmoniformes: Salmonidae: *Coregonus artedii*	Formaldehyde fixation, alcohol preservation	Decrease in length and weight
Vitt et al. 1985	Lizard eggs (teiids and iguanids) Sqamata: Iguanidae and Teiidae	10% formaldehyde	Weight gain of 42.3–64.0% of initial fresh weights
Williams 1962	Lizards Squamata: Iguanidae	Formaldehyde	"reduces the varied shades [of color] present in life to an undistinguished and unpleasant brown and destroys all but the most prominent patterns"
		Alcohol	"loses much the colors of life and may retain one only . . . of a repertoire of patterns"

Winokur and Hillyard 1992	Toads Anura: Bufonidae: *Bufo* sp.	Fresh and "preserved museum specimens" treated with Bouin's fluid	"Postfixing of preserved museum specimens using Bouin's fluid resulted in poor muscle definition; however Lugol's iodine solution . . . was effective in staining the muscles of both fresh and museum specimens." (Winokur and Hillyard 1992, 760)
Yeh and Hodson 1975	Fish Perciformes: Centrarchidae: *Lepomis macrochirus* and *Pomoxis annularis*	10% formaldehyde	Both species shrank slightly in length (98.70–99.85% of live length) and increased significantly in weight (107.89–113.09% of live weight), resulting in a pronounced effect on values for the length-weight relationship

Table 16. Summary of Fixative and Preservative-Induced Changes in Vertebrates by Taxonomic Group (See Table 15 for Details)

Group	*Organisms*	*Change*	*Reference*
Fish	Various freshwater fish	Shrinkage	Leslie and Moore 1986
		Various changes caused by freezing	Mills 1975
Actinoplterygii	Myctophiformes	Variable concentrations of minerals	Gibbs et al. 1974
Atheriniformes	Atherinopsidae	Shrinkage	Bayer and Counihan
		Shrinkage	Cunningham et al. 2000
Clupeiformes	Clupedae	Shrinkage in length and body height	Blaxter 1971
		Shrinkage in length	Davis 1978
		Significant shrinkage	Farris 1963
		Shrinkage in length	Hay 1981
		Shrinkage in length	Hay 1982
		Weight loss	Hay 1984
		Weight loss	Sameoto 1971
		Shrinkage	Schnack and Rosenthal 1978
		Shrinkage in length	Stevenson 1962
	Engraulidae	Shrinkage in length	Theilacker 1980
Cypriniformes	Sparidae	Shrinkage in length	Rosenthal et al. 1978
	Cyprinidae	Weight increase	Amosov 1960
		Length and weight loss	DiStefano et al. 1994
		Shrinkage	Lai 1963
		Shrinkage and lengthening	Leslie 1983
		Shrinkage	Nowak and Popek 2012
		Shrinkage in length and body depth	Paulet and Kaiser 2004
		Appearance of vermillion spot on abdomen in alcohol	Bliss 1872
Esociformes	Esocidae	Shrinkage and lengthening	Leslie 1983
Gadiformes	Gadidae	Shrinkage of embryo and yolk sac	Geldmacher and Wieland 1999
		No significant shrinkage	Radtke 1989
		Significant shrinkage	Radtke and Waiwood 1980

Myctophiformes	Myctophidae	Significant shrinkage	Moku et al. 2004
		Shrinkage and weight loss	Kristoffersen and Salvanes 1998
Osmeriformes	Osmeridae	Shrinkage	Kruse and Dalley 1990
		Shrinkage and lengthening	Leslie 1983
	Plecoglossidae	Shrinkage and weight loss	Takizawa et al. 1994
Perciformes	Carangidae	Shrinkage	Jawad 2003
	Centrarchidae	Significant shrinkage	Hile 1941
		Otolith damage	McMahon and Tash 1979
		Slight shrinkage in length and significant increase in weight	Yeh and Hodson 1975
	Chiclidae	Slight increase in length; return to live length	Billy 1982
		Shrinkage	Mabee et al. 1998
	Moronidae	Significant shrinkage	Jennings 1991
		Shrikage; correction factor	Sigler 1949
	Mullidae	Variable change in body measurements	Al-Hassan et al. 2000
		Decrease in length	Fowler and Smith 1983
	Percidae	Shrinkage and weight gain	Engel 1974
		Shrinkage, weight increase, then decrease	Stobo 1972
		Shrinkage	Treasurer 1992
		No change in egg size	Serns 1982
		No significant shrinkage in eggs	Sterns 1982
	Siganidae	No significant change	Rosenthal and von Westernhagen 1976
	Sparidae	Shrinkage and weight loss	Takizawa et al. 1994
Petromyzontiformes	Petromyzontidae	Shrinkage	Morkert and Bergstedt 1990
Pleuronectiformes	Paralichthyidae	Shrinkage and weight loss	Takizawa et al. 1994
	Pleuronectidae	Shrinkage in length	Lockwood 1973
		Shrinkage in length and weight	Lockwood and Daly 1975
		Shrinkage in length	Pearcy 1962
		Significant shrinkage	Tucker and Chester 1984

(continued)

Table 16. ***(Continued)***

Group	*Organisms*	*Change*	*Reference*
Salmoniformes	Salmonidae	Decrease in length; gain in weight	Engel 1974
		Loss of length	Hile 1936
		Shrinkage in length and weight	Clutter and Whitesel 1956
		Significant shrinkage	Dabrowski and Baradega 1982
		Shrinkage	Armstrong and Stewart 1997
		Significant shrinkage	Burgner 1962
		Length and weight loss	Engel 1974
		Weight increase	Heming and Preston 1981
		Decrease in length, increase in weight	Hoar 1939
		Shrinkage	Parker 1963
		Variable decrease in body length and increase in body widths	Sagnes 1997
		Shrinkage; correction factor	Shetter 1936
		Decrease in length and weight	Van Oosten 1929
Scorpaeniformes	Cottidae	Shrinkage	Bailey 1952
Squaliformes	Squalidae	Weight and volume increase, then shrinkage	Jones and Geen 1977
Stomiiformes	Sternoptychidae	Length and weight loss	Kristoffersen and Salvanes 1998
Anura	Bufonidae and Hylidae	Variable shrinkage	Deichmann et al. 2009
	Bufonidae	Mixed increase and decrease in measurements	Lee 1982
		Muscle definition	Winokur and Hillyard 1992
	Microhylidae	Skin pustularity	Nelson 1971
	Ranidae	Softening after freezing	Scott and Aquino-Shuster 1989
Squamata	Iguanidae	Weight increase (eggs)	Vitt et al. 1958
		Color loss	Williams 1962
	Phrynosomatidae	Eggs increase in weight in formaldehyde; decrease in ethanol	Guillette et al. 1988

		Weight increase	Vitt et al. 1985
	Phrynosomatidae	Significant shrinkage	Martin 1978
	Polychrotidae	Shrinkage in length	Lazell 1972
	Teiidae	Darkening in fixative	Stuart 1995
		Weight increase (eggs)	Vitt et al. 1958
	Colubridae	Differential shrinkage	Klauber 1943
		Darkening, shrinkage, and softening after freezing	Scott and Aquino-Shuster 1989
		Significant color loss	Smith 1955
	Various snakes	Significant shrinkage in length but not mass	Reed 2001
Coraciiformes	Meropidae	Loss of color in feathers in alcohol	Fry 1985
Passeriformes	Philepittidae	Color change of facial skin in formaldehyde fixative	Prum et al. 1994
Artiodactyla	Suidae	Shrinkage	Patten and Philpott 1921
	Bovidae	Shrinkage	Stickland 1975
Carnivora	Felidae	Variable shrinkage and swelling in fixatives	Tarkhan 1931
Chiroptera	Vespertilionidae	Variation in measurements	Bininda-Edmonds and Russell 1992
Rodentia	Muridae	Variable weight change	Bernard and Wynn 1964
Human	Brain stem	Shrinkage	Quester and Schröder 1997

Table 17. Proprietary Preservatives (Based on Manufacturer's MSDS, Advertisements, and Published Analyses)

Brand Name	*Manufacturer*	*Comments*
Carosafe or Perfect Solution	Carolina Biological Supply	Sold as a shipping and holding fluid; contains propylene glycol; 2-amino-2-ethyl-1, 3-propanediol; ethylene glycol phenyl ester; some residual formaldehyde; and other ingredients.
DMDM-Hydantoin (Glydant)	Lonza	Contains 1,3-Bis(hydroxymethyl)-5,5-dimethylimidazolidine-2,4-dione. A pH neutral, non-polymerizing biocide, formaldehyde releasing agent; proposed as a possible fluid preservative by van Dam (2003).
Nascoguard	Nasco	Contains propylene glycol (1,2-propanediol), water, 0.3% or less formaldehyde, and a trace of phenol. The product literature states that "Most Nasco-Guard Specimens are initially fixed in a 3.7% formaldehyde solution, which is still the best animal tissue fixative known. In the Nasco-Guard process, skilled Nasco technicians remove the formaldehyde fixative in a series of water baths then painstakingly perfuse the specimens with ethylene glycol until the tissue fluids contain at least 25% glycol."
Novec HFE-7100	3M	Sold as a holding solution; contains methoxy-nonafluorobutane ($C_4F_9OCH_3$), a product developed to clean and degrease electronics; clear, colorless, and odorless, high-boiling point, low-surface tension, density of 1.52 g/cm^3.
Opresol	Philip Harris (UK)	Contains 2-phenoxyethanol, diethylene glycol, and proprietary ingredients.
Perfect Solution	Carolina Biological Supply	Proprietary ingredients, but apparently contains some formaldehyde and phenol.
RNA*later*	Ambion Products	MSDS lists ingredients as "Trade secret."
Ward-Safe	Wards Scientific	Contains methyl alcohol (2.87%); 1,2 propanediol [propylene glycol] (1.44%); a proprietary ingredient (0.68%); and water (95.01%). The proprietary ingredient is most likely glutaraldehyde, based on published analyses.
2-Phenoxyethanol	various	Contains phenoxetol or ethylene glycol monophenyl ether, which is a poor biocide and more acidic than ethanol.

Table 18. Summary of Published Fluid Concentration and pH Testing of Fluid-Preserved Collections

Reference	*Percent Alcohol (Range)*	*Percent Formaldehyde (Range)*	*Average pH of Alcohol Preservative (Range)*	*Average pH of Formaldehyde Preservative (Range)*
Cato 1990	64 (40–71)	Not measured	6.2 (5.0–7.0)	Not reported
Kotrba and Golbig 2009	Unknown	Not measured	4.5–9.5	Not reported
Palmer 1996	39–94	Not measured	4.8–8.9	Not reported
Pickering 1997	63.1 invertebrates; 63.4 vertebrates	Not measured	Not reported	Not reported
Thede 1996	n/a	> 10	4.2	4.2
Von Endt 1994	Not measured	Not measured	6.5 (5.7–7.3)	Not measured
Waller and Simmons 2003	59.3 (40.1–74.1)	13.02 (9.7–17.8)	6.38 (5.19–7.55)	6.32 (4.26–8)

Table 19. Characteristics of Containers for Fluid-Preserved Specimens (Based on Simmons 2002)

Type of Container	*Relative Cost*	*Availability*	*Quality of Seal*	*Longevity of Seal*	*Rating*
Glass jar with glass stopper, antique glass	Very high	Very poor	May be very good if the stopper was ground to match the jar; many do not seal well	Life of the container	Very good to poor
Glass jar with glass stopper, borosilicate (modern)	Very high	Fair	Excellent (due to precision machining of modern glassware)	Life of the container	Excellent
Glass jar with glass lid clamped by wire bail	High	Poor	Fair to good (compressible gaskets deteriorate rapidly with exposure to preservatives)	< 5 years	Good
Glass jar with thermoplastic threaded lid (e.g., polypropylene)	Medium	Very good	Very good, may be improved by using lids with liners and wrapping threads of jar with Teflon tape	>20 years	Very good
Glass jar with threaded metal lid	Medium	Good	Fair (lids subject to oxidation); use only lids with liners; wrap jar threads with Teflon tape	< 10 years	Poor
Glass jar with snap-on thermoplastic lid (polyethylene)	Medium	Poor	Fair (due to deterioration of the thermoplastic)	5–10 years	Poor
Glass jar with thermoset threaded lid (phenol, melamine Bakelite)	Medium	Good	Fair (rigid lids tend to unscrew themselves with slight changes in temperature and are susceptible to stress cracking)	< 5 years	Poor
Glass vial with compressible stopper	Low	Good	Fair (due to deterioration of compressible stopper; deterioration products may contaminate vial contents)	< 10 years	Poor
Polyethylene terephthalate (PET) container with polypropylene closure	Low	Good	Unknown	Unknown	Unknown
Polycarbonate containers with polycarbonate snap-on lids	Medium	Fair	Good (polycarbonate will discolor and crack if exposed to UV radiation)	>10 years	Good
Stainless steel with compressible gasket	Very high	Fair	Fair to good (limitation is the life of the compressible gasket)	< 10 years (gasket)	Fair
High-density polyethylene (HDPE) bucket or barrel	Low	Good	Good (HDPE will crack if exposed to UV radiation)	> 15 years	Good

Table 20. Oxygen Permeability of Container Materials

Table 20a. Data from van Dam et al. 2000

Material	*Permeability Constant for Oxygen*
HDPE	34.0
PET	3.0
PP	151.0
glass	0.0

Table 20b. The Permeability of Polymer Foils to Oxygen in cm^3 of O_2 x m^{-2} area x 24 hr^{-1} x bar^{-1} (after van Dam et al. 2000)

Material	*°C*	*µm*	*Permeability*
HDPE	25	40	1890
PET	23	25	80/110
PP	25	40	1900

Table 21. Published Recommendations for Label Substrates and Inks

Reference	*Label Substrate*	*Ink*	*Label Production*
Anon 1905	White cotton cloth	Lead pencil	Handwritten
Anon 1938b	Fiber waterproof tags (Turtox)	Higgins Eternal Ink or pencil	Handwritten
Anon n.d., AMNH	Tough white paper	Soft pencil	Handwritten
Bentley 2004	Spun-bond polyester	Wax-based carbon ink	Thermal printer
Carter 1996	Found Resistall better than goatskin parchment	Found PermaDri superior to Indian Ink	Handwritten
Child 1994	High quality cellulose paper such as Byron-Weston Resistall or Atlantis Archival Copysafe	Rotring 17	Handwritten
Cook 1965	100% cotton rag-content paper (e.g., Dennison or Resistall)	Waterproof India ink or soft pencil	Handwritten
Cristobo et al. 1992	Polyester paper or parchment	Soft pencil or indelible ink	Handwritten
Doolittle 1925	Paper waterproofed by melting paraffin over the lettering	"Anything opaque," e.g., typewriter, lead pencil, red or blue pencil, any kind of ink	Handwritten
Fink et al. 1979	Byron Weston Resistall Linen Record, Dennison tag stock, parchment, Byron Weston Defiance 100% Index Bristol, Nalgene Polypaper, K&E Albanene tracing paper, Blackstone ledger paper	Pelikan No. 17, Higgins Engrossing No. 893, Koh-I-Noor 3080F, Higgins Eternal No. 813, Higgins Black Magic	Handwritten
Hangay and Dingley 1985a	Heavy rag base paper	HB pencil or waterproof India ink	Handwritten
Köhler 2001	Waterproof fiber or Dymo plastic tags	No recommendation	Handwritten or machine punched
Lincoln and Sheals 1979	Preprinted on fine quality paper, no pulp paper	Waterproof Indian ink	Handwritten
Moore 2002	Goatskin Parchment (Arjo Wiggins)	Indian ink	Handwritten
Pettitt 1976	Tyvek (spun-bonded polyethylene)	Pencil	Handwritten
	Goatskin parchment		

Pitkin 1995	100% rag paper	Indelible, permanent ink	Laserjet printers for nonsubmersible labels only, deskjet printers with alcohol resistant ink, dot matrix printers with alcohol resistant ink
	Goatskin parchment		
	Resistall Paper		
Ross 1961	Goatskin parchment	Good quality Indian ink	Handwritten
Schmidt 1932	Pure tin or Dennison's fiber-proof paper	No recommendation	Machine printed or stamped
Simmons 1999	100% cotton stock or spun-bonded polyethylene (e.g., Tyvek or Polypaper)	No recommendation	Impact printer
Slevin 1927	Pure tin tags	Machine numbered or inscribed with a steel point	Machine punched
	Lead strip	Inscribed with a steel point	Inscribed
	Thin strip of wood	Lead pencil	Handwritten
Snyder 1999	Forbon White Tag (vulcanized fiber paper, Allen-Bailey Tag & Label)	No recommendation	
	Curtis Parchment Parchkin	Koh-i-noor 3080-F India Ink	Printed on a plotter
	Domtar Wet Strength Laundry Tag	Toner	Laser jet printer
	Byron Weston Resistall	Printer ribbon cartridges from Automated Office Products	Dot matrix printer
Stejneger 1891	Parchment paper	No. 2 lead pencil or India ink	Handwritten
Van Guelpen et al. 1987	Wet Strength Laundry Tag (Domtar) or Suture Label (Domtar)	India ink or graphite pencil	Handwritten
	Tyvek	India Ink	Handwritten

Table 22. Rehydration Techniques for Fluid-Preserved Specimens

Technique	*Reference*	*Comments*
Decon 90 or TEEPOL in cold or hot bath (40°C) or by injection, followed by refixation in formaldehyde solution; remove air bubbles with vacuum pump to prevent specimen from floating	Waterhouse and Graner 2009	Decon 90 in warm water produces fair to good results, depending on the tissue type and the degree of dehydration.
30% ammonia and warm water	Traditional technique	Poor results; ammonia damages proteinaceous tissues
2:1 solution of hydrogen peroxide and water	Traditional technique	Poor results; hydrogen peroxide causes significant tissue damage
3:1 solution of isopropyl alcohol and water, warmed in microwave; rinse in 99% isopropyl	Traditional technique	May initially soften specimens, but alcohol will quickly cause further dehydration of tissues
Surfactant (detergent) in water	Banks and Williams 1972	Generally poor results on badly dehydrated specimens; surfactant chemicals will contaminate specimens and may cause problems with future chemical analysis
0.25 to 0.50 % solution of commercial grade trisodium phosphate and water.	van Cleave and Ross 1947	Usually unsatisfactory results due to tissue damage from the trisodium phosphate
0.5% trisodium phosphate and water for 24 hr	Marhue 1983	Usually unsatisfactory results due to tissue damage from the trisodium phosphate
Stage through solutions of 10%, 5%, and 0.5% acetic acid, followed by soaking in trisodium phosphate solution	Vogt 1991	Reported as a successful system for some species, but acetic acid and trisodiumphosphate may cause significant damage to protein
50% propylene glycol or ethylene glycol solution	Marhue 1983; Thompson et al. 1955	May be useful for some small, soft-bodied specimens; glycols are unsatisfactory preservatives and difficult to remove from specimens posttreatment
10% solution of enzymatic drain cleaner (e.g., Draino or Liquid Plumber) for 3 wks	Vogt 1998	Extreme risk of tissue damage to specimens; use of proprietary products with unknown ingredients is not recommended
Use of high pH solutions (produced by pH buffer tablets in water)	Vogt 2001	Success depends on previous preservation history of the specimen and short dehydration time
Exposure of specimen to vapors from distilled or deionized water (treated with thymol) until specimen begins to soften, followed by staged immersion by 20% steps from 100% deionized or distilled water to desired preservative strength	Simmons et al. (*forthcoming*); Singer (2014)	A slow rehydration technique during which the specimen and the rehydrating fluids must be closely monitored for mold or bacterial growth; produces satisfactory results when tissue can be softened sufficiently to rehydrate

Table 23. Treatments and Practices That Are Not Recommended for Fluid-Preserved Specimens

Treatment or Practice	*Comment*
Exposing specimens to visible light and UV radiation.	Photochemical reactions cause deterioration of specimens and some containers; UV radiation is particularly damaging.
Diluting fixatives and preservatives with tap water.	Adds contaminants to the solution; use distilled or deionized water.
Not rinsing after formaldehyde fixation.	Results in small amounts of formaldehyde in alcohol which may pose health and pH problems.
Packing containers completely full of specimens.	Use a minimum ratio of 2:1 of fluid-preservative volume to specimen volume.
Repeated topping up with storage strength alcohol solutions.	Dilutes preservative concentration (alcohol evaporates faster than water).
Replacement of discolored preservative solutions.	Accelerates the processes of leaching of proteins and lipids from specimens.
Filling containers as full of preservative as possible.	Leaves no air space for compression when rise in temperature increases internal jar pressure; leave 10 percent air space for alcohol-based preservatives, 5 percent air space for water-based preservatives (e.g., formaldehyde).
Keeping storage areas warm.	Preferred storage temperature is 18°C with minimal fluctuations (preferably a few degrees cooler than laboratory areas to improve container seals).
Leaving containers open while specimens are being examined.	Causes loss of preservative quality as alcohol evaporates faster than water.
Submerging alcohol-preserved specimens in water for examination.	Alcohol specimens are hygroscopic, alcohol is a dehydrating agent. The water the specimens absorb will dilute the alcohol in the storage container.
Working with formaldehyde without splash goggles.	Formaldehyde will rapidly fix the mucus membranes of the face and is a nasal carcinogen. Soft contact lenses may absorb formaldehyde vapors.
Working with formaldehyde and wearing latex gloves.	Formaldehyde penetrates latex; use nitrile or neoprene gloves.
Rehydration with damaging chemicals.	Ammonia damages proteinaceous tissue; trisodium phosphate may clear tissues.
Rehydrating with proprietary compounds.	Never use unknown chemicals with scientific specimens.
Using rapid rehydration techniques.	Rapid rehydration is more likely to rupture cell membranes. Rehydrate specimens very slowly.
Use of anecdotal, untried, or unproven techniques and materials.	Use tested and reliable materials and preservation technology when working with specimens of historic significance. Even proven techniques may produce variable results. Document all techniques and materials used, and monitor all results.

INDEX

About the Author

John E. Simmons holds a BA in systematics and ecology and an MA in historical administration and museum studies. In 1987, he completed the Collections Care Pilot Training Program (funded by the Bay Foundation) to become one of thirty people in the country to receive specialized training in natural history conservation and collections care. He has spent a total of thirty years as collections manager in two of the largest collections of fluid-preserved specimens in the United States (the California Academy of Sciences and the Biodiversity Research Center at the University of Kansas). He has published extensively on collections care topics and conducted seminars, workshops, and training programs in the United States, Latin America, Asia, the Middle East, and Europe on the care of natural history collections (his previous publications include the AAM standard reference on collections management policies).

He runs an international consulting company (Museologica) dealing with museum collections and related issues, and teaches museum studies courses for Kent State University, Juniata College, the Universidad Nacional de Colombia (Bogotá), and the Northern States Conservation Center. His consulting work includes presentation of workshops and seminars, collections assessments, and hands-on work with collections. He also serves as Adjunct Curator of Collections at the Earth and Mineral Sciences Museum and Art Gallery at Penn State University.